# EXOTIC STATES IN QUANTUM NANOSTRUCTURES

# Exotic States in Quantum Nanostructures

*Edited by*

SARBEN SARKAR
*King's College London,
London, United Kingdom*

KLUWER ACADEMIC PUBLISHERS
DORDRECHT / BOSTON / LONDON

A C.I.P. Catalogue record for this book is available from the Library of Congress.

ISBN 1-4020-1030-3

Published by Kluwer Academic Publishers,
P.O. Box 17, 3300 AA Dordrecht, The Netherlands.

Sold and distributed in North, Central and South America
by Kluwer Academic Publishers,
101 Philip Drive, Norwell, MA 02061, U.S.A.

In all other countries, sold and distributed
by Kluwer Academic Publishers,
P.O. Box 322, 3300 AH Dordrecht, The Netherlands.

*Printed on acid-free paper*

All Rights Reserved
© 2002 Kluwer Academic Publishers
No part of this work may be reproduced, stored in a retrieval system, or transmitted
in any form or by any means, electronic, mechanical, photocopying, microfilming, recording
or otherwise, without written permission from the Publisher, with the exception
of any material supplied specifically for the purpose of being entered
and executed on a computer system, for exclusive use by the purchaser of the work.

Printed in the Netherlands.

# Table of Contents

| | |
|---|---|
| Preface | vii |

1. Transport in Single Channel Quantum Wires  1
   *H. Grabert*

2. Lectures on Conformal Field Theory  53
   *A.M. Tsvelik*

3. Magnetization and Orbital Properties of the Two-Dimensional  99
   Electro Gas in the Quantum Limit
   *S. Wiegers, E. Bibow, L.P. Lévy, V. Bayot, M. Simmons
   and M. Shayegan*

4. Skyrmion Excitations of the Quantum Hall Ferromagnet  139
   *R.J. Nicholas*

5. Superconducting Devices for Quantum Computation  165
   *J.F. Annett, B.L. Gyorffy and T.P. Spiller*

6. Giant Magnetoresistance and Layered Magnetic Structures  213
   *D.M. Edwards*

7. From the Fermi Liquid towards the Wigner Solid in Two Dimensions  263
   *J.-L. Pichard, G. Benenti, G. Katomeris, F. Selva
   and X. Waintal*

Index  309

# Preface

Over the last few years molecular beam epitaxy has allowed semiconductor heterostructures to be built up by adding individual atomic layers. The density of electrons can be controlled by means of electric fields leading to wavelengths of the electrons of the order of tens of nanometres . The electron mean free path can be made large (thousands of nanometres). Consequently, below about one degree Kelvin, the quantum mechanical phase coherence typical of microscopic atomic systems can be achieved. Electrons can be made to move under the influence of engineered electrostatic potentials and, even in the presence of impurities, provided there is no dissipation of energy, the wave nature of quantum mechanics plays a crucial role. One manifestation of this engineering is the control of the effective dimension of the electron system. This leads to quantum wires which are one-dimensional and quantum dots which are zero-dimensional, since the electrons are confined in all three directions. Quantum nanostructures encompass such systems and provide an example of so called mesoscopic systems since they are neither microscopic nor macroscopic but lie somewhere in between.

In the presence of perpendicular magnetic fields, two-dimensional systems show that the Hall resistance has plateaux, as a function of magnetic field, superposed on an otherwise linear increase with magnetic field. The resistance is inversely proportional to the fundamental unit of conductance which is determined by a ratio of the square of the charge of the electron and Planck's constant. This phenomenon, the quantum Hall effect, is accompanied by electron transport with almost no dissipation. The effect is universal in the sense that it depends neither on the details of the semiconducting materials used nor on the exact value of the magnetic field. The quantum Hall effect continues to be a rich source of interesting physics.

Molecular beam epitaxy has enabled the construction of magnetic multilayers consisting of alternate magnetic and non-magnetic layers. Magnetic layers are made of ferromagnetic transition metals. The spin directions in the neighbouring magnetic layers can be parallel or anti-parallel depending on the separation of the magnetic layers. The effects of spin have been used in a non-static way in the area of spintronics. This has been exploited commercially in disc drives using magnetoresistive read heads. In the phenomenon of magnetoresistance the resistance of electrical current is controlled by a magnetic field.

Owing to advanced lithography techniques, circuits consisting of Josephson junctions can be engineered with a great deal of flexibility. It is for this reason that they are of some interest as possible candidates for quantum computers.

Analogous to quantum dots and the Coulomb blockade, a similar phenomenon can be considered with a superconducting island connected via a tunnel junction to a superconducting gate. The island is connected capacitively to a gate voltage. Instead of single electrons tunnelling into the island, the Cooper electron pairs which are responsible for superconductivity tunnel in as a pair provided that the charging energy and the energy level spacings are much smaller than the superconducting energy gap. At certain (periodic) values of the gate voltage in the absence of tunnelling there is a degeneracy in energy between two configurations where the number of Cooper pairs differ by one. This degeneracy is lifted by the tunnelling out of the island and leads to coherent superpositions of the two states. When the Josephson energy is much larger than the charging energy, the role of the Cooper pairs is played by the flux quanta threading the ring. Such systems are relevant to the increasingly important subject of quantum information.

Quantum wires can be made from heterostructures as well as from carbon nanotubes. The theoretical description of transport along these wires requires new techniques which go under the name of bosonization. These wires are connected to leads which are described by reservoirs of Fermi liquids (i.e. electronic liquids which are effectively described by a collection of non-interacting electrons). Since wires are one-dimensional, the effects of impurities are particularly strong at low temperatures. The methods of bosonization and conformal field theory lead to exact results.

Although the subject of mesoscopics and quantum nanostructures continues to develop, it has attained a certain maturity and it was thought appropriate to help a new generation of researchers to obtain a pedagogical introduction to the frontiers in the area. This book is based on lectures on the above topics at a recent Summer School on "Exotic States in Quantum Nanostructures" held at Cumberland Lodge Windsor. The chapter on Superconducting Devices for Quantum Computation represents a subject covered at the School but not lectured on by the authors.

I wish to thank Vladimir Falko, John Jefferson and Cohn Lambert for help in organizing the School. The funding from the European Commision and the Office of Naval Research USA is gratefully acknowledged.

Sarben Sarkar

# Chapter 1

# TRANSPORT IN SINGLE CHANNEL QUANTUM WIRES

Hermann Grabert
*Fakultät für Physik*
*Albert-Ludwigs-Universität*

**Abstract**  This tutorial article gives an introduction to the methods needed to treat interacting electrons in a quantum wire with a single occupied band. Since one–dimensional Fermions cannot be described in terms of noninteracting quasiparticles, the Tomonaga–Luttinger model is presented in some detail with an emphasis on transport properties. To achieve a self–contained presentation, the Bosonization technique for one–dimensional Fermions is developed, accentuating features relevant for nonequilibrium systems. The screening of an impurity in the wire is discussed, and the insight gained on the electrostatics of a quantum wire is used to describe the coupling to Fermi–liquid reservoirs. These parts of the article should be readily accessible to students with a background in quantum mechanics including second quantization. To illustrate the usefulness of the methods presented, the current–voltage relation is determined exactly for a spin–polarized quantum wire with a particular value of the interaction parameter. This part requires familiarity with path integral techniques and connects with the current literature.

**Keywords:** Tomonaga–Luttinger liquid, Bosonization, Electronic transport properties in one dimension, impurity scattering, current–voltage relation.

## 1. Introduction

Within the last decade there has been increased interest in the behavior of quasi one–dimensional Fermionic systems, due to significant advances in the fabrication of single channel quantum wires [1, 2, 3] based on semiconductor heterostructures and the observation of non–Fermi liquid behavior in carbon nanotubes [4, 5, 6]. While the unusual equilibrium properties of Fermions in one dimension have been studied since many decades and are well documented in review articles [7, 8], nonequilibrium quantum wires are an area of active research with many important question remaining to be answered.

In this article, we give a rather elementary introduction to the theoretical framework underlying much of the present studies on transport properties of one–dimensional Fermions. While we do not review extensively features of the Tomonaga–Luttinger model [9, 10] upon which these studies are based, we give an elementary introduction to the Bosonization technique which is an essential ingredient of current theoretical methods. We do not review the rather long history of Bosonization starting with the work by Schotte and Schotte [11] in 1969. Some important articles are contained in a book of reprints collected by Stone [12]. Our approach is based on Haldane's algebraic Bosonization [13], which can be understood with the usual graduate level background in physics. For a more in–depth discussion of the method, we refer to a recent review by von Delft and Schoeller [14]. The field theoretical approach to Bosonization, which is probably harder to learn but easier to apply, has lately been expounded by Gogolin, Nersesyan and Tsvelik [15].

We employ the Bosonization technique to describe a quantum wire coupled to Fermi–liquid reservoirs. In this connection the electrostatic properties of the wire play an important role. Landauer's approach [16] to transport in mesoscopic systems, which is based on Fermi liquid theory, is generalized to take the electronic correlations in a single channel quantum wire into account. In the Bosonized version of the model the coupling to reservoirs is shown to be described in terms of radiative boundary conditions [17]. This allows us to use the powerful Bosonization method also for nonequilibrium wires. About a decade ago, Kane and Fisher [18] have noted that transport properties of one–dimensional Fermions are strongly affected by impurities. Even weak impurities have a dramatic effect at sufficiently low temperatures leading to a zero bias anomaly of the conductance. It is the aim of the article to present the theoretical background necessary to study the recent literature on this subject. Again, we do not provide a review of transport properties of the Tomonaga–Luttinger model. Rather, the methods developed are illustrated by treating a particular case.

## 1.1. Noninteracting electrons in one dimension

Let us start by considering first noninteracting electrons of mass $m$ moving along a one–dimensional wire with a scatterer at $x = 0$. For simplicity, the scattering potential is taken as a $\delta$–potential

$$V_{\rm sc}(x) = \frac{\hbar^2}{m} \Lambda\, \delta(x) \qquad (1)$$

where $\Lambda$ characterizes the strength. The Schrödinger equation

$$-\frac{\hbar^2}{2m} \psi''(x) + \frac{\hbar^2}{m} \Lambda\, \delta(x)\, \psi(x) = \varepsilon\, \psi(x) \qquad (2)$$

has for all positive energies

$$\varepsilon_k = \frac{\hbar^2 k^2}{2m} \qquad (3)$$

a solution ($k > 0$)

$$\psi_k(x) = \frac{1}{\sqrt{2\pi}} \begin{cases} e^{ikx} + r_k e^{-ikx} &, x < 0 \\ t_k\, e^{ikx} &, x > 0 \end{cases} \qquad (4)$$

describing a wave incident from the left that is partially transmitted and partially reflected. The transmission amplitude

$$t_k = \frac{1}{1 + i\Lambda/k} = \sqrt{T_k}\, e^{i\eta_k} \qquad (5)$$

determines the transmission coefficient $T_k$ and the phase shift $\eta_k$. Likewise, there is a solution ($k > 0$)

$$\psi_{-k}(x) = \psi_k(-x) \qquad (6)$$

describing a wave incident from the right.

When the ends of the wire are connected to electrodes with electrical potentials $\mu_L$ and $\mu_R$, a voltage

$$U = (\mu_L - \mu_R)/e \qquad (7)$$

is applied to the wire, where $e$ is the electron charge, and an electrical current

$$I = e \int_{-\infty}^{+\infty} dk\, f_k j_k \qquad (8)$$

flows. Here

$$j_k(x) = {\rm Im}\, \psi_k^*(x) \frac{\hbar}{m} \frac{\partial}{\partial x} \psi_k(x) = \frac{\hbar k}{2\pi m} T_k \qquad (9)$$

is the particle current in state $\psi_k$ which is independent of $x$, and the

$$f_k = \begin{cases} f(\varepsilon_k - \mu_L) &, k > 0 \\ f(\varepsilon_k - \mu_R) &, k < 0 \end{cases} \qquad (10)$$

with $f(\varepsilon) = 1/(e^{\beta\varepsilon} + 1)$ are state occupation probabilities determined by the Fermi function of the electrode from which the particles come. Both electrodes are assumed to be at the same inverse temperature $\beta$. Putting $\mu_{L,R} = \varepsilon_F \pm \frac{1}{2}eU$, we readily find for small voltages $U$

$$I = GU \qquad (11)$$

with the conductance

$$\begin{aligned} G &= e^2 \int_0^\infty dk\, \frac{\hbar k}{2\pi m}\, T_k \frac{\partial}{\partial \varepsilon_F} f(\varepsilon_k - \varepsilon_F) \\ &= \frac{e^2}{h} \int_0^\infty d\varepsilon\, T(\varepsilon) \left[ -\frac{\partial}{\partial \varepsilon} f(\varepsilon - \varepsilon_F) \right]. \end{aligned} \qquad (12)$$

Provided $\beta\varepsilon_F$ is large, this yields the Landauer formula for a single transport channel

$$G = \frac{e^2}{h} T_F, \qquad (13)$$

where $T_F$ is the transmission coefficient at the Fermi energy $\varepsilon_F$. If we take into account the spin degeneracy of real electrons, the conductance becomes multiplied by 2.

## 1.2. Fano–Anderson Model

Of course, to describe electrons in one dimension, we may also start from a tight binding model with localized electronic states at positions $x_j = aj$ where $a$ is the lattice constant. The Hamiltonian in the presence of an impurity at $x = 0$ then takes the form of the Fano-Anderson model [19, 20]

$$H = \varepsilon_0 a_0^\dagger a_0 - t \sum_{j=-\infty}^{+\infty} (a_{j+1}^\dagger a_j + \text{h.c.}), \qquad (14)$$

where $t$ is the hopping matrix element and $\varepsilon_0 > 0$ is the extra energy needed to occupy the perturbed site at $x = 0$. The $a_j$ are Fermi operators obeying the usual anti–commutation relations. This model is also exactly solvable [21] with eigenstates in the energy band

$$\varepsilon_k = -2t\cos(ka)\, , |k| \leq \frac{\pi}{a}, \qquad (15)$$

and the transmission amplitude takes the form

$$t_k = \frac{1}{1 + i/\lambda_k}, \qquad (16)$$

where

$$\lambda_k = \frac{2t}{\varepsilon_0} \sin(ka) = \frac{\hbar v_k}{\varepsilon_0 a}. \qquad (17)$$

The last relation follows by virtue of $v_k = (1/\hbar)\partial\varepsilon_k/\partial k$. It is now easily checked that this model leads to the same result for the linear conductance than the free electron model examined previously, provided we match parameters of the unperturbed models such that the Fermi velocities $v_F$ at the two Fermi points coincide, and we adjust the strength of the $\delta$-function in Eq. (1) such that

$$\frac{\hbar^2}{m} \Lambda = \varepsilon_0 a. \qquad (18)$$

Then, we end up with the same transition coefficient $T_F$ at the Fermi energy.

## 1.3. Noninteracting Tomonaga–Luttinger model

Apparently, for low temperatures and small applied voltages, the conductance only depends on properties of states in the vicinity of the Fermi energy $\varepsilon_F$. We can take advantage of this fact by introducing still another model, the noninteracting Tomonaga–Luttinger (TL) model, which has the same properties near the two Fermi points $\pm k_F$ but is more convenient once we introduce electron interactions. Let us formally decompose the true energy dispersion curve $\varepsilon_k$ into two branches $+$ and $-$ of right-moving and left-moving electrons, respectively, where these branches comprise states in the energy interval $[\varepsilon_F - \Delta, \varepsilon_F + \Delta]$. (cf. Fig. 1.1). If $\Delta$ is chosen large enough, these branches should suffice to describe the low energy physics of the true physical model, since for low temperatures and small applied voltages states with energy below $\varepsilon_F - \Delta$ are always occupied while states above $\varepsilon_F + \Delta$ are empty. The Hamiltonian of the noninteracting TL model reads

$$H = \sum_p \sum_k \varepsilon_{p,k} \left[ c^\dagger_{p,k} c_{p,k} - \langle c^\dagger_{p,k} c_{p,k} \rangle_0 \right]. \qquad (19)$$

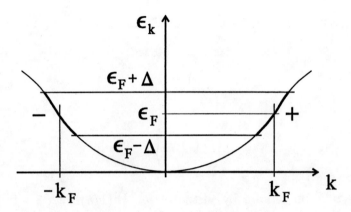

*Figure 1.1.* Physical energy dispersion curve $\epsilon_k$ (thin line) and the two branches $+$, $-$ (thick lines) of the TL model

Here $p = \pm$ labels the two branches. We have introduced a quantization length $L$ such that wave vectors are discrete[1]

$$k = \frac{2\pi}{L} n_k \;, \; n_k \text{ integer.} \qquad (20)$$

Further, the $\varepsilon_\pm(k)$ are single particle energies measured relative to the Fermi energy $\varepsilon_F$, i.e., $\varepsilon_+(k_F) = \varepsilon_-(-k_F) = 0$. The operators $c^\dagger_{p,k}$ and $c_{p,k}$ are Fermi creation and annihilation operators obeying anti-commutation relations, in particular

$$\left[ c_{p,k} , c^\dagger_{p',k'} \right]_+ = \delta_{p,p'} \delta_{k,k'} . \qquad (21)$$

The sum over $k$ states in Eq. (19) is restricted to $k$ values near $\pm k_F$ such that $\varepsilon_\pm(k) \in [-\Delta, \Delta]$. Finally, the ground state energy is subtracted in Eq. (19) where the ground state $|0,0\rangle_0$ is defined by

$$c_{+,k}|0,0\rangle_0 = c_{-,-k}|0,0\rangle_0 = 0 \text{ for } k > k_F,$$
$$c^\dagger_{+,k}|0,0\rangle_0 = c^\dagger_{-,-k}|0,0\rangle_0 = 0 \text{ for } k \leq k_F. \qquad (22)$$

Frequently, the spectra $\varepsilon_{\pm,k}$ are linearized about the Fermi points, i.e.,

$$\varepsilon_{\pm,k} = \pm \hbar v_F (k \mp k_F) , \qquad (23)$$

---

[1] We use periodic boundary conditions here and consider the limit of large $L$ in the sequel. The same techniques can also be used for other boundary conditions [14, 22].

and then the cutoff energy $\Delta$ is used as a large energy scale regularizing divergent expressions. Since for low energy systems only inert empty or occupied states are added, an increase of $\Delta$ is admissible. However, the linearization of the spectra is only realistic in the close vicinity of the Fermi points. In fact, some phenomena not discussed here, e.g. the thermopower [23], depend on band curvature. We shall come back to the limitations of the linearization (23) below.

Before we discuss transport properties of the TL model, we first introduce methods, the advantage of which becomes apparent only when we pass on to the case of interacting electrons. These methods are independent of the precise dispersion law as long as the $\varepsilon_{p,k}$ are monotonous functions of $k$.

## 2. Bosonization

The noninteracting TL model allows for a formulation in terms of Bose operators. We discuss the bosonization technique here for spinless Fermions first, and then extend it to the spinful case.

### 2.1. Density operators and their algebra

While the physical problem of one-dimensional Fermions is described by a single energy dispersion curve $\varepsilon_k$ with empty states at both ends of the range of $k$ values, the TL model introduces two branches with empty states at one end but occupied states at the other end of the $k$ range. This leads to unusual algebraic properties we will discuss now. Since the range of allowed $k$ values plays an important role in this discussion, we keep track of it in detail by introducing

$$W_{p,k} = \begin{cases} 1 & \text{for } \varepsilon_{p,k} \in [-\Delta, \Delta] \\ 0 & \text{else} \end{cases} . \quad (24)$$

Let us define Fourier components of the densities of $p$–movers ($p = \pm$) by

$$\tilde{\rho}_{p,q} = \sum_k W_{p,k} W_{p,k+q} c^\dagger_{p,k} c_{p,k+q} . \quad (25)$$

Since Fermi operators on different branches anti-commute, we have

$$[\tilde{\rho}_{p,q}, \tilde{\rho}_{p',q'}]_- = 0 \text{ for } p \neq p' . \quad (26)$$

On the other hand, for the commutator on the same, say the + branch, we find

$$[\tilde{\rho}_{+,q}, \tilde{\rho}_{+,q'}]_- = \sum_{kk'} W_{+,k} W_{+,k+q} W_{+,k'} W_{+,k'+q'} \quad (27)$$

$$\times \left[ c^\dagger_{+,k} c_{+,k+q} c^\dagger_{+,k'} c_{k'+q'} - c^\dagger_{+,k'} c_{+,k'+q'} c^\dagger_{+,k} c_{+,k+q} \right] .$$

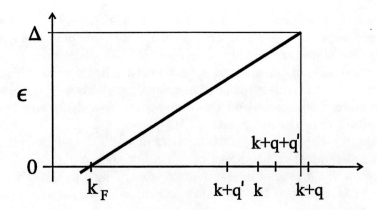

*Figure 1.2.* Wave vectors in the sum (28) for $q > 0$ and $q' < 0$ near the upper cutoff energy of the + branch.

We now use anti–commutation relations for the second and third operator in each of the two products of four Fermi operators, e.g., $c_{+,k+q}c^{\dagger}_{+,k'} = \delta_{k+q,k'} - c^{\dagger}_{+,k'}c_{+,k+q}$. Then, the remaining products of four Fermi operators are easily seen to cancel by virtue of the anti–commutation relations, and the two terms with two Fermi operators can be written as

$$[\tilde{\rho}_{+,q}, \tilde{\rho}_{+,q'}]_- \qquad (28)$$
$$= \sum_k W_{+,k}W_{+,k+q+q'}\left[W^2_{+,k+q} - W^2_{+,k+q'}\right]c^{\dagger}_{+,k}c_{+,k+q+q'}.$$

The terms of the sum are finite only for $W_{+,k} = W_{+,k+q+q'} = 1$. But then $W_{+,k+q}$ and $W_{+,k+q'}$ are also equal to 1 if $q$ and $q'$ have the same sign. In this case all terms of the sum (28) vanish. Hence, a nontrivial commutator can only arise if $q$ and $q'$ have different signs, say $q > 0$ and $q' < 0$. Then non–vanishing terms of the sum (28) may occur near the upper cutoff where $\varepsilon_{+,k} \approx \Delta$, in particular, when

$$W_{+,k} = W_{+,k+q+q'} = W_{+,k+q'} = 1, \quad \text{but } W_{+,k+q} = 0,$$

as illustrated in Fig. 1.2. However, in this case the operator $c^{\dagger}_{+,k}c_{+,k+q+q'}$ tries to annihilate a particle in the state $k + q + q'$ with an energy near $\Delta$. In the low energy sector of the model these states are always empty and $c_{+,k+q+q'}$ can be replaced by zero. Hence, for a low energy system we get no contribution to the commutator from states near the upper cutoff energy.

A contribution near the lower cutoff energy may arise from terms with $W_{+,k} = W_{+,k+q+q'} = W_{+,k+q} = 1$ but $W_{+,k+q'} = 0$ since $\varepsilon_{+,k+q'} < -\Delta$.

In this case we use the anti–commutation relations to write Eq. (28) as

$$[\tilde{\rho}_{+,q}, \tilde{\rho}_{+,q'}]_- = \sum_k W_{+,k} W_{+,k+q+q'} \left[ W^2_{+,k+q} - W^2_{+,k+q'} \right]$$
$$\times \left[ \delta_{q,-q'} - c_{+,k+q+q'} c^\dagger_{+,k} \right]. \quad (29)$$

With a similar argument as above, the operator $c^\dagger_{+,k}$ may now be replaced by zero since near the lower cutoff all states are occupied. We thus obtain

$$[\tilde{\rho}_{+,q}, \tilde{\rho}_{+,q'}]_- = \delta_{q,-q'} \sum_{k<k_F} W^2_{+,k} \left[ W^2_{+,k+q} - W^2_{+,k-q} \right], \quad (30)$$

where the condition $k < k_F$ ensures that we do not get contributions from the region near the upper cutoff where the Fermi operators in Eq. (29) cannot be dropped. Finally, since

$$\sum_{k<k_F} W^2_{+,k} W^2_{+,k+q} - \sum_{k<k_F} W^2_{+,k-q} W^2_{+,k} = n_q, \quad (31)$$

where $q = \frac{2\pi}{L} n_q$ determines the number $n_q$ of additional non–vanishing terms in the first sum, and with a similar reasoning for the $-$ branch and other signs of $q$, we get

$$[\tilde{\rho}_{p,q}, \tilde{\rho}_{p',q'}]_- = p\, \delta_{p,p'} \delta_{q,-q'}\, n_q. \quad (32)$$

As we have seen this nontrivial commutator only arises since we have introduced two branches with empty states at one end and occupied states at the other end. The result (32) can also be derived in the same way when the sharp cutoff functions $W_{\pm,k}$ defined in Eq. (24) are replaced by a smooth cutoff. Furthermore, the relation (32) is not an exact operator relation but holds only in the low energy sector of the Fock space. However, for technical convenience, we may use the linearized spectrum (23) and send the cutoff $\Delta$ to infinity. We then obtain a model where the relations (32) hold as formally exact commutation relations but have to remember that only the low energy properties of this model are related to the physical problem of one–dimensional electrons.

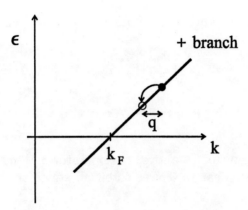

*Figure 1.3.* Action of $b_q$ for $q > 0$ on states of the + branch.

## 2.2. Bose operators and basis vectors

With the result (32) it is now straightforward to introduce Bose annihilation and creation operators[2]

$$b_q = \frac{-i}{\sqrt{|n_q|}} \sum_p \theta(pq) \tilde{\rho}_{p,q},$$

$$b_q^\dagger = \frac{i}{\sqrt{|n_q|}} \sum_p \theta(pq) \tilde{\rho}_{p,-q}, \tag{33}$$

where $q \neq 0$. Inserting (25), we see that for $q > 0$ the operator $b_q$ lowers the wave vector $k$ of right–movers by $q$, provided the state $k - q$ is empty, cf. Fig. 1.3, while $b_{-q}$ acts accordingly on left–movers. By virtue of Eq. (32), the $b_q$ and $b_q^\dagger$ satisfy Bose commutation relations, in particular

$$[b_q, b_{q'}^\dagger]_- = \delta_{q,q'}. \tag{34}$$

When acting on a $N$–particle state, the operators $b_q^\dagger$ and $b_q$ create and annihilate electron–hole pairs, respectively, but they conserve of course the particle numbers $N_p$ of each branch. Consider now $N_+$–particle ground states of right–movers where $|N_+ = 0\rangle_0$ is the many-body state with all single particle states $k \leq k_F$ occupied and all states $k > k_F$ empty, while in the state $|N_+ = 1\rangle_0$ also the level $k = k_F + \frac{2\pi}{L}$ is occupied, and so on, as illustrated in Fig. 1.4. One can then show that the states

---

[2] As usual, $\theta(x)$ denotes the step function: $\theta(x) = 1$ for $x > 0$, $\theta(x) = 0$ for $x < 0$.

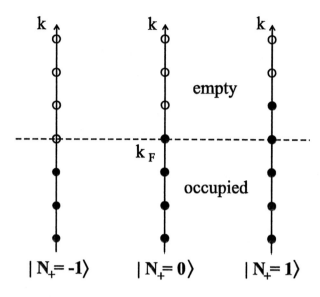

*Figure 1.4.* Graphical representation of $N_+$–particle ground states of right–movers.

$$|N_+, \{m_q\}_{q>0}\rangle = \prod_{q>0} \frac{\left(b_q^\dagger\right)^{m_q}}{\sqrt{m_q!}} |N_+\rangle_0 , \qquad (35)$$

with $N_+ = 0, \pm 1, \pm 2, \ldots$ and $m_q = 0, 1, 2, \ldots$ form a complete basis in the Fock space of right–movers. It is obvious that the states (35) are all within the Fock space spanned by the standard basis vectors $|\{N_{+,k}\}\rangle$, where the $N_{+,k} = 0, 1$ are the usual Fermi occupation numbers, that is eigenvalues of $c_{+,k}^\dagger c_{+,k}$. Haldane [13] has demonstrated completeness of the basis (35) with the help of the model Hamiltonian

$$h_+ = \sum_k (n_k - \frac{1}{2}) : c_{+,k}^\dagger c_{+,k} : , \qquad (36)$$

where

$$: c_{+,k}^\dagger c_{+,k} := \begin{cases} c_{+,k}^\dagger c_{+,k} & \text{for } k > 0 \\ c_{+,k} c_{+,k}^\dagger & \text{for } k \leq 0 . \end{cases} \qquad (37)$$

This Hamiltonian assigns positive energies to all states and allows for an explicit evaluation of the partition function $Z = \sum_\alpha \exp(-\beta E_\alpha)$ for the two sets $\{\alpha\}$ of basis vectors. Since $Z$ is a sum of positive terms, and the sum over the states (35) gives the same result as the sum over the states $|\{N_{+,k}\}\rangle$, the basis $|N_+, \{m_q\}_{q>0}\rangle$ spans the entire Fock space of right–movers.

Likewise we may introduce a basis

$$|N_-, \{m_q\}_{q<0}\rangle = \prod_{q<0} \frac{\left(b_q^\dagger\right)^{m_q}}{\sqrt{m_q!}} |N_-\rangle_0 \qquad (38)$$

in the Fock space of left–movers and combine both to the basis

$$|N_+, N_-, \{m_q\}\rangle = \prod_{q\neq 0} \frac{\left(b_q^\dagger\right)^{m_q}}{\sqrt{m_q!}} |N_+, N_-\rangle_0 \qquad (39)$$

in the Fock space of one–dimensional particles.

## 2.3. Ladder and particle number operators

It is clear that the Bose operators $b_q$, $b_q^\dagger$ cannot generate the whole algebra of operators in the Fock space since they preserve the particle numbers $N_p$. Hence, the Bose operators need to be supplemented by ladder operators removing or adding a particle of branch $p$. The lowering operators $U_p$ are defined by

$$U_p|N_p, N_{-p}\rangle_0 = p^{N_+}|N_p - 1, N_{-p}\rangle_0 \,, \qquad (40)$$

where the sign factor $p^{N_+}$ is one possible choice assuring anti–commutation relations for Fermi operators on different branches, and

$$[U_p, b_q]_- = [U_p, b_q^\dagger]_- = 0 \,. \qquad (41)$$

The adjoint raising operator obeys $U_p^\dagger = U_p^{-1}$. The action of the ladder operator $U_+$ on a state in the Fock space of right–movers is illustrated in Fig. 1.5. Since the ladder operators commute with the Bose operators $b_q$, $b_q^\dagger$, they preserve the electron–hole pair excitations present in a state.

Now, the particle number $N_p$ is an eigenvalue of the particle operator

$$\hat{N}_p = \sum_k \left( c_{p,k}^\dagger c_{p,k} - \langle c_{p,k}^\dagger c_{p,k} \rangle_0 \right) \,, \qquad (42)$$

where the average $\langle \ \rangle_0$ is over the $|N_p = 0\rangle_0$ ground state. The ladder operators and the particle numbers satisfy the commutation relations

$$\begin{aligned}
\left[\hat{N}_p, U_{p'}\right]_- &= -\delta_{p,p'} U_p \,, \\
\left[\hat{N}_p, U_{p'}^\dagger\right]_- &= \delta_{p,p'} U_p^\dagger \,,
\end{aligned} \qquad (43)$$

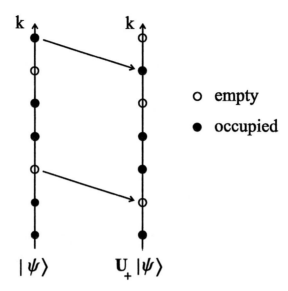

*Figure 1.5.* Illustration of the action of $U_+$ on a many–body state $|\psi\rangle$ of right-movers.

that are easily shown by applying them to an arbitrary vector from the basis set (39). Furthermore, the ladder operators obey

$$[U_+, U_-]_+ = [U_+^\dagger, U_-^\dagger]_+ = 0,$$
$$[U_p, U_{p'}^\dagger]_+ = 2\delta_{p,p'}, \tag{44}$$

which may be demonstrated in the same way taking the sign factor in Eq. (40) into account.

### 2.4. Bosonic phase fields

With the Bose and ladder operators we should be able to represent all operators in the Fock space of the TL model. To demonstrate this one usually constructs explicitly the Fermi annihilation operators

$$\Psi_p(x) = \frac{1}{\sqrt{L}} \sum_k e^{ikx} c_{p,k}. \tag{45}$$

It is convenient to first introduce Bosonic phase fields

$$\varphi_p(x) = \frac{1}{2\pi} \sum_{q \neq 0} \theta(pq) \frac{1}{\sqrt{|n_q|}} e^{iqx} b_q, \tag{46}$$

and associated Hermitian phase fields

$$\Phi_p(x) = \varphi_p(x) + \varphi_p^\dagger(x). \tag{47}$$

These are related to the densities

$$\rho_p(x) = \Psi_p^\dagger(x)\Psi_p(x) - \langle \Psi_p^\dagger(x)\Psi_p(x)\rangle_0 = \frac{1}{L}\left[\sum_{q\neq 0} e^{iqx}\tilde{\rho}_{p,q} + \hat{N}_p\right], \quad (48)$$

where the $\tilde{\rho}_{p,q} = \sum_k c_{p,k}^\dagger c_{p,k+q}$ are the Fourier components studied above. Using Eq. (33), the phase field (46) can also be written as

$$\varphi_p(x) = -\frac{i}{2\pi}\sum_{q\neq 0}\theta(pq)\frac{1}{|n_q|}e^{iqx}\tilde{\rho}_{p,q}. \quad (49)$$

Because of the factor $\theta(pq)$, we may replace $|n_q|$ by $pn_q = \frac{L}{2\pi}pq$. Then, adding the Hermitian conjugate field, we readily find for the gradient

$$\frac{\partial}{\partial x}\Phi_p(x) = \frac{p}{L}\sum_{q\neq 0}e^{iqx}\tilde{\rho}_{p,q} = p\left(\rho_p(x) - \frac{1}{L}\hat{N}_p\right). \quad (50)$$

The Bosonic phase fields $\varphi(x)$, $\varphi^\dagger(x)$ obey the commutation relations

$$[\varphi_p(x), \varphi_{p'}(x')]_- = [\varphi_p^\dagger(x), \varphi_{p'}^\dagger(x')]_- = 0 \quad (51)$$

while by virtue of Eqs. (34) and (46) the commutator

$$[\varphi_p(x), \varphi_{p'}^\dagger(x')]_- = \frac{\delta_{p,p'}}{4\pi^2}\sum_{n=1}^\infty \frac{1}{n}e^{ip\frac{2\pi}{L}(x-x')n}. \quad (52)$$

Since

$$\sum_{n=1}^\infty \frac{y^n}{n} = -\ln(1-y), \quad (53)$$

the right hand side of Eq. (52) is logarithmically divergent for $x = x'$. However, we have to remember that the sum over $n$ originates from a sum over wave vectors $q = (2\pi/L)n$ which due to the energy cutoff are restricted to small $q$. The effect of a cutoff is seen when we multiply the terms of the sum (52) by an exponential cutoff function $\exp(-aq) = \exp(-\frac{2\pi}{L}an)$ which limits the sum to wave vectors $q \lesssim 1/a$, where $a$ is a cutoff length related to a cutoff energy $\Delta \approx \hbar v_F/a$. We then obtain

$$\begin{aligned}[\varphi_p(x), \varphi_{p'}^\dagger(x')]_- &= -\frac{\delta_{p,p'}}{4\pi^2}\ln\left(1 - \exp\left\{-\frac{2\pi}{L}[a - ip(x-x')]\right\}\right) \\ &= -\frac{\delta_{p,p'}}{4\pi^2}\ln\left(\frac{2\pi}{L}[a - ip(x-x')]\right) \quad (54)\end{aligned}$$

where the second equality holds for $L \gg a, |x-x'|$. For later convenience we note that in particular the commutator for vanishing distance

$$\left[\varphi_p(x), \varphi_{p'}^\dagger(x)\right]_- = -\frac{\delta_{p,p'}}{4\pi^2} \ln\frac{2\pi a}{L} \tag{55}$$

is cutoff and size dependent.

Simpler algebraic properties are found for the Hermitian phase field (47). From Eqs. (51) and (54) we readily find

$$\left[\Phi_p(x), \Phi_{p'}(x')\right]_- = -\frac{\delta_{p,p'}}{4\pi^2} \ln\frac{a - ip(x-x')}{a + ip(x-x')}, \tag{56}$$

where $L$ has dropped out, and we have for $|x-x'| \gg a$

$$\left[\Phi_p(x), \Phi_{p'}(x')\right]_- = \frac{ip\,\delta_{p,p'}}{4\pi} \text{sign}(x-x'). \tag{57}$$

Here $\text{sign}(x-x')$ is the sign of $x-x'$ which produces a step at $x=x'$ that in the presence of a cutoff is smeared over a length of order $a$.

Instead of the fields $\Phi_\pm(x)$ we shall mainly use the linear combinations

$$\begin{aligned}\phi(x) &= \sqrt{\pi}\left[\Phi_+(x) + \Phi_-(x)\right] \\ \vartheta(x) &= \sqrt{\pi}\left[\Phi_+(x) - \Phi_-(x)\right]\end{aligned} \tag{58}$$

that are readily seen to obey the commutation relations

$$\left[\phi(x), \phi(x')\right]_- = \left[\vartheta(x), \vartheta(x')\right]_- = 0 \tag{59}$$

and

$$\left[\phi(x), \vartheta(x')\right]_- = \frac{i}{2}\text{sign}(x-x'), \tag{60}$$

which shows that

$$\Pi_\vartheta(x) = -\hbar\frac{\partial}{\partial x}\phi(x) \tag{61}$$

is the field conjugate to $\vartheta(x)$ with the canonical commutator

$$\left[\vartheta(x), \Pi_\vartheta(x')\right] = i\hbar\delta(x-x'). \tag{62}$$

Likewise $\Pi_\phi(x) = -\hbar\frac{\partial}{\partial x}\vartheta(x)$ is the field canonically conjugate to $\phi(x)$. Finally, we note that by virtue of Eq. (50)

$$\frac{1}{\sqrt{\pi}}\frac{\partial}{\partial x}\vartheta(x) = \rho_+(x) + \rho_-(x) - \frac{\hat{N}_+ + \hat{N}_-}{L} \tag{63}$$

describes the fluctuations of the total density of right- and left-movers.

## 2.5. Bose representation of Fermi operators

To construct the Fermi operators $\Psi_p(x)$ in terms of the Bosonic fields, we start by evaluating the commutator

$$[b_q, \Psi_p(x)]_- = \alpha_{p,q}(x)\,\Psi_p(x), \qquad (64)$$

where

$$\alpha_{p,q}(x) = \frac{i}{\sqrt{|n_q|}}\,\theta(pq)\,e^{-iqx}. \qquad (65)$$

This follows readily by inserting Eqs. (33) and (45) into (64) with the help of the Fermi commutator $[\tilde{\rho}_{p',q}, c_{p,k}]_- = -\delta_{p,p'}\,c_{p,k+q}$. In the same way we find

$$\left[b_q^\dagger, \Psi_p(x)\right]_- = \alpha^*_{p,q}(x)\,\Psi_p(x). \qquad (66)$$

Operating with $b_q\Psi_p(x)$ onto a $N$–particle ground state $|N_+, N_-\rangle_0$ we have

$$b_q\Psi_p(x)|N_+, N_-\rangle_0 = [b_q, \Psi_p(x)]_-\,|N_+, N_-\rangle_0$$
$$= \alpha_{p,q}(x)\Psi_p(x)|N_+, N_-\rangle_0, \qquad (67)$$

where the first relation holds because $b_q|N_+, N_-\rangle_0 = 0$. Now, Eq. (67) shows that $\Psi_p(x)|N_+, N_-\rangle_0$ is an eigenstate of $b_q$ with eigenvalue $\alpha_{p,q}(x)$. Eigenstates of Bose annihilation operators are known as coherent states, and from their familiar properties it follows that $\Psi_p(x)|N_+, N_-\rangle_0$ is of the form

$$\Psi_p(x)|N_+, N_-\rangle = \lambda_p(x)\exp\left(\sum_{q\neq 0}\alpha_{p,q}(x)\,b_q^\dagger\right)|N_p - 1, N_{-p}\rangle_0, \qquad (68)$$

where $\lambda_p(x)$ is an as yet undetermined normalization factor, and we have taken into account that $\Psi_p(x)$ reduces the number $N_p$ of $p$-movers by 1. With the help of the Bose commutator (34), which implies

$$\left[b_q, \exp\left(\sum_{q'\neq 0}\alpha_{p,q'}(x)\,b_{q'}^\dagger\right)\right]_- = \alpha_{p,q}(x)\exp\left(\sum_{q'\neq 0}\alpha_{p,q'}(x)\,b_{q'}^\dagger\right), \qquad (69)$$

it is readily seen that the ansatz (68) indeed satisfies Eq. (67). From the definition (46) of the Bosonic phase field $\varphi_p(x)$ and Eq. (65) we find

$$\sum_{q\neq 0}\alpha_{p,q}(x)\,b_q^\dagger = 2\pi i\,\varphi_p^\dagger(x), \qquad (70)$$

and hence from Eq. (68)

$$_0\langle N_p - 1, N_{-p}| \Psi_p(x) |N_p, N_{-p}\rangle_0 = \lambda_p(x), \qquad (71)$$

where we have made use of

$$_0\langle N_p - 1, N_{-p}| \exp\left[2\pi i \varphi_p^\dagger(x)\right] |N_p - 1, N_{-p}\rangle_0 = 1, \qquad (72)$$

which is easily seen by expanding the exponential. We now insert the representation (45) of $\Psi_p(x)$ into Eq. (71) and use

$$_0\langle N_p - 1, N_{-p}| c_{p,k} |N_p, N_{-p}\rangle_0 = p^{N_+} \delta_{k,k_p}, \qquad (73)$$

where

$$k_p = p\left(k_F + \frac{2\pi}{L} N_p\right). \qquad (74)$$

The sign factor $p^{N_+}$ arises from the anti-commutation relations of Fermi operators on different branches in accordance with Eq. (40). This yields

$$\lambda_p(x) = \frac{1}{\sqrt{L}} p^{N_+} e^{ip(k_F + \frac{2\pi}{L} N_p)x}, \qquad (75)$$

which can now be combined with the definition (40) of the lowering operator to obtain from Eq. (68)

$$\Psi_p(x)|N_+, N_-\rangle_0 = \frac{1}{\sqrt{L}} e^{ip(k_F + \frac{2\pi}{L} N_p) + 2\pi i \varphi_p^\dagger(x)} U_p |N_+, N_-\rangle_0. \qquad (76)$$

This represents the action of a Fermi annihilation operator on a $N$-particle ground state in terms of Bose and ladder operators. Since particle-hole excitations are created by Bose operators, it is straightforward to generalize Eq. (76) for arbitrary states in the Fock space. For a basis vector (39) we have as a consequence of the commutator (66)

$$\Psi_p(x)|N_+, N_-, \{m_q\}\rangle$$
$$= \sum_{q \neq 0} \frac{\left[b_q^\dagger - \alpha_{p,q}^*(x)\right]^{m_q}}{\sqrt{m_q!}} \Psi_p(x)|N_+, N_-\rangle_0. \qquad (77)$$

Using Eq. (76), we find

$$\Psi_p(x)|N_+, N_-, \{m_q\}\rangle \qquad (78)$$
$$= \frac{1}{\sqrt{L}} U_p e^{ip(k_F + \frac{2\pi}{L} N_p)x + 2\pi i \varphi_p^\dagger(x)} \sum_{q \neq 0} \frac{\left[b_q^\dagger - \alpha_{p,q}^*(x)\right]^{m_q}}{\sqrt{m_q!}} |N_+, N_-\rangle_0,$$

since the Bose creation operators commute with $\varphi_p^\dagger(x)$ and $U_p$. In view of

$$\sum_{q\neq 0} \alpha_{p,q}^*(x) b_q = -2\pi i\, \varphi_p(x) \tag{79}$$

and the familiar Bose relation $e^{\alpha b}\, b^\dagger\, e^{-\alpha b} = b^\dagger + \alpha$ we have

$$e^{2\pi i\varphi_p(x)}\, b_q^\dagger\, e^{-2\pi i\varphi_p(x)} = b_q^\dagger - \alpha_{p,q}^*(x), \tag{80}$$

and Eq. (78) may be written as

$$\Psi_p(x)|N_+, N_-, \{m_q\}\rangle \tag{81}$$
$$= \frac{1}{\sqrt{L}} U_p\, e^{ip(k_F + \frac{2\pi}{L}\hat{N}_p)x}\, e^{2\pi i\varphi_p^\dagger(x)}\, e^{2\pi i\varphi_p(x)}\, |N_+, N_-, \{m_q\}\rangle.$$

Here we have made use of $\varphi_p(x)|N_+, N_-\rangle_0 = 0$, and have replaced in the exponent $N_p$ by the operator $\hat{N}_p$, which is appropriate since $\hat{N}_p$ commutes with the Bose operators and thus acts on an eigenstate with eigenvalue $N_p$. Now, the operator acting on the basis vector on the right hand side of Eq. (81) has the same form for any vector, and we thus obtain the operator identity

$$\Psi_p(x) = \frac{1}{\sqrt{L}} U_p\, e^{ip(k_F + \frac{2\pi}{L}\hat{N}_p)x}\, e^{2\pi i\varphi_p^\dagger(x)}\, e^{2\pi i\varphi_p(x)}. \tag{82}$$

This is the desired representation of the Fermi operators $\Psi_p(x)$ in terms of Bose and ladder operators. The result (82) is in normal ordered form with all Bose creation operators to the left of the annihilation operators. It should be noted that this remarkable relation is independent of the form of the Hamiltonian, it just relates the form (45) of the Fermi operator with an obvious meaning in the standard occupation number basis $|\{N_{+,k}\}, \{N_{-,k}\}\rangle$ of the Fock space to a representation with an obvious interpretation in the basis $|N_+, N_-, \{m_q\}\rangle$.

For much of the following discussions another form of $\Psi_p(x)$ is often more convenient. Using the commutator (55) and the operator relation

$$e^A\, e^B = e^{A+B}\, e^{\frac{1}{2}[A,B]_-}, \tag{83}$$

which holds if $[A, B]_-$ commutes with $A$ and $B$, we may transform the relation (82) to read

$$\Psi_p(x) = \frac{1}{\sqrt{2\pi a}} U_p\, e^{ip(k_F + \frac{2\pi}{L}\hat{N}_p)x + 2\pi i\Phi_p(x)}, \tag{84}$$

where $\Phi_p(x)$ was introduced in Eq. (47). Finally, in terms of the phase fields $\phi(x)$ and $\vartheta(x)$ defined in Eq. (58), we have

$$\Psi_p(x) = \frac{1}{\sqrt{2\pi a}} U_p\, e^{ip[k_F x + \frac{2\pi}{L}\hat{N}_p x + \sqrt{\pi}\vartheta(x)] + i\sqrt{\pi}\phi(x)}, \tag{85}$$

which is the form employed mostly in the literature.[3]

## 2.6. Bose representation of the Hamiltonian

In the following, we shall make use explicitly of the Hamiltonian (19) with the linearized energy dispersion curves (23). From Eq. (33) we see that the Bose creation operators may be written as

$$b_q^\dagger = \frac{i}{\sqrt{|n_q|}} \sum_{pk} \theta(pq)\, c_{p,k+q}^\dagger\, c_{p,k}. \tag{86}$$

With the help of the Fermi anti–commutation relations we then find

$$\left[H_0, b_q^\dagger\right]_- = \hbar v_F |q|\, b_q^\dagger, \tag{87}$$

where we have also made use of $\varepsilon_{p,k+q} - \varepsilon_{p,k} = p\hbar v_F q$ which holds for the linear spectrum (23). Now, a $N$–particle ground state $|N_+, N_-\rangle_0$ is an eigenstate of $H_0$ with the energy

$$E_0(N_+, N_-) = \hbar v_F \frac{\pi}{L} \sum_p N_p(N_p + 1). \tag{88}$$

Since $|0,0\rangle_0$ has zero energy by definition and the single particle energies (23) are counted relative to the Fermi energy, the result (88) is obtained readily by adding for $N_p > 0$ the single particle energies of the states occupied additionally in the $N$–particle ground state $|N_+, N_-\rangle_0$, while for $N_p < 0$ we have to subtract the (negative) energies of the particles removed. Using the commutator (87) we obtain for an arbitrary basis vector (39)

$$H_0 \prod_{q\neq 0} \frac{\left(b_q^\dagger\right)^{m_q}}{\sqrt{m_q!}} |N_+, N_-\rangle_0$$

$$= \prod_{q\neq 0} \frac{\left(b_q^\dagger\right)^{m_q}}{\sqrt{m_q!}} \left(H_0 + \sum_{q\neq 0} \hbar v_F |q| m_q\right) |N_+, N_-\rangle_0, \tag{89}$$

which shows that $|N_+, N_-, \{m_q\}\rangle$ is an eigenstate of $H_0$ with eigenvalue

$$E_0(N_+, N_-, \{m_q\}) = E_0(N_+, N_-) + \hbar v_F \sum_{q\neq 0} |q| m_q. \tag{90}$$

---

[3] Some authors use non–standard definitions of the Fermi annihilation operator fields (45). To compare with our notation, one has to make proper replacements, e.g., replace $x$ by $-x$. In addition, the fields $\vartheta(x)$ and $\phi(x)$ are sometimes defined the other way round.

Since the basis (39) is complete and the basis vectors are clearly eigenstates of

$$H_0 = \hbar v_F \left[ \sum_{q \neq 0} |q| b_q^\dagger b_q + \frac{\pi}{L} \sum_p \hat{N}_p(\hat{N}_p + 1) \right] \quad (91)$$

with the proper eigenvalues (90), we see that Eq. (91) gives indeed a Bose representation of the Hamiltonian (19).

Combining Eqs. (33) and (50), we may write the gradient of the Bosonic phase field $\Phi_p(x)$ as

$$\frac{\partial}{\partial x} \Phi_p(x) = \frac{p}{\sqrt{2\pi L}} \sum_{q \neq 0} \theta(pq) \sqrt{|q|} \left[ i e^{iqx} b_q + \text{h.c.} \right], \quad (92)$$

which gives

$$\sum_p \int_{-L/2}^{L/2} dx \left( \frac{\partial \Phi_p(x)}{\partial x} \right)^2 = \frac{1}{2\pi} \sum_{q \neq 0} |q| \left( b_q b_q^\dagger + b_q^\dagger b_q \right). \quad (93)$$

The Hamiltonian (91) may thus be written as

$$H_0 = \pi \hbar v_F \int_{-L/2}^{L/2} dx \sum_p \left[ : \left( \frac{\partial \Phi_p}{\partial x} \right)^2 : + \frac{1}{L^2} \hat{N}_p(\hat{N}_p + 1) \right], \quad (94)$$

where : : puts the Bose operators in normal order. Further, in terms of the fields (58) this reads

$$H_0 = \frac{\hbar v_F}{2} \int_{-L/2}^{L/2} dx \left[ : \left( \frac{\partial \phi}{\partial x} \right)^2 + \left( \frac{\partial \vartheta}{\partial x} \right)^2 : + \frac{2\pi}{L^2} \sum_p \hat{N}_p(\hat{N}_p + 1) \right]. \quad (95)$$

Since $\Pi_\vartheta = -\hbar \partial \phi / \partial x$ is the conjugate density to the phase field $\vartheta$ with the canonical commutator (62), we finally obtain the $\vartheta$–representation of the Hamiltonian

$$H_0 = \frac{\hbar v_F}{2} \int_{-L/2}^{L/2} dx \left[ : \frac{1}{\hbar^2} \Pi_\vartheta^2 + \left( \frac{\partial \vartheta}{\partial x} \right)^2 : + \frac{2\pi}{L^2} \sum_p \hat{N}_p(\hat{N}_p + 1) \right]. \quad (96)$$

Likewise, using $\Pi_\phi = -\hbar \partial \vartheta / \partial x$, we can readily write down a $\phi$–representation of $H_0$.

## 2.7. Action functional

In the usual way, we may introduce a Lagrangian

$$L_0 = \int_{-L/2}^{L/2} dx \, \Pi_\vartheta \frac{\partial \vartheta}{\partial t} - H_0, \qquad (97)$$

where the time rate of change of the $\vartheta$–field reads

$$\frac{\partial}{\partial t}\vartheta = \frac{i}{\hbar}[H_0, \vartheta]_- = -v_F \frac{\partial}{\partial x}\phi = \frac{v_F}{\hbar}\Pi_\vartheta, \qquad (98)$$

which follows from the commutation relations (60). In the limit $L \to \infty$ this gives

$$L_0 = \hbar \int dx \left[\frac{1}{2v_F}\left(\frac{\partial \vartheta}{\partial t}\right)^2 - \frac{v_F}{2}\left(\frac{\partial \vartheta}{\partial x}\right)^2 - \pi v_F \left(\bar{\rho}_+^2 + \bar{\rho}_-^2\right)\right] \qquad (99)$$

where

$$\bar{\rho}_p = \frac{N_p}{L} \qquad (100)$$

is the average density of $p$–movers. Note that by definition the densities $\bar{\rho}_p$ vanish in the ground state $|0,0\rangle_0$, where the Fermi levels are at $\pm k_F$. The $\bar{\rho}_p$ determine a shift of the Fermi points, while the density fluctuations $\rho_p(x) - \bar{\rho}_p$ arising from electron–hole pair excitations are described by the phase field $\vartheta$. However, we may define a shifted phase field

$$\vartheta' = \vartheta + \sqrt{\pi}\left(\bar{\rho}_+ + \bar{\rho}_-\right) x - \sqrt{\pi}\left(\bar{\rho}_+ - \bar{\rho}_-\right) v_F t \qquad (101)$$

with the properties

$$\frac{1}{\sqrt{\pi}}\frac{\partial \vartheta'}{\partial x} = \rho_+ + \rho_-$$

$$\frac{1}{\sqrt{\pi}}\frac{\partial \vartheta'}{\partial t} = -v_F(\rho_+ - \rho_-). \qquad (102)$$

Now, the gradient determines the total density of right– and left–movers including the ground state density $\bar{\rho}_+ + \bar{\rho}_-$, while the time rate of change is proportional to the particle current. The Lagrangian then takes the simple form

$$L_0 = \frac{\hbar}{2}\int dx \left[\frac{1}{v_F}\left(\frac{\partial \vartheta}{\partial t}\right)^2 - v_F \left(\frac{\partial \vartheta}{\partial x}\right)^2\right], \qquad (103)$$

where we have omitted the prime. The noninteracting TL model with dispersionless spectrum (23) can thus be characterized by the classical action functional

$$S_0 = \frac{\hbar}{2} \int dt \int dx \left[ \frac{1}{v_F} \left( \frac{\partial \vartheta}{\partial t} \right)^2 - v_F \left( \frac{\partial \vartheta}{\partial x} \right)^2 \right], \qquad (104)$$

which is the action of a harmonic string with wave velocity $v_F$ and a dimensionless displacement field $\vartheta(x,t)$ measured in units of $(\hbar/v_F \mu)^{1/2}$ where $\mu$ is the mass density in the string. From this mechanical analogue it is obvious that Bosonization provides an alternative description of one–dimensional Fermions in terms of charge density oscillations rather than electron–hole pair excitations.

We remark that in the general case of an arbitrary dispersion law $\varepsilon_{p,k}$ the Bosonization identities, in particular the representation (85) of the Fermi operators remain valid, however, the Hamiltonian $H_0$ is no longer quadratic in the Bose operators $b_q$, $b_q^\dagger$. As a consequence, the mechanical analogue will be an anharmonic string which may, of course, be treated in the harmonic approximation when we restrict ourselves to low energy excitations. In combination with the Feynman path integral representation the classical action functional (104) can be a convenient starting point for quantum mechanical calculations.

## 2.8. Electron density operator

When we use the TL model to make predictions for one–dimensional fermions, we have to remember that the model provides a local approximation to the physical model in the vicinity of the Fermi points, while there is only a single energy dispersion curve for real electrons, cf. Fig. 1.1. This has consequences for observables like the density operator

$$\rho(x) = \frac{1}{L} \sum_{k,k'} e^{-i(k-k')x} \left( a_k^\dagger a_{k'} - \langle a_k^\dagger a_{k'} \rangle_0 \right), \qquad (105)$$

where the $a_k$, $a_{k'}^\dagger$ are Fermi operators for the underlying physical model with a single branch, and we have subtracted the constant density of the ground state with the Fermi points at $\pm k_F$. When we restrict ourselves to low energy states, the operators $a_k^\dagger a_{k'}$ can give a nonvanishing contribution only if both wave vectors $k$ and $k'$ are in the vicinity of the two Fermi points $\pm k_F$. However, for $k \approx p k_F$, $k' \approx p' k_F$, $(p, p' = \pm)$, the operators $a_k^\dagger$, $a_{k'}$ may be replaced by the Fermi operators $c_{p,k}^\dagger$, $c_{p',k'}$ of the TL model, and Eq. (105) splits into four terms

$$\rho(x) = \sum_{pp'} \frac{1}{L} \sum_{kk'} e^{-i(k-k')x} \left( c_{p,k}^\dagger c_{p',k'} - \langle c_{p,k}^\dagger c_{p',k'} \rangle_0 \right)$$

$$= \sum_{pp'} \left( \Psi_p^\dagger(x)\Psi_{p'}(x) - \langle \Psi_p^\dagger(x)\Psi_{p'}(x)\rangle_0 \right), \tag{106}$$

where we have used Eq. (45) to obtain the second line. Now, the diagonal terms ($p = p'$) just give the densities (48) of $p$–movers, while for the non–diagonal terms we employ the representation (85) to find

$$\Psi_p^\dagger(x)\Psi_{-p}(x) = \frac{1}{2\pi a} e^{-2ipk_F x} e^{-2ip\sqrt{\pi}\vartheta(x)} e^{-ip\frac{2\pi}{L}(\hat{N}_p+\hat{N}_{-p}+1)} U_p^\dagger U_{-p}, \tag{107}$$

where we have used the commutators (44) and the fact that $\vartheta(x)$ and $\phi(x)$ commute at the same position $x$, which is seen from Eq. (60). Note that in the presence of a cutoff the sign function in Eq. (60) takes the form of the right hand side of Eq. (56) and therefore vanishes for vanishing argument. The density (106) may thus be written as

$$\rho(x) = \rho_+(x) + \rho_-(x) + \rho_{2k_F}(x), \tag{108}$$

where

$$\rho_{2k_F}(x) = \frac{1}{2\pi a} \left( e^{-2i[k_F x + \sqrt{\pi}\vartheta(x) + \frac{\pi}{L}(\hat{N}_p+\hat{N}_{-p}+1)]} U_+^\dagger U_- + \text{h.c.} \right). \tag{109}$$

Hence, the density operator for real electrons is not just the sum of the densities in the two branches of the TL model, but there is an additional $2k_F$-component $\rho_{2k_F}(x)$ which comes from the fact that right– and left–movers are propagating in the same channel and interfere.

In the limit $L \to \infty$ with constant average densities (100), we may introduce the shifted phase field (101) with $t = 0$ since the expression (108) gives the operator in the Schrödinger picture. We then have in view of Eq. (102)

$$\rho(x) = \frac{1}{\sqrt{\pi}} \frac{\partial}{\partial x} \vartheta(x) + \rho_{2k_F}(x), \tag{110}$$

where we have again suppressed the prime on $\vartheta$ which now contains the terms in the exponent of Eq. (109) that depend on the particle numbers. The operators $U_p^\dagger U_{-p}$ in Eq. (109) associated with the scattering of an electron from the $(-p)$– into the $p$–branch can often be suppressed since changes of the particle numbers $N_p$ by 1 can be neglected for $L \to \infty$. Then $\rho_{2k_F}(x)$ takes the simple form

$$\rho_{2k_F}(x) = \frac{k_F}{\pi} \cos\left[2k_F x + 2\sqrt{\pi}\vartheta(x)\right], \tag{111}$$

where we have chosen a cutoff length $a = k_F^{-1}$ of order a typical microscopic length.

## 2.9. Fermions with spin

We now briefly summarize the modifications necessary to include the electron spin. Then, the electron spectrum has two branches $s = \uparrow, \downarrow$ for spin up and spin down particles. For each species $s$, we may proceed exactly as for spinless fermions and define Bosonic phase fields $\vartheta_s(x)$ and $\phi_s(x)$. There are now four branches of the TL model and correspondingly four particle number operators $\hat{N}_{p,s}$. When defining the ladder operators, we have to include appropriate sign factors in the generalization of Eq. (40) to assure anti-commutation relations

$$[U_{p,s}, U_{p',s'}]_+ = 2\delta_{p,p'}\delta_{s,s'}(U_{p,s})^2$$

$$[U^\dagger_{p,s}, U^\dagger_{p',s'}]_+ = 2\delta_{p,p'}\delta_{s,s'}\left(U^\dagger_{p,s}\right)^2 \quad (112)$$

$$[U_{p,s}, U^\dagger_{p',s'}]_+ = 2\delta_{p,p'}\delta_{s,s'}$$

that extend the relations (44) to the case of four branches. The Bose representation (85) of the Fermi operators takes again the same form for each species $s$, i.e.,

$$\Psi_{p,s}(x) = \frac{1}{\sqrt{2\pi a}} U_{p,s} e^{ip[k_F x + \frac{2\pi}{L}\hat{N}_{p,s} x + \sqrt{\pi}\vartheta_s(x)] + i\sqrt{\pi}\phi_s(x)}. \quad (113)$$

It is often convenient to transform to the phase fields

$$\vartheta_\rho(x) = \frac{1}{\sqrt{2}}[\vartheta_\uparrow(x) + \vartheta_\downarrow(x)]$$

$$\phi_\rho(x) = \frac{1}{\sqrt{2}}[\phi_\uparrow(x) + \phi_\downarrow(x)] \quad (114)$$

and

$$\vartheta_\sigma(x) = \frac{1}{\sqrt{2}}[\vartheta_\uparrow(x) - \vartheta_\downarrow(x)]$$

$$\phi_\sigma(x) = \frac{1}{\sqrt{2}}[\phi_\uparrow(x) - \phi_\downarrow(x)] \quad (115)$$

that satisfy the commutation relations

$$[\vartheta_\rho(x), \vartheta_\rho(x')]_- = [\vartheta_\sigma(x), \vartheta_\sigma(x')]_- = 0$$

$$[\phi_\rho(x), \phi_\rho(x')]_- = [\phi_\sigma(x), \phi_\sigma(x')]_- = 0$$

$$[\vartheta_\rho(x), \vartheta_\sigma(x')]_- = [\phi_\rho(x), \phi_\sigma(x')]_- = 0 \qquad (116)$$

$$[\vartheta_\rho(x), \phi_\sigma(x')]_- = 0$$

and

$$[\vartheta_\rho(x), \phi_\rho(x')] = [\vartheta_\sigma(x), \phi_\sigma(x')] = \frac{i}{2}\,\text{sign}(x - x'). \qquad (117)$$

These fields describe charge and spin density excitations. In particular, instead of Eq. (63) we now have

$$\sqrt{\frac{2}{\pi}}\frac{\partial}{\partial x}\vartheta_\rho(x) = \rho_{+,\uparrow}(x) + \rho_{+,\downarrow}(x) + \rho_{-,\uparrow}(x) + \rho_{-,\downarrow}(x) - \frac{\hat{N}_{+,\uparrow} + \hat{N}_{+,\downarrow} + \hat{N}_{-,\uparrow} + \hat{N}_{-,\downarrow}}{L}, \qquad (118)$$

which gives the fluctuations of the total particle density of right– and left–movers, while

$$\sqrt{\frac{2}{\pi}}\frac{\partial}{\partial x}\vartheta_\sigma(x) = \rho_{+,\uparrow}(x) + \rho_{-,\uparrow}(x) - \rho_{+,\downarrow}(x) - \rho_{-,\downarrow}(x) - \frac{\hat{N}_{+,\uparrow} + \hat{N}_{-,\uparrow} - \hat{N}_{+,\downarrow} - \hat{N}_{-,\downarrow}}{L} \qquad (119)$$

determines fluctuations of the spin density. In the limit $L \to \infty$ with given average densities

$$\bar{\rho}_{p,s} = \frac{N_{p,s}}{L} \qquad (120)$$

we may again introduce shifted phase fields $\vartheta'_s$ for each species $s$ according to Eq. (101) and then obtain the classical action functional of the noninteracting model with dispersionless energy spectrum as

$$S_0 = \frac{\hbar}{2}\sum_\alpha \int dt \int dx \left[\frac{1}{v_F}\left(\frac{\partial \vartheta_\alpha}{\partial t}\right)^2 - v_F\left(\frac{\partial \vartheta_\alpha}{\partial x}\right)^2\right], \qquad (121)$$

where the sum is over the two spin directions $\alpha = \uparrow, \downarrow$. Then, $S_0$ is the sum of the actions (104) for each species $s$. The expression (121) remains, however, also valid if the sum is over $\alpha = \rho, \sigma$ with the phase fields introduced in Eqs. (114) and (115). Hence, the action functional for spinful electrons can be split in a charge and spin contribution

$$S_0 = S_{0,\rho} + S_{0,\sigma}, \qquad (122)$$

where each term has the form (104) of the action of a harmonic string.

The representation of the true electron density operator (105) in terms of the Bosonic phase fields has also a straightforward extension to the spinful case. Here we give explicitly only the generalization of the expression (110) valid in the limit $L \to \infty$. One now obtains

$$\rho(x) = \sqrt{\frac{2}{\pi}} \frac{\partial}{\partial x} \vartheta_\rho(x) + \rho_{2k_F}(x), \qquad (123)$$

where

$$\rho_{2k_F}(x) = \frac{2k_F}{\pi} \cos\left[2k_F x + \sqrt{2\pi}\vartheta_\rho(x)\right] \cos\left[\sqrt{2\pi}\vartheta_\sigma(x)\right] \qquad (124)$$

is the $2k_F$-contribution from the interference between right- and left-movers. Again we have suppressed the ladder operators that need to be taken into account in general.

## 3.  Interaction, Voltage Bias, and Impurities in the Tomonaga–Luttinger Model

In the previous section we have investigated noninteracting Fermions in a single channel quantum wire and seen that the low energy physics of the Fermi gas can be described either in terms of occupation numbers of single electron states or as excitations of Bosonic density waves. In higher dimensions noninteracting quasiparticles, supplemented by an electroneutrality constraint, give a rather accurate description also of the system in presence of Coulomb interaction, provided some parameters are replaced by effective parameters [24]. It has been known since quite some time that in one dimension the Fermi liquid description breaks down. Remarkably, the interaction leads to rather moderate modifications of the action in the Bose representation. The real advantages of Bosonization will thus only emerge in this section where we first present the real, interacting TL model and then extend it to describe a quantum wire with a scatterer in presence of an applied voltage.

### 3.1.  Electron-electron interaction and the Tomonaga–Luttinger model

We now take the interaction into account but restrict ourselves to spinless Fermions first. The electronic charge density in a quantum wire is largely compensated by a homogeneous positive background charge density. Therefore, we now fix the Fermi wave number $k_F$ so that the state $|N_+ = 0, N_- = 0\rangle_0$, where all single particle states of $p$-movers with $pk \leq k_F$ are occupied, is electrically neutral. The density $\rho(x)$

introduced in Eq. (105) multiplied by the electron change $e$ is then the excess charge density in the wire, and the interaction may be written

$$H_{\text{int}} = \frac{1}{2} \int dx \int dy \, \rho(x) \, U(x-y) \, \rho(y) \,. \tag{125}$$

The electron–electron interaction potential [4]

$$U(x) = \frac{1}{L} \sum_q U_q \, e^{iqx} \tag{126}$$

has real Fourier components $U_q$ since $U(-x) = U(x)$. Usually, the potential $U(x)$ deviates from a simple Coulomb potential both at small and large distances. At small distances of order the lateral dimensions of the quantum wire, the wave function for the transversal motion of the strictly speaking three–dimensional particles becomes relevant. When these transversal components are integrated out, the resulting effective potential $U(x)$ in the one–dimensional model remains finite for $x \to 0$. On the other hand, at large distances one has to take into account the effect of gate electrodes and other nearby conductors that screen the long–range part of the Coulomb interaction. The effective potential $U(x)$ then has a finite range $R$, which implies that the logarithmic increase for small $q$ of the Fourier transformed Coulomb potential is cut off at $q$–values of order $R^{-1}$. In particular, $U_{q=0}$ then remains finite, with the precise value depending on the geometry of the problem. We assume that this externally screened potential is still sufficiently long ranged so that $U_0 \gg U_{2k_F}$. Then the $2k_F$–component of the electronic density (110) will give a negligible contribution when inserted in the interaction (125), and we find for the Bosonized interaction energy

$$H_{\text{int}} = \frac{1}{2\pi} \int dx \int dy \, \frac{\partial \vartheta(x)}{\partial x} \, U(x-y) \, \frac{\partial \vartheta(y)}{\partial y} \,. \tag{127}$$

When we restrict ourselves to low energy excitations with wavelengths large compared to the range $R$ of the interaction potential, we may replace $U(x)$ by a local interaction $U_0 \, \delta(x)$, and then obtain the interaction term of the TL model

$$H_{\text{int}} = \frac{U_0}{2\pi} \int dx \left( \frac{\partial \vartheta}{\partial x} \right)^2 \,. \tag{128}$$

We remark that a microscopic local interaction $\sim \delta(x)$ would of course have no effect on spinless Fermions as a consequence of the anti–commutation rules. Essentially, in Eq. (128) one neglects the wave number

---

[4] In the limit $L \to \infty$ the sum $\frac{1}{L} \sum_q$ is replaced by $\frac{1}{2\pi} \int dq$.

dependence of $U_q$ for small $q$, but still $U_{2k_F} \ll U_0$. The interaction term (128) can readily be put in the action functional (104), which for the interacting model takes the form

$$S_\rho = \frac{\hbar}{2} \int dt \int dx \left[ \frac{1}{v_F} \left( \frac{\partial \vartheta}{\partial t} \right)^2 - v_F \left( 1 + \frac{U_0}{\pi \hbar v_F} \right) \left( \frac{\partial \vartheta}{\partial x} \right)^2 \right] \qquad (129)$$

and thus remains of the form of the action of a harmonic string. This clearly shows the great advantage of the Bose representation: We still have a model of free Bosonic charge density excitations, only the wave velocity is altered by the interaction, while in the Fermi representation the interaction leads to quartic terms in the Fermi operators.

It is customary to introduce the coupling constant

$$g = \left( 1 + \frac{U_0}{\pi \hbar v_F} \right)^{-1/2} \qquad (130)$$

and the charge density wave velocity

$$v = \frac{v_F}{g}. \qquad (131)$$

In terms of these quantities the action functional of the TL model reads

$$S_\rho = \frac{\hbar}{2g} \int dt \int dx \left[ \frac{1}{v} \left( \frac{\partial \vartheta}{\partial t} \right)^2 - v \left( \frac{\partial \vartheta}{\partial x} \right)^2 \right]. \qquad (132)$$

It would go beyond the scope of this article to demonstrate that the action (132) indeed describes the low energy properties of spinless one-dimensional Fermions correctly. Here we refer to the literature [8, 15]. We would, however, like to point out that the matter is in fact more complex than what our plausible "derivation" of Eq. (132) might suggest. The Coulomb interaction is strong and affects all states not only those near the two Fermi points. Hence, it needs to be introduced in the underlying physical model with a single energy dispersion curve $\varepsilon_k$. Afterwards, one may integrate out states far from the Fermi points until one reaches energy scales sufficiently close to the Fermi energy to allow for a linearization of the spectrum. At this point the Hamiltonian may be re-written in terms of the right- and left-movers of the TL model, but the parameters of the model are then already renormalized by the aforementioned elimination of high energy excitations and additional interaction vertices are generated. One can, however, conclude from a renormalization group study [7] that the action (132) is indeed a low energy fixpoint of a (spin-polarized) quantum wire. From these remarks it is clear that the parameter $U_0$ in Eq. (128) does not necessarily coincide with the Fourier coefficient at $q = 0$ of the interaction potential. This latter quantity can be considered an estimate of $U_0$ which becomes more accurate for large electron densities at which the effect of the Coulomb interaction is weaker. In the sequel we use $g$ and $v$ as fundamental parameters of the model. For repulsive interaction $U_0 > 0$ and thus $g < 1$.

## 3.2. Screening of external charges

Let us assume that we perturb the quantum wire by an external charge density

$$\rho_{\text{ext}}(x,t) = eQ(x,t), \tag{133}$$

which interacts with the electronic charge density $e\rho(x,t)$ via the same effective potential $U(x)$ introduced in the previous section. We are interested in the long wavelength response of the quantum wire and may thus disregard the $2k_F$-component of the electronic charge density. With the local approximation $U_0\,\delta(x)$, the action (132) is then modified to read

$$S = S_\rho - \frac{U_0}{\sqrt{\pi}} \int dt \int dx\, Q(x,t) \frac{\partial}{\partial x} \vartheta(x,t). \tag{134}$$

Since the action has only terms linear and quadratic in $\vartheta$, the average electronic density $\langle \rho(x,t) \rangle$ caused by $Q(x,t)$ can be determined from the phase field $\bar{\vartheta}(x,t)$ minimizing the action (134). The equation of motion

$$\left( \frac{\partial^2}{\partial t^2} - v^2 \frac{\partial^2}{\partial x^2} \right) \bar{\vartheta}(x,t) = \frac{gvU_0}{\sqrt{\pi}\hbar} \frac{\partial}{\partial x} Q(x,t) \tag{135}$$

obeyed by the minimal action field is readily solved in terms of the Fourier representation

$$\bar{\vartheta}(x,t) = \frac{1}{(2\pi)^2} \int dq \int d\omega\, \tilde{\vartheta}(q,\omega)\, e^{iqx - i\omega t}. \tag{136}$$

We find

$$\tilde{\vartheta}(q,\omega) = -\frac{gU_0}{\sqrt{\pi}\hbar} \frac{ivq\,\tilde{Q}(q,\omega)}{\omega^2 - v^2 q^2}, \tag{137}$$

which yields for the electronic density $\langle \rho(x,t) \rangle = \frac{1}{\sqrt{\pi}} \frac{\partial}{\partial x} \bar{\vartheta}(x,t)$ in Fourier space

$$\langle \tilde{\rho}(q,\omega) \rangle = (1 - g^2) \frac{v^2 q^2 \tilde{Q}(q,\omega)}{\omega^2 - v^2 q^2}, \tag{138}$$

where we have expressed $U_0$ in terms of $g$ and $v$ by means of Eqs. (130) and (131).

Now, the relation between the external charge density $eQ(x,t)$ and the resulting screening charge density $e\langle\rho(x,t)\rangle$ is governed by the dielectric function

$$\varepsilon(q,\omega) = \frac{\tilde{Q}(q,\omega)}{\tilde{Q}(q,\omega) + \langle\tilde{\rho}(q,\omega)\rangle}. \tag{139}$$

Combining this with Eq. (138) we find

$$\varepsilon(q,\omega) = \frac{\omega^2 - v^2 q^2}{\omega^2 - g^2 v^2 q^2}. \tag{140}$$

In particular, in the static case $\omega = 0$ we have

$$\varepsilon = \frac{1}{g^2} = 1 + \frac{U_0}{\pi \hbar v_F}, \tag{141}$$

which shows that the interaction parameter $g$ is directly related to the dielectric constant of the quantum wire. In a metallic system the dielectric function has a pole for $\omega \to 0$, $q \to 0$ associated with the perfect screening of static charges leading to electroneutrality. However, in the TL model the long range part of the Coulomb interaction is assumed to be screened by other conductors as explained above. Then $U_0$ is finite and there is a finite dielectric constant in the zero frequency and long wavelength limits.

This is in accordance with the fact that the total screening charge in units of $e$

$$Q_s = \int dx \, \langle\rho(x)\rangle = \langle\tilde{\rho}(q=0)\rangle \tag{142}$$

accumulated near a static impurity charge $Q$ follows from Eq. (138) as [25]

$$Q_s = -(1-g^2)Q. \tag{143}$$

Hence a fraction $g^2 Q$ of the external charge remains unscreened, and the quantum wire is in general not electroneutral. As we will discuss in greater detail in the next section, the charge $g^2 Q$ is screened by the electrode responsible for the finite range of the interaction. Formally, the limit of long range Coulomb interaction corresponds to $g \to 0$ which implies electroneutrality of the wire.

We mention that apart from the long wavelength response of the quantum wire to an impurity charge there is also a $2k_F$-response leading to Friedel oscillations of the charge density. We will not discuss this here but refer to the recent literature [26, 27].

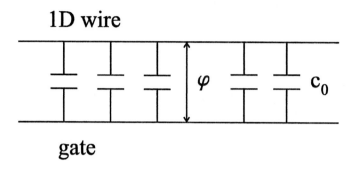

*Figure 1.6.* Electrostatic model of a quantum wire coupled by a capacitance per unit length $c_0$ to a gate. In a nonequilibrium wire the band bottom is shifted by an electric potential $\varphi$.

### 3.3. Electrostatics of a quantum wire

As we have seen in the preceeding section, the electrode screening the long range part of the Coulomb interaction plays an important role in the electrostatic response of a quantum wire to external charges. We can visualize the TL model as a one–dimensional quantum wire screened by a gate coupled to the wire by a distributed capacitance as depicted in Fig. 1.6. The interaction energy (128) can then be interpreted as the charging energy of the wire–gate capacitance

$$H_{\text{int}} = \frac{U_0}{2} \int dx\, \rho(x)^2 = \int dx\, \frac{e^2 \rho(x)^2}{2c_0}, \qquad (144)$$

where the capacitance per unit length $c_0$ is determined by

$$\frac{e^2}{c_0} = U_0 = \pi \hbar v_F \left(\frac{1}{g^2} - 1\right). \qquad (145)$$

An electronic charge density $e\rho(x)$ in the wire polarizes the capacitance, and the resulting electric potential

$$\varphi(x) = \frac{e\rho(x)}{c_0} \qquad (146)$$

shifts the band bottom of the quantum wire. This is directly related to the underscreening of external charges discussed previously as can be seen from the following consideration. In a noninteracting system an increase of the Fermi energy by $\Delta E$ shifts the wave number of the Fermi points from $\pm k_F$ to $\pm(k_F + \Delta k)$ where $\Delta k = \Delta E/\hbar v_F$. There are $(L/2\pi)\Delta k$ single particle states in the wave number interval $\Delta k$ and thus the density of $p$-movers increases by

$$\Delta\rho_p = \frac{\Delta k}{2\pi} = \frac{\Delta E}{2\pi\hbar v_F}. \tag{147}$$

Here the factor $1/2\pi\hbar v_F$ gives the density of states of $p$-movers at the Fermi energy. Accordingly, the electronic density increases by

$$\Delta\rho = \Delta\rho_+ + \Delta\rho_- = \frac{\Delta E}{\pi\hbar v_F}. \tag{148}$$

On the other hand, in an interacting, electroneutral system the shift of the Fermi energy by $\Delta E$ is accompanied by a shift of the band bottom by the same amount and the electronic density remains unchanged. In the TL model the situation is in between these two extremal cases. The band bottom is shifted by $e\varphi$ where $\varphi$ is the electric potential difference between the wire and the gate electrode. Then, the change of the electronic density is determined by

$$\Delta\rho = \frac{\Delta E - e\varphi}{\pi\hbar v_F}. \tag{149}$$

In view of the relations (145) and (146) we have

$$e\varphi = \frac{e^2}{c_0}\Delta\rho = \pi\hbar v_F\left(\frac{1}{g^2} - 1\right)\Delta\rho, \tag{150}$$

which gives

$$\Delta\rho = \frac{g^2\Delta E}{\pi\hbar v_F}. \tag{151}$$

Hence, we see again that a fraction $g^2$ of the "bare" charge density (148) caused by the shift of the Fermi level persists as true electronic charge density in the wire.

## 3.4. Voltage bias and boundary conditions

It is not difficult to generalize the preceeding considerations to a nonequilibrium quantum wire in presence of an applied voltage. Let us consider a quantum wire which is attached at the ends to two- or three-dimensional Fermi liquid reservoirs. We assume that the contacts between the wire and the reservoirs are adiabatic, which means that at the ends the quantum wire widens sufficiently slowly to avoid any backscattering of outgoing particles into the wire. This is the usual assumption underlying Landauer's approach [16] to the conductance of mesoscopic wires. If the contacts are not adiabatic, there will be an additional resistance depending on the precise realization of the contacts and not only on intrinsic properties of the quantum wire.

In equilibrium there is an equal amount of right– and left–movers in the wire and the electrochemical potential is constant. When we attach the wire to reservoirs, the influx of right–movers at the left end of the wire will depend on the electrochemical potential of the left electrode which we assume to be $eU_L$ above the Fermi energy of the equilibrium quantum wire. $U_L$ is then the voltage between the left reservoir and the gate electrode screening the wire. In the absence of interactions, the shift of the Fermi energy by $eU_L$ would increase the density of right–movers near the left end of the wire by

$$\rho_+^{\text{bare}} = \frac{eU_L}{2\pi\hbar v_F} \tag{152}$$

as depicted schematically in Fig. (1.7)a. Note that below we will consider quantum wires with impurities. Then, the reservoir determines the density of incoming particles only in the clean section of the wire near the end, where the incoming particles have not yet interacted with the impurities. The density of outflowing particles, on the other hand, will be affected by impurities.

In the presence of Coulomb interaction the excess charge density caused by the reservoirs will charge the distributed wire–gate capacitance leading to a shift of the band bottom by

$$e\varphi = \frac{e^2}{c_0}\rho = \pi\hbar v_F \left(\frac{1}{g^2} - 1\right)\rho, \tag{153}$$

where $e\rho$ is the true charge density in the wire that has to be determined selfconsistently. The true density of right–movers near the left end of the wire is then, cf. Fig. (1.7)b,

$$\rho_+ = \frac{e(U_L - \varphi)}{2\pi\hbar v_F}, \tag{154}$$

which means that the bare charge density (152) is partially screened.

In terms of the Bosonic phase field $\vartheta$, we have by virtue of the relations (102)

$$\rho_+ = \frac{1}{2\sqrt{\pi}}\left(\frac{\partial\vartheta}{\partial x} - \frac{1}{v_F}\frac{\partial\vartheta}{\partial t}\right), \tag{155}$$

where we have omitted the prime on $\vartheta$ as in all equations following (102). We remark that in the relation

$$\frac{1}{\sqrt{\pi}}\frac{\partial\vartheta}{\partial t} = -v_F(\rho_+ - \rho_-) \tag{156}$$

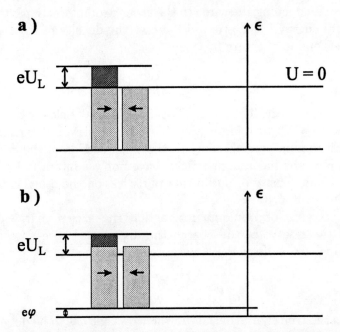

*Figure 1.7.* Density of right–movers in a quantum wire attached at the left end to a reservoir with voltage $U_L$ against the gate electrode. The equilibrium density is represented as light-grey area and the nonequilibrium excess density as dark-grey area. a) Sketch of the situation in the absence of Coulomb effects and b) with interaction taking the shift $e\varphi$ of the band bottom into account.

the Fermi velocity $v_F$ is not altered by the Coulomb interaction such as the velocity of charge density waves, because the microscopic expression of the particle current operator in terms of Fermi operators is independent of the interaction in the absence of a vector potential [28]. For the electric potential $\varphi$ we obtain from Eqs. (153) and (102)

$$e\varphi = \sqrt{\pi}\hbar v_F \left(\frac{1}{g^2} - 1\right) \frac{\partial \vartheta}{\partial x}, \tag{157}$$

which can be combined with the relations (154) and (155) to give for $x$ near the left end of the wire

$$\frac{1}{g^2}\frac{\partial \vartheta}{\partial x} - \frac{1}{v_F}\frac{\partial \vartheta}{\partial t} = \frac{eU_L}{\sqrt{\pi}\hbar v_F}. \tag{158}$$

Similar considerations hold for the density of left–movers near the right end of the wire attached to an electrode at potential $U_R$. Strictly speaking, all of the above considerations hold for the average densities of right– and left–movers. Hence, we find that the coupling to reservoirs can be described in the Bosonized TL model in terms of radiative boundary conditions for the phase field [17]

$$\left(\frac{1}{g^2}\frac{\partial}{\partial x} - \frac{1}{v_F}\frac{\partial}{\partial t}\right) \langle \vartheta(x,t) \rangle_{x=-\frac{L}{2}} = \frac{eU_L}{\sqrt{\pi}\hbar v_F},$$

$$\left(\frac{1}{g^2}\frac{\partial}{\partial x} + \frac{1}{v_F}\frac{\partial}{\partial t}\right) \langle \vartheta(x,t) \rangle_{x=\frac{L}{2}} = \frac{eU_R}{\sqrt{\pi}\hbar v_F}. \tag{159}$$

If we impose these conditions at a point near the reservoirs, they will be seen to be obeyed at any point in the impurity-free clean sections at the ends of the wire.

## 3.5. Impurity potential

Nontrivial dc transport properties of the quantum wire do not arise from the Coulomb interaction and the resulting non–Fermi liquid behavior alone but only in connection with impurities causing backscattering of electrons. A nonmagnetic impurity at position $x_0$ couples to the charge density and gives rise to an energy

$$H_W = \int dx\, W(x - x_0)\, \rho(x), \tag{160}$$

where $W(x)$ is the impurity potential. When we want to add this coupling term to the TL model, we have to note that the perturbation affects all states, also those far from the Fermi points, and we should again start from the underlying physical model with a single branch. Applying essentially the same line of reasoning used to include electron–electron interactions in the TL model, we obtain for the case of an interaction with an impurity at position $x_0$

$$H_W = \frac{W_0}{\sqrt{\pi}} \frac{\partial \vartheta}{\partial x}\bigg|_{x=x_0} + \frac{W_{2k_F} k_F}{\pi} \cos\left[2k_F x_0 + 2\sqrt{\pi}\vartheta(x_0)\right]. \quad (161)$$

Of course, this form results when we insert the representation (110) of the electronic density into Eq. (160) and then replace the Fourier coefficients $W_q$ of the impurity potential by constants for $q \approx 0$ and $|q| \approx 2k_F$, respectively. Since an elimination of states far from the Fermi points is necessary before we can rewrite the Hamiltonian in terms of the Fermi operators $c_{p,k}, c^\dagger_{p,k}$ of the TL model, the coefficients $W_0$ and $W_{2k_F}$ must be interpreted as effective parameters that are not directly related to Fourier coefficients of the microscopic interaction potential. In addition, the elimination of states will generate higher order processes involving several electrons with momentum transfer $4k_F$, $6k_F$, and so on. It turns out that the $2k_F$-processes dominate at low temperatures [18].

Although an impurity is typically laterally displaced from the center of the wire, the potential $W(x)$ will essentially have the same behavior as the electron–electron potential $U(x)$ discussed previously. Thus, from the bare Fourier components we would conclude $W_{2k_F} \ll W_0$. But, as we shall see, the forward scattering $W_0$, which scatters a $p$-mover into another state of the same branch, has no effect on the transport properties of the wire, while even a small backscattering term $W_{2k_F}$ has a dramatic effect at low energy scales [18].

To show that forward scattering is unimportant, we write the TL Hamiltonian in presence of an impurity in the form

$$H = \frac{\hbar v_F}{2} \int dx \left[\left(\frac{\partial \phi}{\partial x}\right)^2 + \left(\frac{\partial \vartheta}{\partial x}\right)^2\right] + \frac{U_0}{2\pi} \int dx \left(\frac{\partial \vartheta}{\partial x}\right)^2$$

$$+ \frac{W_0}{\sqrt{\pi}} \frac{\partial \vartheta}{\partial x}\bigg|_{x=x_0} + \frac{W_{2k_F} k_F}{\pi} \cos\left[2k_F x_0 + 2\sqrt{\pi}\vartheta(x_0)\right], \quad (162)$$

which includes the Hamiltonian (95) of the noninteracting model, the interaction (128), and the impurity term (161). Now, the unitary transformation

$$\mathcal{U} = \exp\left[-i\sqrt{\pi} \int dx\, \alpha(x)\, \phi(x)\right] \quad (163)$$

shifts the charge density, since

$$\mathcal{U} \frac{\partial \vartheta}{\partial x} \mathcal{U}^{-1} = \frac{\partial \vartheta}{\partial x} + \sqrt{\pi}\, \alpha(x). \quad (164)$$

This is readily shown by considering the auxiliary function

$$F(s) = e^{isA} \frac{\partial \vartheta}{\partial x} e^{-isA}, \tag{165}$$

where $A = \sqrt{\pi} \int dx\, \alpha(x)\phi(x)$ is the exponent of $\mathcal{U}$. One then finds

$$\frac{\partial}{\partial s} F(s) = e^{isA}\, i\, [A,\, \partial\vartheta/\partial x]_-\, e^{-isA} = \sqrt{\pi}\,\alpha(x), \tag{166}$$

where the commutator is evaluated by means of Eq. (60). Since $F(0) = \partial\vartheta/\partial x$, we find $F(1) = \partial\vartheta/\partial x + \sqrt{\pi}\,\alpha(x)$ which is just the relation (164).

Based on the same commutator, one also finds

$$\mathcal{U}\, e^{2i\sqrt{\pi}\vartheta(x)}\, \mathcal{U}^{-1} = e^{2i\sqrt{\pi}\vartheta(x) + i\eta(x)}, \tag{167}$$

where

$$\eta(x) = \pi \int dy\, \text{sign}(x-y)\,\alpha(y). \tag{168}$$

To see this, we write the left hand side of Eq. (167) as $e^{iA}\, e^{iB}\, e^{-iA}$ with the operator $A$ introduced previously and $B = 2\sqrt{\pi}\vartheta(x)$. Then, using twice the relation (83), we find

$$e^{iA}\, e^{iB}\, e^{-iA} = e^{iA}\, e^{i(B-A)}\, e^{-\frac{1}{2}[A,B]_-} = e^{iB - [A,B]_-}, \tag{169}$$

which gives the transformation (167).

With the help of the relations (164) and (167), the Hamiltonian (162) is transformed into

$$\begin{aligned}
\mathcal{U} H \mathcal{U}^{-1} = &\frac{\hbar v_F}{2} \int dx \left[ \left(\frac{\partial \phi}{\partial x}\right)^2 + \frac{1}{g^2}\left(\frac{\partial \vartheta}{\partial x}\right)^2 \right] \\
&+ \frac{W_0}{\sqrt{\pi}} \frac{\partial \vartheta}{\partial x}\bigg|_{x=x_0} + \frac{\sqrt{\pi}\hbar v_F}{g^2} \int dx\, \alpha(x) \frac{\partial \vartheta}{\partial x} \\
&+ \frac{W_{2k_F} k_F}{\pi} \cos\left[2k_F x_0 + 2\sqrt{\pi}\vartheta(x_0) + \eta(x_0)\right],
\end{aligned} \tag{170}$$

where we have omitted terms that depend only on $\alpha(x)$ but not on the phase fields. Further, we have introduced the interaction parameter (130). Now, with the choice

$$\alpha(x) = -\frac{g^2 W_0}{\pi \hbar v_F}\, \delta(x - x_0) \tag{171}$$

the terms in the second line of Eq. (170) cancel, and we are left with the Hamiltonian

$$H' = \frac{\hbar v_F}{2} \int dx \left[ \left(\frac{\partial \phi}{\partial x}\right)^2 + \frac{1}{g^2}\left(\frac{\partial \vartheta}{\partial x}\right)^2 \right] \quad (172)$$

$$+ \frac{W_{2k_F} k_F}{\pi} \cos\left[2k_F x_0 + 2\sqrt{\pi}\vartheta(x_0)\right],$$

which contains only a backscattering term. Note that for an impurity with charge $eQ$ giving rise to the Coulomb potential $W(x) = QU(x)$, the quantity

$$\alpha(x) = -\frac{g^2 U_0}{\pi \hbar v_F} Q\, \delta(x - x_0) = -(1 - g^2) Q\, \delta(x - x_0) \quad (173)$$

is just the screening charge density (138) caused by the static impurity charge. The unitary transformation thus removes the screening cloud, which is the main effect of the forward scattering term. Below we will determine the current–voltage relation of a quantum wire in presence of a single impurity at position $x = 0$. This study will be based on the action

$$S = \frac{\hbar}{2g} \int dt \int dx \left[\frac{1}{v}\left(\frac{\partial \vartheta}{\partial t}\right)^2 - v\left(\frac{\partial \vartheta}{\partial x}\right)^2\right]$$

$$- \lambda \int dt\, \cos\left[2\sqrt{\pi}\vartheta(x=0, t)\right] \quad (174)$$

associated with the Hamiltonian (172), where $\lambda$ characterizes the impurity strength.

## 3.6. Interacting electrons with spin

Here we briefly summarize the changes necessary to include the electron spin. For noninteracting electrons we found that the action of spinful electrons can be split into a charge and a spin contribution. Since the Coulomb interaction couples only to the charge density, we might argue that the charge part of the action is modified by the interaction in the same way as for spinless electrons while the spin part remains unchanged. However, this would ignore the fact that the Coulomb interaction must be introduced in the underlying physical model and the transcription to TL Fermions can only be made after an elimination of high energy excitations. This has consequences also for the spin density waves, in particular, the spin wave velocity $v_\sigma$ becomes smaller than the

Fermi velocity $v_F$ ([29]). The low energy physics is then governed by the action

$$S = \frac{\hbar}{2g} \int dt \int dx \left[ \frac{1}{v} \left( \frac{\partial \vartheta_\rho}{\partial t} \right)^2 - v \left( \frac{\partial \vartheta_\rho}{\partial x} \right)^2 \right]$$
$$+ \frac{\hbar}{2} \int dt \int dx \left[ \frac{1}{v_\sigma} \left( \frac{\partial \vartheta_\sigma}{\partial t} \right)^2 - v_\sigma \left( \frac{\partial \vartheta_\sigma}{\partial x} \right)^2 \right].$$
(175)

The TL model is thus characterized by three parameters $v$, $v_\sigma$ and $g$. There is no coupling constant $g_\sigma$ for the spin sector, which can be traced back to spin rotation invariance [8, 15]. A detailed discussion of these issues would go beyond the scope of this article and we refer to the literature cited. As in the noninteracting case, the action splits into a charge and a spin part. However, the difference between the charge and spin wave velocities in an interacting wire has important consequences and leads to the notable phenomenon of spin–charge separation. When an electron is injected into a wire it causes a charge and a spin pulse propagating with different velocities.

In case an impurity is added at position $x = 0$ to the model we see from the $2k_F$–part of the electron density in the spinful case (124) that the important backscattering term now has the form

$$S_\lambda = -\lambda \int dt \, \cos\left[\sqrt{2\pi}\vartheta_\rho(x=0,t)\right] \cos\left[\sqrt{2\pi}\vartheta_\sigma(x=0,t)\right]. \quad (176)$$

Hence, the impurity couples the charge and spin sectors making the theory of dirty quantum wires in the absence of a spin–polarizing magnetic field more involved.

## 4. Current–voltage relation of a quantum wire

In this section we apply the theory developed so far and determine the current in a quantum wire with an impurity as a function of the applied voltage and the temperature. Rather than giving an overview of the results available in the literature, we treat a special case in some detail to illustrate how the formalism explained in the previous sections can be employed to obtain concrete results. With this background readers should then be prepared to embark on reading the recent original literature on the subject.

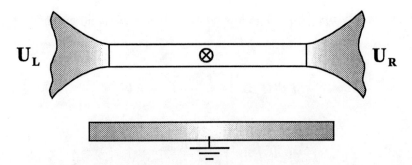

*Figure 1.8.* Gated quantum wire adiabatically connected to reservoirs at voltages $U_L$ and $U_R$. A backscatterer is located in the center of the wire.

## 4.1. Particular solution and four–terminal voltage

We study a single channel quantum wire with an impurity at position $x = 0$. The conductor is attached to reservoirs at voltages $U_L$ and $U_R$ relative to the gate electrode screening the wire as sketched in Fig. (1.8). The applied voltage

$$U = U_L - U_R \tag{177}$$

will then drive a current $I$ through the wire. For simplicity, we shall restrict ourselves to the case of spinless electrons. We can then base the consideration on the action functional (174) and the boundary conditions (159).

As discussed previously, the phase field $\vartheta(x,t)$ describes also the mean particle densities (100) due to a shift introduced in Eq. (101). To take care of the boundary conditions, we look for a particular field $\theta(x,t)$ that satisfies the wave equation in the clean parts of the wire and the boundary conditions. This solution is of the form of the shift in Eq. (101)

$$\theta(x,t) = \frac{g^2 e}{2\sqrt{\pi}\hbar v_F}[(U_L + U_R)x - V|x|] - \frac{e}{2\sqrt{\pi}\hbar}(U_L - U_R - V)t, \tag{178}$$

where $V$ is an arbitrary parameter. We now split the phase field $\vartheta(x,t)$ into

$$\vartheta(x,t) = \theta(x,t) + \varphi(x,t). \tag{179}$$

The deviation $\varphi(x,t)$ from the particular solution (178) will then obey equilibrium boundary conditions. Further, we fix the parameter $V$ by the requirement that in the steady nonequilibrium state

$$\frac{\partial}{\partial t}\langle \varphi(x,t)\rangle = 0. \tag{180}$$

Now, the average current $I$ may be written as

$$I = ev_F \langle \rho_+ - \rho_- \rangle = -\frac{e}{\sqrt{\pi}} \frac{\partial}{\partial t} \langle \vartheta \rangle, \qquad (181)$$

where we have used Eq. (102) to obtain the second equality. Hence, the condition (180) means that the average current is determined solely by the particular solution (178), and we have

$$I = \frac{e^2}{h}(U - V). \qquad (182)$$

Since in a steady state the current $I$ is independent of $x$, we may impose the condition (180) fixing the voltage $V$ at any point $x$.

The particular solution (178) not only determines the average current but also the average charge density in the nonequilibrium quantum wire. Since $\langle \varphi \rangle$ obeys equilibrium boundary conditions as well as the condition (180), it does not contribute to the average density (110)

$$\langle \rho \rangle = \frac{1}{\sqrt{\pi}} \frac{\partial}{\partial x} \langle \vartheta \rangle, \qquad (183)$$

where we have omitted the $2k_F$-component which gives an additional oscillatory contribution near the impurity. This Friedel oscillation component is not seen in a density smoothed over length scales of order $\lambda_F = 2\pi/k_F$. From the particular solution (178) we obtain for the average density

$$\langle \rho \rangle = \frac{g^2 e(U_L + U_R)}{2\pi \hbar v_F} - \frac{g^2 eV}{2\pi \hbar v_F} \text{sign}(x). \qquad (184)$$

The first term just describes the change (151) of the average electronic density as a consequence of the average shift $e(U_L + U_R)/2$ of the Fermi energy. This term is absent if the voltage $U$ is applied asymmetrically, i.e., for $U_L = -U_R = U/2$. The second term gives an asymmetric component of the charge density in presence of an impurity. The density drop

$$\Delta \rho = \frac{g^2 eV}{\pi \hbar v_F} \qquad (185)$$

across the impurity site is associated with a difference

$$\Delta \mu = g^2 eV \qquad (186)$$

of the effective chemical potential on both sides of the impurity. Furthermore, the drop of the charge density across the impurity site is also associated with a change of the electric potential (153) by

$$\Delta \varphi = \frac{\pi \hbar v_F}{e} \left( \frac{1}{g^2} - 1 \right) \Delta \rho = (1 - g^2) V. \qquad (187)$$

In a hypothetical ideal measurement of the voltage drop across the impurity site one would observe the difference of the electrochemical potential [16]

$$\Delta\varphi + \frac{1}{e}\Delta\mu = V. \qquad (188)$$

Hence, the parameter $V$ introduced above coincides with the average four-terminal voltage $V$, which is the part of the applied voltage $U$ dropping across the scatterer.[5]

In view of Eq. (182) the determination of the current–voltage relation corresponds to a calculation of the four-terminal voltage. Two limiting cases are evident from physical grounds. In the absence of a backscatterer ($\lambda \to 0$) we have $V = 0$ and obtain from Eq. (182)

$$I = G_0 U, \qquad (189)$$

where $G_0 = e^2/h$ is the conductance of a clean wire. This is the same result as obtained previously for noninteracting electrons. Hence, for a clean quantum wire with adiabatic contacts to the reservoirs the interaction has no effect on the conductance [30, 31, 32, 17]. On the other hand, for a very strong backscatterer ($\lambda \to \infty$) we have $V = U$ and the current $I$ vanishes. In the remainder we shall discuss how we get from one limit to the other.

## 4.2. Path integral on the Keldysh contour

To treat the nonequilibrium quantum wire with the action functional (174) quantum mechanically, we have to evaluate a Feynman path integral on the Keldysh contour. For an introduction to the Keldysh technique we refer to the review article [33], however, the basic idea can be understood in the following way. Assume that at time $t_0$ the system is described by the density matrix $W(t_0)$ and let $H$ be the Hamiltonian including the coupling to the reservoirs. The density matrix at a later time $t_f$ is then given by

$$W(t_f) = e^{-\frac{i}{\hbar}H(t_f-t_0))} W(t_0) e^{\frac{i}{\hbar}H(t_f-t_0)}. \qquad (190)$$

Each of the two time evolution operators $e^{\pm\frac{i}{\hbar}H(t_f-t_0)}$ may be written as a Feynman path integral. Since we are interested in steady state properties independent of the initial state $W(t_0)$, we take the limit $t_0 \to -\infty$. The

---

[5]The readers should be aware of the fact that in some early treatments of transport properties of the TL model the discrimination between $U$ and $V$ was not made.

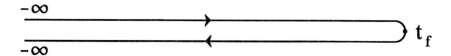

*Figure 1.9.* The Keldysh contour runs from $-\infty$ to $t_f$ along the real time axis and back to $-\infty$. The two branches of the contour correspond to the forward and backward propagators in Eq. (190).

trace over such a time propagated operator leads to a path integral of the form

$$Z_0 = \int \mathcal{D}[\vartheta]\, e^{\frac{i}{\hbar}S[\vartheta]}, \tag{191}$$

where $S[\vartheta]$ is the action functional (174) with the time integration $\int dt$ running along the Keldysh contour depicted in Fig. (1.9). In order to employ this path integral for the calculation of expectation values, we first decompose the phase field according to Eq. (179). The action (174) then takes the form

$$\begin{aligned}S = {} & \frac{\hbar}{2g}\int dt \int dx \left[\frac{1}{v}\left(\frac{\partial\varphi}{\partial t}\right)^2 - v\left(\frac{\partial\varphi}{\partial x}\right)^2\right] \\ & - \frac{eV}{\sqrt{\pi}}\int dt\, \varphi(0,t) \\ & - \lambda \int dt\, \cos\left[2\sqrt{\pi}\,\varphi(0,t) - \frac{e}{\hbar}(U-V)t\right],\end{aligned} \tag{192}$$

where we have omitted terms independent of $\varphi(x,t)$. Further, we have taken into account that

$$\int dt\, \frac{\partial\varphi}{\partial t} = 0 \tag{193}$$

for a time integral along the Keldysh contour and

$$\int dx\, \mathrm{sign}(x)\frac{\partial}{\partial x}\varphi(x,t) = -2\varphi(0,t). \tag{194}$$

For the phase field $\varphi(x,t)$, which obeys equilibrium boundary conditions, the nonequilibrium situation becomes apparent in two modifications of the action (192). The term in the second line comes from the voltage drop across the impurity site. It describes the potential energy

$$\int dx \left[-\frac{V}{2}\,\mathrm{sign}(x)\right]\frac{e}{\sqrt{\pi}}\frac{\partial\varphi(x,t)}{\partial x} = \frac{eV}{\sqrt{\pi}}\varphi(0,t) \tag{195}$$

of a charge density fluctuation $(e/\sqrt{\pi})\frac{\partial\varphi}{\partial x}$ in presence of an electrochemical potential $-(V/2)\,\mathrm{sign}(x)$. The remainder $U-V$ of the applied voltage

$U$ shows up in the third line of Eq. (192) as a Josephson–type phase shift in the pinning potential caused by the impurity.

Instead of the path integral (191) we now study the functional

$$Z[\eta] = \int \mathcal{D}[\varphi]\, e^{\frac{i}{\hbar}S[\varphi]+i\sqrt{\pi}\int dt\, \eta(t)\frac{\partial}{\partial t}\varphi(0,t)}, \qquad (196)$$

where $S[\varphi]$ is the action (192), and where we have introduced an auxiliary field $\eta(t)$ on the Keldysh contour. $Z[\eta]$ is a generating functional for expectation values of $\frac{\partial}{\partial t}\varphi(0,t)$. In particular, the condition (180) is now equivalent to

$$\left.\frac{\delta Z[\eta]}{\delta \eta(t)}\right|_{\eta=0} = 0, \qquad (197)$$

which needs to be evaluated to determine the four–terminal voltage $V$.

## 4.3. Coulomb gas representation

In the sequel we outline one of the methods available to determine the current–voltage relation. While for the one impurity problem considered here, an approach based on the thermodynamic Bethe ansatz is most powerful [34], we present here a technique which remains useful also for multi impurity problems. First, we split the Keldysh contour explicitly into the two branches and denote the phase field on the branch from $-\infty$ to $t_f$ by $\varphi(x,t)$ and the field on the branch from $t_f$ back to $-\infty$ by $\varphi'(x,t)$. Further, we split the action (192) into two terms

$$S_0 = \frac{\hbar}{2g}\int_{-\infty}^{t_f} dt \int dx \left[\frac{1}{v}\left(\frac{\partial\varphi}{\partial t}\right)^2 - v\left(\frac{\partial\varphi}{\partial x}\right)^2 - \frac{1}{v}\left(\frac{\partial\varphi'}{\partial t}\right)^2 + v\left(\frac{\partial\varphi'}{\partial x}\right)^2\right]$$
$$\qquad (198)$$
$$-\frac{eV}{\sqrt{\pi}}\int_{-\infty}^{t_f} dt\, [\varphi(0,t) - \varphi'(0,t)]$$

and

$$S_\lambda = -\lambda \int_{-\infty}^{t_f} dt\, \left\{\cos\left[2\sqrt{\pi}\varphi(0,t) - \frac{e}{\hbar}(U-V)t\right]\right.$$
$$\left. - \cos\left[2\sqrt{\pi}\varphi'(0,t) - \frac{e}{\hbar}(U-V)t\right]\right\}. \qquad (199)$$

With the help of the trigonometric relation

$$\cos\alpha - \cos\beta = -2\sin\frac{\alpha+\beta}{2}\sin\frac{\alpha-\beta}{2} \qquad (200)$$

the action $S_\lambda$ may be written as

$$S_\lambda = 2\lambda \int_{-\infty}^{t_f} dt \, \cos A(t) \sin B(t), \tag{201}$$

where

$$\begin{aligned}
A(t) &= \sqrt{\pi}\,[\varphi(0,t) + \varphi'(0,t)] - \frac{e}{\hbar}(U-V)t + \delta \\
B(t) &= \sqrt{\pi}\,[\varphi(0,t) - \varphi'(0,t)].
\end{aligned} \tag{202}$$

From Eqs. (199) and (200) we obtain $\delta = -\frac{\pi}{2}$, however, the precise value of this phase must be irrelevant, since we can always add a constant phase to the particular solution (178), e.g., by replacing $t$ by $t-t_0$. Such an additional phase of $\theta(x,t)$ leads to a shift of $\delta$.

Using Eq. (201) we find by expanding in powers of $\lambda$

$$e^{\frac{i}{\hbar}S_\lambda} = 1 + \sum_{n=1}^{\infty} \left(\frac{2i\lambda}{\hbar}\right)^n \int \mathcal{D}_n t \prod_{j=1}^{n} \cos A(t_j) \sin B(t_j), \tag{203}$$

where we have introduced the abbreviation

$$\int \mathcal{D}_n t = \int_{-\infty}^{t_f} dt_n \int_{-\infty}^{t_n} dt_{n-1} \cdots \int_{-\infty}^{t_2} dt_1. \tag{204}$$

Next, we write the trigonometric functions as

$$\begin{aligned}
\cos A(t_j) &= \frac{1}{2} \sum_{u_j = \pm} e^{i u_j A(t_j)}, \\
\sin B(t_j) &= \frac{1}{2i} \sum_{v_j = \pm} v_j \, e^{i v_j B(t_j)},
\end{aligned} \tag{205}$$

which gives

$$e^{\frac{i}{\hbar}S_\lambda} = 1 + \sum_{n=1}^{\infty} \sum_{\{u_j,v_j\}} \left(\prod_{j=1}^{n} \frac{\lambda v_j}{2\hbar}\right) \int \mathcal{D}_n t \, e^{\frac{i}{\hbar}S_n}, \tag{206}$$

where

$$\frac{1}{\hbar} S_n = \sum_{j=1}^{n} [u_j A(t_j) + v_j B(t_j)] \tag{207}$$

is linear in the phase fields $\varphi$, $\varphi'$ by virtue of Eq. (202). The benefit of this expansion is that now the path integral (196) is Gaussian order by order.

Therefore, we can integrate out the $\varphi$ and $\varphi'$ fields. Essentially, the calculation of these path integrals amounts to an explicit computation of the fields minimizing the action.

On a formal level we may consider the auxiliary variables $u_j$, $v_j$ as charges on the time axis that are coupled to a harmonic string described by the phase fields. The integration over the fields $\varphi$, $\varphi'$ then corresponds to the elimination of a harmonic bath in the context of dissipative quantum mechanics. This problem is well studied in the literature [35, 36, 37]. It would go beyond the scope of this article to present these methods explicitly here. Once the Gaussian integrals over the phase fields are carried out, the sum over the charges $v_j$ can be done straightforwardly, and the generating functional (196) is obtained in the form

$$Z[\eta] = \exp\left\{-\int dt \int^t dt'\, \eta(t)\, \ddot{C}(t-t')\, \eta(t') - ig\frac{e}{\hbar}V \int dt\, \eta(t)\right\} \quad (208)$$

$$\times \left(1 + \sum_{m=1}^{\infty} Z_m[\eta]\right),$$

where

$$Z_m[\eta] = \left(\frac{i\lambda}{\hbar}\right)^{2m} \int \mathcal{D}_{2m} t \sum_{\{u_j\}'} \exp\left(\sum_{j>k=1}^{2m} u_j u_k C(t_j - t_k) \right.$$

$$\left. + \sum_{j=1}^{2m} u_j \left[\int dt\, \eta(t)\dot{C}(t-t_j) - i\frac{e}{\hbar}(U - V + gV)t_j\right]\right)$$

$$\times \sin[\pi g \eta(t_{2m})] \prod_{j=1}^{2m-1} \sin\left(\pi g \left[\eta(t_j) + \sum_{k=j+1}^{2m} u_k\right]\right) \quad (209)$$

Here the sine functions arise from the sum over the charges $v_j$ by means of Eq. (205). The function

$$C(t) = 2g \ln\left[\frac{\beta \Delta}{\pi} \sinh\left(\frac{\pi |t|}{\hbar \beta}\right)\right] \quad (210)$$

describes an effective interaction between the charges $u_j$ whereby $\beta = 1/k_B T$ is the inverse temperature of the quantum wire and $\Delta$ is the cutoff energy of the harmonic string. The temperature emerges from the asymptotic conditions on the fluctuations of the phase field for $t \to -\infty$. To simplify notation we have chosen the same auxiliary field $\eta(t)$ on both branches of the Keldysh contour so that the second term in the exponent of (196) reads $\sqrt{\pi} \int_{-\infty}^{t_f} dt\, \eta(t) \frac{\partial}{\partial t}[\varphi(0,t)+\varphi'(0,t)]$. Furthermore, we have taken the limit $t_f \to \infty$.

In the series (208) only even terms in $\lambda$ survive, because $C(t)$ is a long range interaction which suppresses all terms that do not satisfy the charge neutrality condition

$$\sum_{j=1}^{n} u_j = 0. \tag{211}$$

Since the charges $u_j = \pm 1$, this condition can only hold for $n = 2m$. The constraint (211) is indicated as a prime at the sum over the $u_j$ in Eq. (209). Because of this condition the phase $\delta$ in Eq. (202) drops out. The representation (208) of the generating functional is known as the Coulomb gas representation of the problem, since some terms allude to the partition function of one–dimensional charges interacting with the "Coulomb potential" (210). Note that in view of the factors $\sin[\pi g \eta(t_{2m})]$ the generating functional (208) obeys the normalization $Z[\eta = 0] = 0$.

The four–terminal voltage $V$ can now be determined from the condition (197). A nonvanishing contribution of $Z_m[\eta]$ only arises if the variational derivative acts upon $\sin[\pi g \eta(t_{2m})]$. Introducing the time–difference variables

$$\tau_m = t_{m+1} - t_m \tag{212}$$

we find

$$\frac{e}{\hbar} V = K\left(\frac{e}{\hbar}[U - V + gV]\right), \tag{213}$$

where

$$K(\Omega) = \pi \operatorname{Im} \sum_{m=1}^{\infty} (-1)^m \left(\frac{\lambda}{\hbar}\right)^{2m} \int_0^{\infty} d\tau_1 \ldots \int_0^{\infty} d\tau_{2m-1} \tag{214}$$

$$\times \sum_{\{u_j\}'} \exp\left[\sum_{j>k=1}^{2m} u_j u_k C\left(\sum_{l=k}^{j-1} \tau_l\right)\right] \prod_{j=1}^{2m-1} \sin(\pi g p_j)\, e^{-i p_j \Omega \tau_j}.$$

Here

$$p_j = \sum_{k=j+1}^{2m} u_k = -\sum_{k=1}^{j} u_k \tag{215}$$

is the accumulated charge in the time interval $t > t_j$. Sums of the form (214) with an interaction (210) are again familiar from dissipative quantum mechanics [38]

## 4.4. Exact solution for $g = \frac{1}{2}$

The explicit evaluation of the sum (214) is greatly simplified in the special case $g = \frac{1}{2}$, since the sine–functions then suppress many terms. The so-called "collapsed blip approximation" [38] of the Coulomb gas problem becomes exact for $g = \frac{1}{2}$ and $K(\Omega)$ can easily be calculated explicitly. One finds

$$K(\Omega) = 2\frac{\lambda_B}{\hbar} \operatorname{Im}\psi\left(\frac{1}{2} + \frac{\beta[\lambda_B + i\hbar\Omega]}{2\pi}\right), \quad (216)$$

where $\psi(z)$ is the digamma function and

$$\lambda_B = \pi\frac{\lambda^2}{\Delta} \quad (217)$$

is a renormalized energy scale of the impurity. The four-terminal voltage $V$ then follows from Eq. (213), which for $g = \frac{1}{2}$ reads

$$\frac{e}{\hbar}V = K\left(\frac{e}{\hbar}[U - V/2]\right). \quad (218)$$

It can be easily seen that the solution is of the form

$$\frac{eV}{\lambda_B} = f\left(\frac{eU}{\lambda_B}, \frac{k_B T}{\lambda_B}\right). \quad (219)$$

This shows that the solution exhibits scaling. When the energies $eU$ and $k_B T$ are measured in units of the renormalized impurity energy scale $\lambda_B$, we find the same four-terminal voltage (in units of $\lambda_B$) for any impurity and hence also the same form of the current–voltage relation (182).

In Fig. (1.10) we depict the current–voltage curve for various temperatures. We see that the interaction has a dramatic effect on the $I$–$V$–characteristics in presence of a scatterer. While in the noninteracting model the conductance is essentially independent of $k_B T$ and $eU$ for small energy scales, there is a strong suppression of the conductance in the interacting system for energies below $\lambda_B$. At zero temperature

$$\frac{I}{I_0} = \frac{1}{24}\left(\frac{eU}{\lambda_B}\right)^2 \quad \text{for } eU \ll \lambda_B \quad (220)$$

and

$$\frac{I}{I_0} = 1 - \pi\frac{\lambda_B}{eU} \quad \text{for } eU \gg \lambda_B, \quad (221)$$

where $I_0 = (e^2/h)U$. Hence, the differential conductance shows a perfect zero bias anomaly and vanishes $\sim U^2$ for $U \to 0$. On the other hand, the

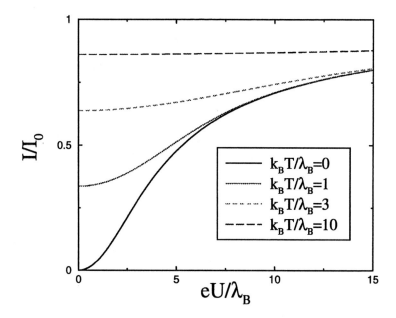

*Figure 1.10.* Current–voltage characteristics for several temperatures $T$. The current is normalized to $I_0 = (e^2/h)U$.

conductance approaches the value $e^2/h$ of a clean wire for large voltages. In practice, this latter limit will not be reached, since for large $U$ the low energy description in terms of the TL model breaks down. On the other hand for $T > 0$ the linear conductance for $U \to 0$ reads

$$G(T) = \frac{e^2}{h} \frac{\Theta - \psi'(\frac{1}{2} + \frac{1}{\Theta})}{\Theta + \psi'(\frac{1}{2} + \frac{1}{\Theta})}, \qquad (222)$$

where $\Theta = 2\pi k_B T/\lambda_B$. Hence, the conductance vanishes $\sim T^2$ for $k_B T \ll \lambda_B$. We mention that the exact solution for $g = \frac{1}{2}$ was first obtained by re–Fermionization techniques [17].

## 4.5. Concluding remarks

The explicit results for the current–voltage characteristics derived in the preceeding section show that for a spin–polarized quantum wire the crossover from the weak impurity problem at higher temperatures or larger applied voltages to the strong coupling problem at low temperatures and small voltages can be solved exactly. The scaling function can also be obtained for arbitrary $g < 1$ [34]. Mostly, one has to be content with more limited results. In the spinful case the current is known only asymptotically for large and small energy scales [18]. Several topics are still under investigation. For instance, in the presence of many impurities additional features such as resonant tunneling may arise. Another area of active research is the current noise in one-dimensional conductors.

We have focussed here on the physics of screened single channel quantum wires at low energy scales where the finite range of the interaction is unimportant. Many results can in fact be extended to finite range interactions. However, in the absence of a gate, the long range nature of the Coulomb interaction requires special attention [39]. There are many similarities between the transport properties of quantum wires and edge currents in fractional quantum Hall bars [40]. Since in these latter devices right- and left-moving branches are spatially separated, they are described by a chiral TL model [41, 42, 43] with somewhat different transport properties. The methods presented here are also useful to model the low energy electronic properties of carbon nanotubes [44, 45]. Again, due to features of the bandstructure, some differences arise leading to four channels where one channel is charged as in a semiconductor quantum wire. We hope that this article will facilitate the study of these and many other exciting new results on transport properties of one-dimensional Fermionic systems.

## Acknowledgments

The author wishes to thank Reinhold Egger for an enjoyable collaboration on the subject of this article in recent years. The manuscript has benefitted from a critical reading by Wolfgang Häusler, Jörg Rollbühler, and Björn Trauzettel. Financial support was provided by the Deutsche Forschungsgemeinschaft (Bonn).

## References

[1] A.R. Goñi et al. Phys. Rev. Lett. **67**, 3298 (1991).

[2] S. Tarucha, T. Honda, T. Saku, Solid State Comm. **94**, 413 (1995).

[3] A. Yacoby et al., Solid State Comm. **101**, 77 (1997); O.M. Auslaender et al., Phys. Rev. Lett. **84**, 1764 (2000).

[4] S.T. Tans et al., Nature **386**, 474 (1997); ibid. **394**, 761 (1998).

[5] M. Bockrath et al., Science, **275**, 1922 (1997); Nature **397**, 598 (1999).

[6] C. Schönenberger et al., Appl. Phys. A **69**, 283 (1999).

[7] J. Solyom, Adv. Phys. **28**, 201 (1979).

[8] J. Voit, Rep. Progr. Phys. **57**, 977 (1995).

[9] S. Tomonaga, Progr. Theor. Phys. **5**, 544 (1950).

[10] J.M. Luttinger, J. Math. Phys. **4**, 1154 (1963).
[11] K.D. Schotte, U. Schotte, Phys. Rev. **182**, 479 (1969).
[12] *Bosonization*, edited by D. Stone, World Scientific, Singapore (1994).
[13] F.D.M. Haldane, J. Phys. C **14**, 2585 (1981).
[14] J. von Delft, H. Schoeller, Ann. Phys. (Leipzig) **4**, 225 (1998).
[15] A.O. Gogolin, A.A. Nersesyan, A.M. Tsvelik, *Bosonization and Strongly Correlated Systems*, Cambridge University Press, Cambridge (1998).
[16] R. Landauer, IBM J. Res. Dev. **1**, 223 (1957); S. Datta, *Electronic Transport in Mesoscopic Systems*, Cambridge University Press, Cambridge (1995).
[17] R. Egger, H. Grabert, Phys. Rev. Lett. **77**, 538 (1996); *ibid.* **80**, 2255 (1998) (Erratum); Phys. Rev. B **58**, 10761 (1998).
[18] C.L. Kane, M.P.A. Fisher, Phys. Rev. Lett. **68**, 1220 (1992); Phys. Rev. B **46**, 15233 (1992).
[19] U. Fano, Phys. Rev. **124**, 1866 (1961).
[20] P.W. Anderson, Phys. Rev. **124**, 41 (1961).
[21] G.D. Mahan *Many-Particle Physics*, Plenum, New York (1990).
[22] M. Fabrizio, A.O. Gogolin, Phys. Rev. B **51**, 17827 (1995).
[23] C.L. Kane, M.P.A. Fisher, Phys. Rev. Lett. **76**, 3192 (1996).
[24] D. Pines, P. Nozières, *The Theory of Quantum Liquids*, W.A. Benjamin, New York (1966).
[25] R. Egger, H. Grabert Phys. Rev. Lett. **79**, 3463 (1997).
[26] R. Egger, H. Grabert Phys. Rev. Lett. **75**, 3505 (1995).
[27] A. Leclair, F. Lesage, H. Saleur, Phys. Rev. B **54**, 13597 (1996).
[28] see e.g. G. Baym, *Lectures on Quantum Mechanics*, Addison–Wesley, Reading (1974).
[29] C.E. Creffield, W. Häusler, A.H. MacDonald, Europhys. Lett. **53**, 221 (2001).
[30] D.L. Maslov, M. Stone, Phys. Rev. B **52**, R5539 (1995).
[31] V.V. Ponomarenko, Phys. Rev. B **52**, R8666 (1995).
[32] I. Safi, H.J. Schulz, Phys. Rev. B **52**, R17040 (1995).
[33] J. Rammer, H. Smith, Rev. Mod. Phys. **58**, 323 (1986).
[34] R. Egger *et al.*, Phys. Rev. Lett. **84**, 3682 (2000).
[35] R.P. Feynman, F.L. Vernon, Ann. Phys. (N.Y.) **24**, 118 (1963).

[36] A.O. Caldeira, A.J. Leggett, Ann. Phys. (N.Y.) **149**, 374 (1983).

[37] H. Grabert, P. Schramm, G.-L. Ingold, Phys. Rep. **168**, 115 (1988).

[38] U. Weiss, *Quantum Dissipative Systems*, World Scientific, Singapore (1999).

[39] H.J. Schulz, Phys. Rev. Lett. **71**, 1864 (1993).

[40] A.M. Chang, L.N. Pfeiffer, K.W. West, Phys. Rev. Lett. **77**, 2538 (1996).

[41] X.G. Wen, Phys. Rev. Lett. **64**, 2206 (1990); Phys. Rev. B **43**, 11025 (1991).

[42] C.L. Kane and M.P.A. Fisher, Phys. Rev. Lett. **72**, 724 (1994).

[43] P. Fendley, A.W.W. Ludwig, H. Saleur, Phys. Rev. B **52**, 8934 (1995).

[44] R. Egger, A.O. Gogolin, Phys. Rev. Lett. **79**, 5082 (1997); Eur. Phys. J. B **3**, 281 (1998).

[45] C.L. Kane, L. Balents, M.P.A. Fisher, Phys. Rev. Lett. **79**, 5086 (1997).

Chapter 2

# LECTURES ON CONFORMAL FIELD THEORY

Alexei M. Tsvelik
*Department of Physics, University of Oxford, and Brasenose College*

**Abstract**  An introduction to conformal field theory in two-dimensions is given.

**Keywords:** Gaussian model, conformal symmetry, finite size effects, conformal anomaly, Virasoro fusion rules, Virasoro algebras, Ward identities, primary fields, descendants, correlation functions, anharmonic ratios, crossing symmetry, monodromy matrix, central charge, Ising model, classical percolation, dense polymers, minimal models, Coulomb gas, Kac-Moody algebras, Sugawara hamiltonian, Wess-Zumino-Novikov-Witten model, Knizhnik-Zamolodchikov equations, non-abelian bosonization

## 1. Gaussian model. Lagrangian formulation

In this Lecture we discuss the theory of the free bosonic massless scalar field in a two-dimensional Euclidean space. Despite of its apparent triviality this model provides a foundation of conformal field theory (CFT). We shall use this model to introduce and illustrate the most important concepts of CFT.

The model we discuss is usually called the Gaussian model. Its action is defined as

$$S = \frac{1}{2}\int_{\mathcal{A}} \mathrm{d}\tau \mathrm{d}x [v^{-1}(\partial_\tau \Phi)^2 + v(\partial_x \Phi)^2] \qquad (1.1)$$

where $\mathcal{A}$ is some area of the infinite $(\tau, x)$-plane. The most important physical case is, of course, when $\mathcal{A}$ is a rectangle:

$$\mathcal{A} = (0 < \tau < \beta,\ 0 < x < L) \qquad (1.2)$$

Here, as usual, $\beta$ is the inverse temperature. It is important, however, to keep an option of arbitrary $\mathcal{A}$ because the model (1.1) possesses a hidden symmetry which makes itself manifest in transformation properties of its correlation functions with respect to the area change. As usual for Lorentz invariant theories one needs to restrict its energy spectrum to make correlation functions nonsingular (regularization). To regularize the theory we introduce the small distance cut-off $a$ assuming this to be the smallest possible *interval* between two points in $(\tau, x)$-space. The Gaussian model (1.1) forms a backbone of the bosonization approach. Its simplicity makes the latter approach a great success story. We shall see, however, that this apparent simplicity hides many highly nontrivial features. The first such feature is that the model (1.1) has quite remarkable correlation functions. The correlation functions of such a simple theory can be most conveniently calculated using the path integral approach. Introduce the generating functional of fields $\Phi$:

$$Z[\eta] = \int D\Phi(\mathbf{x}) e^{-S - \int d\tau dx \eta(\mathbf{x}) \Phi(\mathbf{x})} \tag{1.3}$$

where $S$ is the action (1.1), and $\mathbf{x} = (\tau, x)$. The generating functional can be calculated explicitly for any $\eta$ which will be very convenient. We can do it for any area $\mathcal{A}$, but to keep the calculations simple, we shall consider a rectangular area (1.2) first. To calculate the path integral one has to write down all functions in terms of their Fourier components. Then we get

$$Z[\eta] = \int \prod_\mathbf{p} d\Phi(\mathbf{p}) \exp\left[-\frac{1}{2}\sum_\mathbf{p} \Phi(-\mathbf{p})(vq^2 + v^{-1}\omega^2)\Phi(\mathbf{p}) + \sum_\mathbf{p} \eta(-\mathbf{p})\Phi(\mathbf{p})\right] \tag{1.4}$$

where
$$\mathbf{p} = (\omega, q), \quad \omega = 2\pi n_0/\beta, \quad q = 2\pi n_1/L$$
and
$$\Phi(\mathbf{p}) = \int d\tau dx e^{i\omega_n \tau} e^{-iqx} \Phi(\tau, x)$$

The integral (1.4) is just a product of simple Gaussian integrals; it can be estimated by the shift of variables, which removes the term linear in $\Phi$ from the exponent:

$$\tilde{\Phi}(\mathbf{p}) = \Phi(\mathbf{p}) + G(\mathbf{p})\eta(\mathbf{p}) \tag{1.5}$$

$$\ln Z[\eta] = \ln Z[0] + \frac{1}{2\beta V} \sum_\mathbf{p}' \eta(-\mathbf{p}) G(\mathbf{p}) \eta(\mathbf{p}) \tag{1.6}$$

The prime in the summation means that the term with $\mathbf{p} = 0$ is excluded and $G(\mathbf{p}) = (vq^2 + v^{-1}\omega^2)^{-1}$ is the Fourier transform of the Green's function of the Laplace operator satisfiyng the equation

$$-(v\partial_x^2 + v^{-1}\partial_\tau^2)G(x,\tau;x',\tau') = \delta(x-x')\delta(\tau-\tau') \qquad (1.7)$$

One can easely generalize this formula for an arbitrary area $\mathcal{A}$ writing it in the following form:

$$Z[\eta(x)]/Z[0] = \exp\left[\frac{1}{2}\eta(\xi)G(\xi,\xi')\eta(\xi')\right] \qquad (1.8)$$

where $G(\xi,\xi')$ is the Green's function of the Laplace operator on the area $\mathcal{A}$. This general expression will turn out very useful for us later.

It will be more convenient for later purposes to use the complex coordinates $z = \tau + ix/v, \bar{z} = \tau - ix/v$ where

$$\Delta = 4v^{-1}\partial_z\partial_{\bar{z}}, \quad \partial_z = \frac{1}{2}(\partial_\tau - vi\partial_x), \quad \partial_{\bar{z}} = \frac{1}{2}(\partial_\tau + vi\partial_x)$$

Let us consider the case of a very large rectangle or a disk of very large radius $R$. The words 'very large' in this context mean that we shall be interested only in correlations very far from the boundaries. The Green's function has the following well known form:

$$G(z,\bar{z}) = \frac{1}{4\pi}\ln\left(\frac{R^2}{z\bar{z}+a^2}\right) \qquad (1.9)$$

(recall that $a$ is the minimal interval in the theory).

Now let us consider a particular choice of $\eta$, namely

$$\eta(\xi) = \eta_0(\xi) \equiv i\sum_{n=1}^{N}\beta_n\delta(\xi-\xi_n) \qquad (1.10)$$

where $\beta_n$ are some numbers.

The generating functional (1.8) for this particular choice of $\eta$ coincides with the correlation function of bosonic exponents:

$$Z[\eta_0]/Z[0] \equiv \langle \exp[i\beta_1\Phi(\xi_1)]...\exp[i\beta_N\Phi(\xi_N)]\rangle \equiv \mathcal{F}(1,2,...N) \qquad (1.11)$$

Substituting Eq. (1.10) into (1.8) we get the expresssion which is, probably, the **most important** formula in the book:

$$F(1,2,...N) = \exp\left[-\sum_{i>j}\beta_i\beta_j G(\xi_i;\xi_j)\right]\exp\left[-\frac{1}{2}\sum_i \beta_i^2 G(\xi_i;\xi_i)\right] \tag{1.12}$$

The terms with the Green's functions of coinciding arguments are singular in the continuous limit. However, since we have regularized the theory, they are finite. Substituting the expression for $G$ from Eq. (1.9) into Eq. (1.12) we get the following result for the correlation function of bosonic exponents:

$$\mathcal{F}(1,2,...N) = \prod_{i>j}\left(\frac{z_{ij}\bar{z}_{ij}}{a^2}\right)^{(\beta_i\beta_j/4\pi)}\left(\frac{R}{a}\right)^{-\left(\sum_n \beta_n\right)^2/4\pi} \tag{1.13}$$

Two simple points need to be mentioned. Usually we are interested in very large systems $R/a \to \infty$. In this case the expression (1.13) is only nonzero provided the *'electroneutrality'* condition is fulfilled:

$$\sum_n \beta_n = 0 \tag{1.14}$$

For a finite system, however, even a single exponent has a nonzero average:

$$\langle\exp[i\beta\Phi(\xi)]\rangle = (R/a)^{-\beta^2/4\pi} \tag{1.15}$$

For a finite system, however, even a single exponent has a nonzero average:

$$\langle\exp[i\beta\Phi(\xi)]\rangle = (R/a)^{-\beta^2/4\pi} \tag{1.16}$$

Thus we have a general expression for correlation functions of bosonic exponents. In fact, as we have already mentioned, Eq. (1.12) holds not only for a plane, but for any area $\mathcal{A}$, provided $G$ is chosen properly. This expression gives us potentially *complete* knowledge of correlation functions of the Gaussian model (1.1). Indeed, since the bosonic exponents form a basis for local functionals of $\Phi$, one can calculate correlation functions of any local functional $F(\Phi)$ by expanding it as the Fourier integral

$$F(\Phi) = \int d\beta \tilde{F}(\beta) e^{i\beta\Phi}$$

*Lectures on ConformalField Theory* 57

and using Eq. (1.13). In fact, even more can be achieved. Let us come back to Eq. (1.13) and rewrite its right-hand side as a product of the analytic and the anti-analytic functions:

$$\langle \exp[i\beta_1 \Phi(\xi_1)] ... \exp[i\beta_N \Phi(\xi_N)] \rangle$$
$$= G(z_1,...,z_N) G(\bar{z}_1,...,\bar{z}_N) \delta_{\sum \beta_n, 0} \qquad (1.17)$$

where

$$G(\{z\}) = \prod_{i>j} \left(\frac{z_{ij}}{a}\right)^{(\beta_i \beta_j / 4\pi)}$$

This factorization garantees that analytic and anti-analytic parts of the correlation functions can be studied **independently**. Since factorization of the correlation functions is a general fact, it can be formally written as factorization of the corresponding fields: *inside the* $\langle ... \rangle$-*sign* one can rewrite $\Phi(z, \bar{z})$ as a sum of independent *analytic* and *anti − analytic* fields (we shall call them *chiral* components of the field $\Phi$):

$$\Phi(z, \bar{z}) = \phi(z) + \bar{\phi}(\bar{z}), \qquad (1.18)$$
$$\exp[i\beta\Phi(z, \bar{z})] = \exp[i\beta\phi(z)] \exp[i\beta\bar{\phi}(\bar{z})] \qquad (1.19)$$

We emphasise that this decomposition should be understood only as a property of correlation functions and *not* as a restriction on the variables in the path integral (1.4). The attentive reader understands that the expression (1.17) is obtained by integration over *all* fields $\Phi$. The meaning of the chiral fields $\phi, \bar{\phi}$ becomes quite transparent in the Hamiltonian formulations of the Gaussian model which is discussed in the next chapter.

For many purposes it is convenient to use the 'dual' field $\Theta(z, \bar{z})$ defined as

$$\Theta(z, \bar{z}) = \phi(z) - \bar{\phi}(\bar{z}) \qquad (1.20)$$

The dual field satisfies the following equations:

$$\partial_\mu \Phi = -i\epsilon_{\mu\nu} \partial_\nu \Theta \qquad (1.21)$$

or, in components

$$\partial_z \Phi = \partial_z \Theta$$
$$\partial_{\bar{z}} \Phi = -\partial_{\bar{z}} \Theta \qquad (1.22)$$

In order to study correlation functions of the analytic and the anti-analytic fields, we define the fields

$$A(\beta, z) \equiv \exp\left\{\frac{i}{2}\beta[\Phi(z,\bar{z}) + \Theta(z,\bar{z})]\right\}$$

$$\bar{A}(\bar{\beta}, \bar{z}) \equiv \exp\left\{\frac{i}{2}\bar{\beta}[\Phi(z,\bar{z}) - \Theta(z,\bar{z})]\right\} \quad (1.23)$$

with, generally speaking, **different** $\beta, \bar{\beta}$. With the operators $A(\beta, z), \bar{A}(\bar{\beta}, \bar{z})$ one can expand local functionals of mutually nonlocal fields $\Phi$ and $\Theta$. Let us construct a complete basis of bosonic exponents for a space of local periodic functionals. Suppose that

$$F(\Phi, \Theta)$$

is a local functional periodic both in $\Phi$ and $\Theta$ with the periods $T_1$ and $T_2$, respectively. This functional can be expanded in terms of the bosonic exponents:

$$F(\Phi, \Theta) = \sum_{n,m} \tilde{F}_{n,m} \exp[(2i\pi n/T_1)\Phi + (2i\pi m/T_2)\Theta]$$

$$\sum_{n,m} \tilde{F}_{n,m} A(\beta_{nm}, z)\bar{A}(\bar{\beta}_{nm}, \bar{z}) \quad (1.24)$$

where

$$\beta_{nm} = 2\pi\left(\frac{n}{T_1} + \frac{m}{T_2}\right)$$

$$\bar{\beta}_{nm} = 2\pi\left(\frac{n}{T_1} - \frac{m}{T_2}\right) \quad (1.25)$$

It turns out that the periods $T_1, T_2$ are not arbitrary, but related to each other. The reason for this lies in the fact that the correlation functions must be uniquely defined on the complex plane. We can see how this argument works using the pair correlation function as an example:

$$\langle A(\beta_{nm}, z_1)\bar{A}(\bar{\beta}_{nm}, \bar{z}_1)A(-\beta_{nm}, z_2)\bar{A}(-\bar{\beta}_{nm}, \bar{z}_2)\rangle$$

$$= (z_{12})^{-\beta_{nm}^2/4\pi}(\bar{z}_{12})^{-\bar{\beta}_{nm}^2/4\pi} = \frac{1}{|z_{12}|^{2d}}\left(\frac{z_{12}}{\bar{z}_{12}}\right)^S \quad (1.26)$$

where we introduce the quantities

$$d = \Delta + \bar{\Delta} = \frac{1}{8\pi}(\beta^2 + \bar{\beta}^2)$$

and
$$S = \Delta - \bar{\Delta} = \frac{1}{8\pi}(\beta^2 - \bar{\beta}^2)$$
which are called the 'scaling dimension' and the 'conformal spin', respectively. The two branch cut singularities in Eq. (1.26) cancel each other and give a uniquely defined function only if

$$2S = \beta_{nm}^2/4\pi - \bar{\beta}_{nm}^2/4\pi = \text{(integer)} \qquad (1.27)$$

i.e., physical fields with uniquely defined correlation functions must have integer or half-integer conformal spins. This equation suggests the relation

$$T_2 = \frac{4\pi}{T_1} \equiv \sqrt{4\pi K} \qquad (1.28)$$

as the minimal solution. The relationship (1.28) specifies the exponents of the multi-point correlation functions. Here we introduce the new notation $K$ for future convenience. The normalization is such that at $K = 1$ the periods for the field $\Phi$ and its dual are equal. The quantities $\Delta, \bar{\Delta}$ are called 'conformal dimensions' or 'conformal weights'. In the case of the Gaussian model (1.1) the conformal dimensions of the basic operators are given by:

$$\Delta_{nm} \equiv \beta_{nm}^2/8\pi = \frac{1}{8}\left(n\sqrt{K} + \frac{m}{\sqrt{K}}\right)^2$$
$$\bar{\Delta}_{nm} \equiv \bar{\beta}_{nm}^2/8\pi = \frac{1}{8}\left(n\sqrt{K} - \frac{m}{\sqrt{K}}\right)^2 \qquad (1.29)$$

## 2. Conformal symmetry and finite size effects

As we have already demonstrated, the apparently trivial Gaussian model has non-trivial correlation functions. We have also seen that these functions can be used in the bosonization procedure. In fact, in order to use this tool effectively, we need to learn more about properties of the Gaussian model. One of the most important properties is that the Gaussian model possesses a special hidden symmetry – the conformal symmetry. This symmetry plays a great role in the modern field theory. We choose the Gaussian model as the simplest example to illustrate the general concept.

Conformal symmetry reveals itself in special properties of the correlation functions under transformations of the area A on which the field $\Phi$ is defined. Let $\mathcal{A}$ be an arbitrary area of the complex plane. As we have said before, the expression (1.17) for multi-point correlation functions of the bosonic exponents is valid for any $\mathcal{A}$:

$$\langle \exp[i\beta_1\Phi(\xi_1)]...\exp[i\beta_N\Phi(\xi_N)]\rangle$$
$$= \exp\left[-\sum_{i>j}\beta_i\beta_j G(\xi_i;\xi_j)\right]\exp\left[-\frac{1}{2}\sum_i \beta_i^2 G(\xi_i;\xi_i)\right] \quad (2.1)$$

where $G$ is the Green's function of the Laplace operator on $\mathcal{A}$. As shown in the theory of the Laplace equation, the Green's function $G(\xi_i;\xi_j)$ can be written explicitly if one knows a transformation $z(\xi)$ which maps $\mathcal{A}$ onto the infinite plane. Then the Green's function of the Laplace operator on $\mathcal{A}$ is given by:

$$G(\xi_1,\xi_2) = -\frac{1}{2\pi}\ln|z(\xi_1) - z(\xi_2)| - \frac{1}{4\pi}\ln[|\partial_{\xi_1}z(\xi_1)\partial_{\xi_2}z(\xi_2)|] \quad (2.2)$$

Let us consider the case of two exponents: $N = 2$. Recall that the correlation functions (2.1) are products of analytic and anti-analytic parts. Substituting expression (2.2) into Eq. (2.1) with $N = 2$ and $\beta_1 = \beta$, $\beta_2 = -\beta$, we get the following expression for the analytic part of the pair correlation function:

$$D(\xi_1;\xi_2) = \frac{1}{[z(\xi_1) - z(\xi_2)]^{2\Delta}}[\partial_{\xi_1}z(\xi_1)\partial_{\xi_2}z(\xi_2)]^{\Delta} \quad (2.3)$$

where $\Delta = \beta^2/8\pi$.

As we see, the correlation functions transform locally under analytic coordinate transformations. Therefore one can assign these transformation to the corresponding operators – the bosonic exponents $A_\Delta(z) \equiv \exp[i\beta\phi(z)]$. As is clear from Eq. (2.3), these operators transform as tensors of rank $(\Delta, 0)$:

$$A_\Delta(\xi) = A_\Delta[z(\xi)](dz/d\xi)^\Delta \quad (2.4)$$

Correspondingly, the antianalytical exponent $\bar{A}_{\bar{\Delta}}(\bar{z})$ transforms as a tensor of rank $(0, \bar{\Delta})$.

There is one transformation which is particularly important for applications:

$$z(\xi) = \exp(2\pi\xi/L) \quad (2.5)$$

This transforms a strip of width $L$ in the $x$-direction into infinite complex plane. For $(1 + 1)$-dimensional systems this transformation (i) relates

*Lectures on ConformalField Theory* 61

correlation functions of finite quantum chains to those of infinite systems, and also (ii) relates correlation functions at $T = 0$ to correlation functions at finite temperatures (in the latter case $T = i/L$). Substituting $z(\xi)$ into Eq. (2.3) we get

$$D(\xi_1; \xi_2) = \langle A_\Delta(z_1) A_\Delta^+(z_2) \rangle = \left\{ \frac{\pi/L}{\sinh[\pi(\xi - \xi')/L]} \right\}^{2\Delta}$$

$$\bar{D}(\xi_1; \xi_2) = \langle \bar{A}_{\bar\Delta}(\bar z_1) \bar{A}_{\bar\Delta}^+(\bar z_2) \rangle = \left\{ \frac{\pi/L}{\sinh[\pi(\bar\xi - \bar\xi')/L]} \right\}^{2\bar\Delta} \quad (2.6)$$

Below we continue to treat our system as a quantum $(1 + 1)$-dimensional one at $T = 0$. Let $T = 0$ and the system be a circle of a finite length $L$. In this case we have

$$\xi = \tau + ix, \quad -\infty < \tau < \infty, \quad 0 < x < L$$

Let us expand the expressions for $D(1,2), \bar{D}(1,2)$ at large $\tau_{12}$. The result is

$$D(\xi_1; \xi_2) \bar{D}(\xi_1; \xi_2)$$
$$= \sum_{n,m=0}^{\infty} C_{nm} (\pi/L)^d \exp\left[-\frac{2\pi}{L}(d+n)\tau_{12}\right] \exp\left[-ix_{12}\frac{2\pi}{L}(S+m)\right] \quad (2.7)$$

where $C_{nm}$ are universal numerical coefficients (note that $n, m \geq 0$, it is important!).

On the other hand this correlation function can be expanded in the Lehmann series:

$$D(\tau_{12}, x_{12}) \bar{D}(\tau_{12}, x_{12})$$
$$= \sum_q |\langle 0| \exp[i\beta\phi(0) + i\bar\beta\bar\phi(0)]|q\rangle|^2 \exp[-E_q \tau_{12} - iP_q x_{12}] \quad (2.8)$$

where $q$ labels eigenstates of the Hamiltonian and $E_q, P_q$ are eigenvalues of energy and momenta of the state $q$. Comparing these two expansions we find:

$$E_q = \frac{2\pi}{L}(d+n) \quad (2.9)$$

$$P_q = \frac{2\pi}{L}(S+m) \quad (2.10)$$

$$\left(\frac{\pi}{L}\right)^{(d+n)} C_{nm} = \sum_q |\langle 0| \exp[i\beta\phi(0) + i\bar\beta\bar\phi(0)]|q\rangle|^2$$
$$\times \delta\left[E_q - \frac{2\pi}{L}(d+n)\right] \delta\left[P_q - \frac{2\pi}{L}(S+m)\right] \quad (2.11)$$

All three expressions are of profound significance. The first two relate conformal dimensions of the correlation functions to the eigenvalues of energy and momentum operators. It is usually much easier to calculate energies than correlation functions. In the latter case the amount of computational work greatly increases since one must calculate also the matrix elements. Therefore it is a tremendous relief that for a quantum system with a gapless spectrum the relationships (2.9) and (2.10) allow us to avoid direct calculations of the correlation functions. Instead one can solve the problem of low-lying energy levels in a finite size system (which can be done numerically or even exactly) and then, using the relationships (2.8), (2.9) and (2.10), restore the correlation functions. From Eqs. (2.9) and (2.10 ) we see that the problem of conformal dimensions can be formulated as an eigenvalue problem for the following operators:

$$\Delta = \frac{L}{4\pi}(\hat{H} + \hat{P}) \equiv \hat{T}_0 \qquad (2.12)$$

$$\bar{\Delta} = \frac{L}{4\pi}(\hat{H} - \hat{P}) \equiv \hat{\bar{T}}_0 \qquad (2.13)$$

It can be shown that the operators $\hat{T}_0$, $\hat{\bar{T}}_0$ are related to analytic and anti-analytic components of the stress energy tensor defined as

$$T_{ab}(x) = \frac{\delta S}{\delta g^{ab}(x)} \qquad (2.14)$$

where $g^{ab}$ is the inverse metric tensor (so it is assumed that the system is in a curved space characterized by the metric tensor $g_{ab}$). Namely,

$$\hat{T}_0 = \int_0^L \mathrm{d}x T_{zz}; \quad \hat{\bar{T}}_0 = \int_0^L \mathrm{d}x T_{\bar{z}\bar{z}}$$

It is not unnatural that the stress-energy tensor appears in the present context. Indeed, the scaling dimensions are related to coordinate transformations and such transformations change the metric on the surface:

$$\mathrm{d}z\mathrm{d}\bar{z} \to (\mathrm{d}z/\mathrm{d}\xi)(\mathrm{d}\bar{z}/\mathrm{d}\bar{\xi})\mathrm{d}\xi\mathrm{d}\bar{\xi}$$

The relations between scaling dimensions and the stress-energy tensor have a general character and hold for all conformal theories. To make the new concept of the stress-energy tensor more familiar, let us calculate $T_{ab}$ for the Gaussian model explicitly. The action on a curved background is given by

$$S = \frac{1}{2}\int \mathrm{d}^2x \sqrt{g} g^{ab}\partial_a\Phi\partial_b\Phi \qquad (2.15)$$

where $g = \det \hat{g}$ and $g^{ab} = (g^{-1})_{ab}$. In the vicinity of a flat space where the metric tensor is equal to $g_{ab} = \delta_{ab}$ we get:

$$T_{ab} = \frac{1}{2} : \left[ \partial_a \Phi \partial_b \Phi - \frac{1}{2} \delta_{ab} (\partial_c \Phi)^2 \right] : \qquad (2.16)$$

(the dots denote the normal ordering, i.e., it is assumed that the vacuum average of the operator is subtracted – $\langle T_{ab} \rangle = 0$).

## 3. Virasoro algebra

In two dimensions the group of conformal transformations is isomorphic to the group of analytic transformations of complex plane. Since the number of analytic functions is infinite, this group is infinite-dimensional, i.e. has infinite number of generators. Infinite number of generators generates infinite number of Ward identities for correlation functions. These identities serve as differential equations on correlation functions. No matter how many operators you have in your correlation function, you always have enough Ward identities to specify it. Thus in conformally invariant theories (for brevity we shall call them 'conformal' theories) one can (at least in principle) calculate all multi-point correlation functions. We have already calculated multi-point correlation functions of the bosonic exponents for the Gaussian model (see Chapter 3), below we shall see less trivial examples.

Let us take for good faith that such models exist, that is we postulate that there are models besides the Gaussian one whose spectrum is linear, whose correlation functions factorize into products of analytic and anti-analytic parts and which have operators transforming under analytic (anti-analytic) transformations as (2.4). Such operators will be called *primary* fields. This is almost a definition of what conformal theory is. To make this definition rigorous we have to add one more property, namely, to postulate that the three-point correlation functions of the primary fields have the following form:

$$\langle A_{\Delta_1}(z_1) A_{\Delta_2}(z_2) A_{\Delta_3}(z_3) \rangle = \frac{C_{123}}{(z_{12})^{\Delta_1 + \Delta_2 - \Delta_3} (z_{13})^{\Delta_1 + \Delta_3 - \Delta_2} (z_{23})^{\Delta_2 + \Delta_3 - \Delta_1}} \qquad (3.1)$$

where $C_{123}$ are some constants.

Since the eigenvalues of the Hamiltonian and momentum are related to the zeroeth components of the stress-energy tensor, it is logical to suggest that the generators of transformations (2.4) are Fourier components of the stress-energy tensor. In order to find commutation relations between these components and the primary fields we have to consider the infinitely small version of transformation (2.4):

$$z' = z + \epsilon(z) \qquad (3.2)$$

According to (2.4) this transformation changes the primary field $A_\Delta(z)$:

$$\begin{aligned}\delta A_\Delta(z) &= A_\Delta(z+\epsilon)(1+\partial\epsilon)^\Delta - A_\Delta(z)\\ &= \epsilon(z)\partial A_\Delta(z) + \partial\epsilon(z)\Delta A_\Delta(z) + O(\epsilon^2)\end{aligned} \qquad (3.3)$$

Let us restrict our correlation functions to the line $\tau = 0$ where $z = ix$. This eliminates the time dependence and therefore we can use the operator language of quantum mechanics. As we know from quantum mechanics, if an operator changes under some transformation, this change can be expressed as an action of some other operator – a generator of this transformation. For our case it means that

$$\delta \hat{A}_\Delta(x) = [\hat{Q}_\epsilon, \hat{A}_\Delta(x)] \qquad (3.4)$$

where $\hat{Q}_\epsilon$ is an element of the group of conformal transformations corresponding to the infinitesimal transformation (3.2). For a general Lie group G with generators $\hat{t}_i$ an infinitesimal transformation is given by $\epsilon_i \hat{t}_i$, where $\epsilon_i$ are infinitely small parameters – coordinates of the transformation. Since conformal transformations in two dimensions are characterized not by a finite set of parameters, but by an entire function $\epsilon(z)$, the conformal algebra is infinite-dimensional. Therefore a general infinitesimal transformation can be written as an integral[1]:

$$\hat{Q}_\epsilon = \frac{1}{2\pi}\int dy \epsilon(y)\hat{T}(y)$$

Therefore we have

$$\frac{i}{2\pi}\int dy \epsilon(y)[\hat{T}(y),\hat{A}_\Delta(x)] = \epsilon(x)\partial_x A_\Delta(x) + \partial_x\epsilon(x)\Delta A_\Delta(x) \qquad (3.5)$$

which corresponds to the following local commutation relations:

$$\frac{i}{2\pi}[\hat{T}(y),\hat{A}_\Delta(x)] = \delta(x-y)\partial_x A_\Delta(x) - \partial_y\delta(y-x)\Delta A_\Delta(x) \qquad (3.6)$$

These commutation relations must be satisfied in any conformal theory.

Following the general rules explained in Chapter 2, we can rewrite these commutation relations as OPE:

$$T(z)A_\Delta(z_1) = \frac{\Delta}{(z-z_1)^2}A_\Delta(z_1) + \frac{1}{(z-z_1)}\partial_{z_1}A_\Delta(z_1) + \ldots \quad (3.7)$$

where dots stand for terms nonsingular at $(z - z_1) \to 0$. Let us now study the algebraic properties of the stress-energy tensor $T(z)$. For this purpose we shall study its correlation functions using the Gaussian model stress energy tensor (2.16) as an example. In order to simplify the results, we change notations and introduce the new components [2]

$$T_{zz} = -\pi(T_{11} - T_{22} - 2iT_{12}) \equiv T = -2\pi :(\partial\phi)^2:$$
$$T_{\bar{z}\bar{z}} = -\pi(T_{11} - T_{22} + 2iT_{12}) \equiv \bar{T} = -2\pi :(\bar{\partial}\bar{\phi})^2:$$
$$\text{Tr}T = T_{11} + T_{22} = 0 \quad (3.8)$$

Now it is obvious that correlation functions of operators $T$ and $\bar{T}$ depend on $z$ or $\bar{z}$ only. Since the two-point correlation function is the simplest one, let us calculate it first. A straightforward calculation yields:

$$\langle T(1)T(2)\rangle = \frac{C}{2(z_1-z_2)^4}$$
$$\langle \bar{T}(1)\bar{T}(2)\rangle = \frac{C}{2(\bar{z}_1-\bar{z}_2)^4} \quad (3.9)$$

where $C = 1$ in the given case. We keep $C$ in the relations (3.9) because in this form they hold for all conformal theories. As far as the correlation functions of Tr$T$ are concerned, it seems that Eq. (3.8) dictates them to be identically zero. The miracle is that it is *wrong*! Here we encounter another example of anomaly which is called *conformal* anomaly.

Let us consider the matter more carefully. Since the stress-energy tensor is a conserved quantity, i.e.,

$$\partial_a T_{ab} = 0$$

which reflects the conservation of energy and momentum, the two-point correlation function must satisfy the identity

$$q_a \langle T_{ab}(-q)T_{cd}(q)\rangle = 0$$

This suggests that the Fourier transformation of Eqs. (3.8) must be modified:

$$\langle T_{ab}(-q)T_{cd}(q)\rangle = \frac{Cq^2}{48\pi}\left(\delta_{ab} - \frac{q_a q_b}{q^2}\right)\left(\delta_{cd} - \frac{q_c q_d}{q^2}\right) \quad (3.10)$$

which gives the nonvanishing correlation function of the traces:

$$\langle \mathrm{Tr}T(-q)\mathrm{Tr}T(q)\rangle = \frac{Cq^2}{48\pi} \qquad (3.11)$$

In real space the pair correlation function of the traces is ultralocal:

$$\langle \mathrm{Tr}T(x)\mathrm{Tr}T(y)\rangle = \frac{C}{48\pi}\nabla^2\delta^{(2)}(\mathbf{x}-\mathbf{y}) \qquad (3.12)$$

Since this correlation function is short-range, it is generated at high energies. It is no wonder, therefore, that we have missed it in the straightforward calculations.

It can be shown that the central charge appears in the expression for the specific heat. Later we shall learn that it also determines the conformal dimensions of the correlation functions. The fact that the specific heat of a conformal theory is proportional to its central charge means that the latter is related to the number of states. Since the conformal dimensions are eigenvalues of the stress-energy tensor components, the relation between $C$ and the correlation functions suggests that $C$ also influences the scaling dimensions. For what follows it is necessary to know the fusion rules for the stress-energy tensor itself. We can find these commutation relations using general properties of the stress-energy tensor. At first, the stress energy tensor is really a tensor, and as such it has definite transformation properties. From this fact it follows that its conformal dimensions are $(2,0)$, which is consistent with Eq. (3.9). Therefore the fusion rules for $T(z)T(z_1)$ must include the right hand side of (3.7) with $\Delta = 2$. This is not all, however; the expansion must contain a term with the identity operator, since otherwise the pair correlation function $\langle T(z)T(z_1)\rangle$ would vanish. This term can be deduced from Eq. (3.9); collecting all these terms together we get the following OPE (the Virasoro fusion rules):

$$T(z)T(z_1) = \frac{C}{2(z-z_1)^4} + \frac{2}{(z-z_1)^2}T(z_1) + \frac{1}{(z-z_1)}\partial_{z_1}T(z_1)+... \qquad (3.13)$$

A comparison of Eq. (3.13) with Eq. (3.7) shows that the stress-energy tensor *is not a primary field*, that is, despite its name, the stress-energy tensor does not transform as a *tensor* under conformal transformations. It can be shown that instead of the tensorial law (2.4) the stress-energy

*Lectures on ConformalField Theory*

tensor transforms under finite conformal transformations as follows:

$$T(\xi) = T[z(\xi)](dz/d\xi)^2 + \frac{C}{12}\{z,\xi\}$$

$$\{z,\xi\} = \frac{z'''}{z'} - \frac{3}{2}\left(\frac{z''}{z'}\right)^2 \qquad (3.14)$$

In particular, it follows from this expression that the stress energy tensor on a strip of the width $L$ in the $x$-direction ($z = \exp(2\pi\xi/L)$) is equal to

$$T_{\text{strip}}(\xi) = (2\pi/L)^2 \left[z^2 T_{\text{plane}}(z) - C/24\right] \qquad (3.15)$$

Since $\langle T_{\text{plane}}(z)\rangle = 0$, we find

$$\langle T_{\text{strip}}(z)\rangle = -\frac{C}{24}(2\pi/L)^2 \qquad (3.16)$$

Now recalling the relationship between stress-energy tensor and Hamiltonian (2.13) we reproduce from the latter equation the $1/L$ correction to the stress-energy tensor eigenvalues

It is also convenient to have the Virasoro fusion rules (3.13) in operator form with commutators:

$$\frac{1}{2i\pi}[\hat{T}(x),\hat{T}(y)] = \delta(x-y)\partial_x T(x) - 2\partial_x\delta(x-y)T(x) + \frac{C}{6}\partial_x^3\delta(x-y) \qquad (3.17)$$

Since the commutation relations (3.17) are local, they do not depend on the global geometry of the problem. It is customary to write the Virasoro fusion rules using the Laurent components of the stress-energy tensor and the operators defined on infinite complex plane:

$$\hat{T}(z) = \sum_{-\infty}^{\infty} \frac{L_n}{z^{n+2}}, \quad \hat{A}_\Delta(z) = \sum_{-\infty}^{\infty} \frac{A_{\Delta,n}}{z^{n+2}}$$

The transformations laws (5.6) and (5.14) establish a connection between this expansion and the Fourier expansion which we use in a strip geometry:

$$A_\Delta^{\text{strip}}(x) = \left(\frac{2\pi}{L}\right)^\Delta \sum_n A_{\Delta,n} e^{-2\pi i x(n-\Delta)/L},$$

$$T^{\text{strip}}(x) = \left(\frac{2\pi}{L}\right)^2 \left(\sum_n L_n e^{-2\pi i n x/L} - C/24\right) \qquad (3.18)$$

Substituting these expansions into the fusion rules (3.13) and (3.5) we get the following commutation relations for the Laurent components:

$$[L_n, L_m] = (n-m)L_{n+m} + \frac{C}{12}n(n^2-1)\delta_{n+m,0} \qquad (3.19)$$

and

$$[L_n, A_{\Delta,m}] = [\Delta(1-n) + m + n]A_{\Delta,m+n} \qquad (3.20)$$

From Eq. (3.19) we see that the set of operators $L_n, I$ ($I$ is the identity operator) is closed with respect to the operation of commutation. Therefore components of the stress-energy tensor in conformal theories together with the identity operator form an algebra (Virasoro algebra). As we have said, this fact opens up the possibility of formal studies of conformal theories as representations of the Virasoro algebra.

## 3.1 Ward identities

The identity (3.5) can be used to derive Ward identities for correlation functions of stress-energy tensor with primary fields. Namely, let us insert this identity into a correlation function:

$$\frac{1}{2\pi i}\int_C dz \epsilon(z) \langle T(z) A_{\Delta_1}(1)...A_{\Delta_N}(N)\rangle$$
$$= \sum_{i=1}^{N}[\epsilon(z_i)\partial_{z_i}\langle A_{\Delta_1}(1)...A_{\Delta_N}(N)\rangle + \partial_{z_i}\epsilon(z_i)\Delta_i\langle A_{\Delta_1}(1)...A_{\Delta_N}(N)\rangle]$$
$$(3.21)$$

and assume that $\epsilon(z)$ is analytic in the domain enclosing the points $z_i$ and the contour $C$ encircles these points. Then Eq. (3.21) suggests that as a function of $z$ the integrand of the left hand side has simple and double poles at $z = z_i$:

$$\langle T(z)A_{\Delta_1}(1)...A_{\Delta_N}(N)\rangle$$
$$= \sum_{i=1}^{N}\left[\frac{\Delta_i}{(z-z_i)^2} + \frac{1}{z-z_i}\partial_{z_i}\right]\langle A_{\Delta_1}(1)...A_{\Delta_N}(N)\rangle \qquad (3.22)$$

This is the Ward identity for $T(z)$ we had in mind.

# Lectures on ConformalField Theory

## 3.2 Subalgebra sl(2)

It follows from Eq. (3.19) that three operators $L_0$, $L_{\pm 1}$ compose a subalgebra of the Virasoro algebra isomorphic to the algebra sl(2):

$$[L_0, L_{\pm 1}] = \mp 2L_{\pm 1}, \quad [L_1, L_{-1}] = 2L_0 \qquad (3.23)$$

The corresponding group is the group of all rational transformations of the complex plane

$$w(z) = \frac{az+b}{cz+d}, \quad ad - bc = 1 \qquad (3.24)$$

Let us show that all correlation functions in critical two dimensional theories are invariant with respect to the SL(2) transformations (3.24). This fact follows from the Ward identity (3.22). Indeed, since $\langle T(z) \rangle = 0$ on infinite plane, the correlation function in the left hand side of Eq. (3.22) must decay at infinity as

$$\langle T(z) A_{\Delta_1}(1)...A_{\Delta_N}(N)\rangle \sim z^{-4} D(z_1,...z_N) \qquad (3.25)$$

Expanding the right-hand side of Eq. (3.21) in powers of $z^{-1}$ we find that the terms containing $z^{-1}, z^{-2}$ and $z^{-3}$ must vanish, which gives rise to the following identities:

$$\sum_i \partial_{z_i} \langle A_{\Delta_1}(1)...A_{\Delta_N}(N)\rangle = 0 \qquad (3.26)$$

$$\sum_i (z_i \partial_{z_i} + \Delta_i) \langle A_{\Delta_1}(1)...A_{\Delta_N}(N)\rangle = 0 \qquad (3.27)$$

$$\sum_i (z_i^2 \partial_{z_i} + 2\Delta_i z_i) \langle A_{\Delta_1}(1)...A_{\Delta_N}(N)\rangle = 0 \qquad (3.28)$$

It is easy to see that the operators

$$L_{-1} = \partial_z, \quad L_0 = (z\partial_z + \Delta), \quad L_1 = (z^2 \partial_z + 2\Delta z) \qquad (3.29)$$

satisfy the commutation relations (3.23) thus composing a representation of the sl(2) algebra.

## 4. Structure of Hilbert space in conformal theories

In this chapter we shall generate eigenvectors of the Hilbert space of conformal theories using the Virasoro operators. One may wonder, why we use Virasoro operators whose commutation relations are so awkward? The answer is that we want to make sure that the conformal symmetry is preserved, and the best way to guarantee it is to use as creation and annihilation operators the generators of the conformal group.

In fact, the procedure we are going to follow is a standard one in group theory, where it used for constructing representations. In order to feel ourselves on a familiar ground, let us recall how one constructs representations of spin operators in quantum mechanics. Consider a three dimensional rotator whose Hamiltonian is the sum of squares of generators of the SU(2) group - components of the spin:

$$\hat{H} = \sum_{i=x,y,z} \hat{S}_i^2 = \hat{S}_z^2 + \frac{1}{2}(\hat{S}_+\hat{S}_- + \hat{S}_-\hat{S}_+)$$

Since $[\hat{H}, \hat{S}^z] = 0$, $\hat{S}^z$ is diagonal in the basis of eigenfunctions of $\hat{H}$. The key point is to write down the operator algebra in the proper form:

$$[\hat{S}_\pm, \hat{S}_z] = \mp\hat{S}_\pm, \quad [\hat{S}_+, \hat{S}_-] = 2\hat{S}_z$$

In doing so we separate the diagonal generator $\hat{S}_z$ from the raising and lowering operators. As follows from the commutation relations, the operators $\hat{S}_+$ ($\hat{S}_-$) increase (decrease) an eigenvalue of $\hat{S}_z$ by one. Let us suppose that there exists an eigenvector of $\hat{S}^z$ which is annihilated by $\hat{S}_-$:

$$\hat{S}_-|-S> = 0; \quad \hat{S}_z|-S> = -S|-S>$$

Then $(\hat{S}_+)^j|-S>$ is also an eigenvector of $\hat{S}_z$ with the eigenvalue $j-S$. As we know, it can also be shown that if $2S$ is a positive integer, the state

$$|\chi\rangle = (\hat{S}_+)^{(2S+1)}|-S> \qquad (4.1)$$

is a null vector, that is representations of the SU(2) group with integer $(2S+1)$ are finite dimensional.

Let us show that in the case of the Virasoro algebra the state similar to $|-S\rangle$ is

$$|\Delta, \bar{\Delta}\rangle = A_{\Delta,0} A_{\bar{\Delta},0}|0\rangle$$

According to Eq. (2.7), $\langle n|A_\Delta(z)|0\rangle = 0$ for all negative $n$, which according to the definition of $A_{\Delta,n}$ is equivalent to

$$A_{\Delta,n}|0\rangle = 0, \quad n > 0 \qquad (4.2)$$

Similar considerations carried out for the two-point correlation function of the stress-energy tensor lead to

$$L_n|0\rangle = 0 \quad (n > 0) \qquad (4.3)$$

## Lectures on ConformalField Theory

Now we can prove that operators $L_n$ ($n > 0$) annihilate the state $|\Delta, \bar{\Delta}\rangle$. Indeed, according to Eq. (3.20)

$$L_n A_{\Delta,0}|0\rangle = [\Delta(1-n) + n] A_{\Delta,n}|0\rangle + A_{\Delta,0} L_n|0\rangle = 0 \qquad (4.4)$$

Therefore we can use $L_n$ with positive $n$ as lowering operators and $L_n$ with negative $n$ as raising operators. Thus we look for the eigenvectors in the following form:

$$|\bar{n}_1, \bar{n}_2, ..., \bar{n}_M; n_1, n_2, ..., n_M; \Delta, \bar{\Delta}\rangle$$
$$= \bar{L}_{-\bar{n}_1}\bar{L}_{-\bar{n}_2}...\bar{L}_{-\bar{n}_M} L_{-n_1} L_{-n_2}...L_{-n_M} |\Delta, \bar{\Delta}\rangle \qquad (4.5)$$

We see that a tower of states is built on each primary field, including the trivial identity operator. It follows from the Virasoro commutation relations (3.19), that these vectors are eigenstates of the operators $L_0$ and $\bar{L}_0$ with eigenvalues

$$\Delta_n = \Delta + \sum_j n_j,$$
$$\bar{\Delta}_n = \bar{\Delta} + \sum_j \bar{n}_j \qquad (4.6)$$

Operators obtained from primary fields by action of the Virasoro generators $L_{-n}$ are called *descendants* of the primary fields. It can be shown that for the Gaussian model all primary fields and their descendants are linearly independent and their Hilbert space is isomorphic to the Hilbert space built up by the bosonic creation operators. This, however, is only one particular representation of the Virasoro algebra. There are many others and different methods suggested for finding nontrivial representations of the Virasoro algebra. Some representations can be constructed by truncation of the Gaussian theory Hilbert space, i.e. by postulating that there are linearly dependent states among the states (4.5).

In this chapter we consider only the simplest case when the truncation occurs on the second level. Suppose that the states

$$L_{-2}|\Delta\rangle, \; L_{-1}^2|\Delta\rangle$$

are linearly dependent, i.e., there is a number $\alpha$ such that

$$|\chi\rangle = (L_{-2} + \alpha L_{-1}^2)|\Delta\rangle = 0 \qquad (4.7)$$

that is $|\chi\rangle$ is a null vector similar to (4.1).

Since all other Virasoro generators $L_{-n}$ with $n > 2$ are generated by these two via the commutation relations (3.20), the fact that $L_{-2}$ is proportional to $L_{-1}^2$ means that the only linearly independent eigenvectors from the set (4.5) are
$$|n, \Delta\rangle = (L_{-1})^n |\Delta\rangle$$

Let us consider conditions for existence of the null state $|\chi\rangle$ (4.7). Due to their mutual independence the left and right degrees of freedom can be considered separately. Since we want to preserve conformal invariance of the theory, condition (4.7) must survive conformal transformations, i.e.,
$$L_n |\chi\rangle = 0 \ (n > 0) \tag{4.8}$$

Using the Virasoro commutation relations (3.20) and the properties
$$L_0 |\Delta\rangle = \Delta |\Delta\rangle, \ L_n |\Delta\rangle = 0 \ (n > 0)$$

we get from Eq. (4.8) the following two equations:
$$3 + 2\alpha + 4\alpha\Delta = 0 \tag{4.9}$$
$$4\Delta(2 + 3\alpha) + C = 0 \tag{4.10}$$

whose solution is
$$\alpha = -\frac{3}{2(1 + 2\Delta)}, \ \Delta = \frac{1}{16}\left(5 - C \pm \sqrt{(5-C)^2 - 16C}\right) \tag{4.11}$$

## 4.1 Differential equations for correlation functions

We can use Ward identities for the stress-energy tensor to derive differential equations for the correlation functions. As the first step let us write down the complete OPE for the stress-energy tensor with a primary fields $A_\Delta(z_1)$ which is a generalization of Eq. (5. 7):
$$T(z) A_\Delta(z_1) = \sum_{n=0}^{\infty} (z - z_1)^{n-2} L_{-n} A_\Delta(z_1) \tag{4.12}$$

The absence in this expansion of the Virasoro generators with positive index follows from the Ward identity (3.22). Comparing this OPE with Eq. (3. 7) or Eq. (3.22) we obtain the following identities:
$$L_0 A_\Delta(z_1) = \Delta A_\Delta(z_1), \ L_{-1} A_\Delta(z_1) = \partial_{z_1} A_\Delta(z_1) \tag{4.13}$$

The next step is to use these identities to extract the action of $L_{-2}$. This we do by substracting the singular part of the stress-energy tensor at $z \to z_1$ and using (3.22):

$$\langle L_{-2} A_{\Delta_1}(1)...A_{\Delta_N}(N) \rangle =$$

$$\langle \left[ T(z) - \frac{L_0}{(z-z_1)^2} - \frac{L_{-1}}{z-z_1} \right] A_{\Delta_1}(1)...A_{\Delta_N}(N) \rangle |_{z \to z_1}$$

$$= \sum_{j \neq 1} \left[ \frac{\Delta_j}{(z_1-z_j)^2} + \frac{1}{(z_1-z_j)} \partial_j \right] \langle A_{\Delta_1}(1)...A_{\Delta_N}(N) \rangle \quad (4.14)$$

According to Eq. (4.7) $L_{-1}^2$ is identified with $-\alpha \partial^2$, and we obtain from (4.14) the linear differential equation for a multi-point correlation function of primary fields $A_j(z_j, \bar{z}_j)$ with analytic conformal dimensions $\Delta_j$:

$$\left\{ \frac{3}{2(1+2\Delta_1)} \partial_1^2 - \sum_{j \neq 1} \left[ \frac{\Delta_j}{(z_1-z_j)^2} + \frac{1}{(z_1-z_j)} \partial_j \right] \right\}$$

$$\times \langle A_1(z_1)...A_N(z_N) \rangle = 0 \quad (4.15)$$

We emphasize that this equation holds only for the simplest representation of the Virasoro algebra, namely the one defined by Eq. (4.7).

For the four point correlation function the presence of the SL(2) symmetry helps to reduce Eq. (4.15) to an ordinary differential equation. In the case when all conformal dimensions are equal the general expression for a four-point correlation function acquires the following form:

$$\langle \Phi(1)...\Phi(4) \rangle = \left( \left| \frac{z_{13} z_{24}}{z_{12} z_{14} z_{23} z_{34}} \right| \right)^{4\Delta} G(x, \bar{x}) \quad (4.16)$$

where

$$x = \frac{z_{12} z_{34}}{z_{13} z_{24}}, \quad 1 - x = \frac{z_{14} z_{23}}{z_{13} z_{24}} \quad (4.17)$$

are the so-called anharmonic ratios.

Substituting Eq. (4.16) into Eq. (4.15) we obtain the following conventional differential equation for $G(x, \bar{x})$:

$$x(1-x) G'' + \frac{2}{3}(1-4\Delta)(1-2x) G' + \frac{4\Delta}{3}(1-4\Delta) G = 0 \quad (4.18)$$

(and the same equation for $\bar{x}$). Equations of this type appear frequently in conformal field theory and therefore we shall discuss the solution in detail. Eq. (4.18) is a particular case of the hypergeometric equation.

Its special property is invariance with respect to the transformation $x \to 1-x$ which reflects the invariance of the correlation function with respect to permutation of the coordinates ($2 \to 4$, for instance). Thus if $\mathcal{F}(x)$ is a solution, $\mathcal{F}(1-x)$ is also a solution. If $\frac{2}{3}(1-4\Delta)$ is not an integer number, two linearly independent solutions of Eq. (4.18) nonsingular in the vicinity of $x = 0$ are given by the hypergeometric functions:

$$F^{(0)}(x) = F(a, b, c; x)$$
$$F^{(1)}(x) = x^{1-c} F(a - c + 1, b - c + 1, 2 - c; x)$$
$$a = \frac{1}{3}(1 - 4\Delta); \quad b = -4\Delta; \quad c = \frac{2}{3}(1 - 4\Delta) \qquad (4.19)$$

When $c$ is integer, the second solution is not a hypergeometric function. We shall discuss this case later. A general solution for $G(x, \bar{x})$ is

$$G(x, \bar{x}) = W_{ab} \mathcal{F}^{(a)}(x) \mathcal{F}^{(b)}(\bar{x}) \qquad (4.20)$$

This solution must satisfy two requirements. The first is that it must be invariant with respect to interchange of $x \to 1 - x$ (*crossing* symmetry) which reflects the fact that the entire correlation function does not change when any two coordinates are interchanged. The second requirement is that the correlation function must be single-valued on the $(x, \bar{x})$ plane. These requirements fix the matrix $W$ (up to a factor).

If $c$ is not an integer, we use the known identities for hypergeometric functions and the fact that $a + b = 2c - 1$, to find

$$F^{(a)}(1 - x) = A_{ab} \mathcal{F}^{(b)}(x)$$
$$A_{00} = \frac{\Gamma(c)\Gamma(1-c)}{\Gamma(c-a)\Gamma(1+a-c)}, \quad A_{01} = \frac{\Gamma(c)\Gamma(c-1)}{\Gamma(a)\Gamma(b)}$$
$$A_{10} = (1 - A_{00}^2)/A_{01}, \quad A_{11} = -A_{00} \qquad (4.21)$$

The crossing symmetry condition gives

$$W_{ab} A_{ac} A_{bd} = W_{cd} \qquad (4.22)$$

The monodromy matrix $\hat{A}$ has eigenvalues $\lambda = \pm 1$. Therefore a general solution of Eq. (4.22) is

$$W_{ab} = C_+ e_a^{(+)} e_b^{(+)} + C_- e_a^{(-)} e_b^{(-)} \qquad (4.23)$$

where $e^{(\pm)}$ are the eigenvectors of $\hat{A}^t$. Substituting the explicit expressions for these eigenvectors into the expression for $G(x, \bar{x})$, we find

$$G(x, \bar{x}) =$$

Lectures on ConformalField Theory

$$C_+[A_{01}\mathcal{F}^{(1)}(x)+(1+A_{00})\mathcal{F}^{(0)}(x)][A_{01}\mathcal{F}^{(1)}(\bar{x})+(1+A_{00})\mathcal{F}^{(0)}(\bar{x})]$$
$$+C_-[A_{01}\mathcal{F}^{(1)}(x)+(A_{00}-1)\mathcal{F}^{(0)}(x)][A_{01}\mathcal{F}^{(1)}(\bar{x})+(A_{00}-1)\mathcal{F}^{(1)}(\bar{x})] \quad (4.24)$$

To determine $C_\pm$ we recall that the correlation function must be a single-valued function. From Eq. (4.19) we see that $\mathcal{F}^{(0)}(x)$ is analytic in the vicinity of $x = 0$, but $\mathcal{F}^{(1)}(x)$ has a branch cut. Therefore there should be no terms containing cross products of these two solutions. Such cross products vanish if

$$C_+(1+A_{00}) = C_-(1-A_{00}) \quad (4.25)$$

Finally we get

$$G(x,\bar{x}) \sim A_{01}^2 \mathcal{F}^{(1)}(x)\mathcal{F}^{(1)}(\bar{x}) + (1-A_{00}^2)\mathcal{F}^{(0)}(x)\mathcal{F}^{(0)}(\bar{x}) \quad (4.26)$$

For the future purposes it is convenient to express the solution in terms of $\mathcal{F}^{(0)}$. Using Eq. (4.21), we get

$$G(x,\bar{x}) \sim [\mathcal{F}^{(0)}(x)\mathcal{F}^{(0)}(1-\bar{x}) + \mathcal{F}^{(0)}(\bar{x})\mathcal{F}^{(0)}(1-x)]$$
$$-A_{00}^{-1}[\mathcal{F}^{(0)}(1-x)\mathcal{F}^{(0)}(1-\bar{x}) + \mathcal{F}^{(0)}(x)\mathcal{F}^{(0)}(\bar{x})] \quad (4.27)$$

(here we have changed the normalization factor). The purpose of the latter exercise is to obtain a limit of integer $c$. In this case, according to Eq. (4.21), $A_{00}^{-1} \to 0$ and the cross-symmetric and single-valued solution is

$$G(x,\bar{x}) = \mathcal{F}(x)\mathcal{F}(1-\bar{x}) + \mathcal{F}(\bar{x})\mathcal{F}(1-x) \quad (4.28)$$

Now we shall consider several particular cases. As we know from Eq. (4.11), there are two possible conformal dimensions for each choice of $C$. For $C = 1/2$ we get
• Ising model

$$\Delta_- = 1/16, \quad \Delta_+ = 1/2 \quad (4.29)$$

which correspond to conformal dimensions of the Ising model primary fields (see Chapter 12 for details). Fields with the conformal dimensions $(1/16, 1/16)$ are called order (disorder) parameter fields and are denoted as $\sigma(z,\bar{z})$ ($\mu(z,\bar{z})$). These two fields are mutually nonlocal and one can choose either of them for the basis of primary fields.

There are two other interesting cases: $C = 0$ and $C = -2$. In the first case we have $\Delta_+ = 5/8$ and $\Delta_- = 0$. In the second case we have $\Delta_- = -1/8$ and $\Delta_+ = 1$. In both cases conformal dimensions of some primary fields are integers, thus being degenerate with degenerate with dimensions of the decendants of the unity operator. This leads to interesting modifications in the OPE (see Caux et al. (1996), (1997) and references therein). It goes without saying that both models are nonunitary. The model $C = 0$ describes classical percolation and the model with $C = -2$ describes dense polymers (Rozansky and Saleur (1992)). In the $\Delta = 5/8$ case we have

$$\mathcal{F}(x) = x^2 F(-1/2, 3/2, 3; \; x)$$

and for $\Delta = -1/8$ we have

$$\mathcal{F}(x) = F(1/2, 1/2, 1; \; x)$$

## 4.2 The Coulomb gas scheme for the minimal models

Let us consider how we can modify the Gaussian model without breaking the conformal symmetry. As we have once mentioned, the Gaussian model can be viewed as a model of a two-dimensional classical Coulomb gas where the bosonic exponents generate charges. One may expect that adding an extra charge on infinity (or spreading it over the background) will not break the conformal symmetry. For the sake of brevity we shall denote

$$V_\alpha = \exp[i\sqrt{4\pi}\alpha\Phi]$$

With the extra charge $-Q$ added to the system we shall adopt a new definition for correlation functions:

$$\langle V_{\alpha_1}(1)...V_{\alpha_N}(N)\rangle_Q = \lim_{R\to\infty} R^{2Q^2} \langle V_{\alpha_1}(1)...V_{\alpha_N}(N)V_{-Q}(R)\rangle \quad (4.30)$$

In particular, the only nonvanishing two-point correlation function is now given by

$$\langle V_\alpha(1) V_{(Q-\alpha)}(2)\rangle_Q$$

$$\equiv \lim_{R\to\infty} R^{2Q^2} \langle \exp[i\sqrt{4\pi}\alpha\Phi(1)] \exp[i\sqrt{4\pi}(Q-\alpha)\Phi(2)] \exp[-i\sqrt{4\pi}Q\Phi(R)]\rangle$$

$$= (|z_{12}|)^{-4\Delta} \quad (4.31)$$

$$\Delta = \frac{1}{2}\alpha(\alpha - Q) \quad (4.32)$$

So, as expected, the correlation function still decays as a power-law, but the exponents are modified thus indicating that the stress-energy tensor is changed.

Notice that the bosonic exponents having the same conformal dimension are now $V_\alpha$ and $V_{Q-\alpha}$. If our deformation generates a realistic critical theory, this theory must have novanishing four-point correlation functions of these exponents. It is obvious, however, that one cannot combine four exponents with charges $\alpha$ and $Q - \alpha$ to satisfy the electroneutrality condition

$$\sum_i \alpha_i = Q \qquad (4.33)$$

Therefore such simple modification of the Gaussian model is not sufficient to obtain a nontrivial theory. The way out of this difficulty was found by Dotsenko and Fateev (1984) (DF), who added marginal operators to the action of the Gaussian model. Following their method, let us consider the modified action

$$S = \int d^2x \left[ \frac{1}{2}(\partial_\mu \Phi)^2 + i\sqrt{4\pi} Q R^{(2)} \Phi + \sum_{\sigma=\pm} \mu_\sigma \exp(i\sqrt{4\pi}\alpha_\sigma \Phi) \right] \qquad (4.34)$$

where the 'charges' $\alpha_\pm$ are chosen in such a way that the dimensions of the corresponding exponents (*screening* operators) given by Eq. (4.32) are equal to one:

$$\alpha_\pm(\alpha_\pm - Q) = 2$$

or

$$\alpha_\pm = Q/2 \pm \sqrt{Q^2/4 + 2} \qquad (4.35)$$

The quantity $R^{(2)}$ represents the Riemann curvature of the surface. In this representation the extra charge $Q$ is spread over the entire surface. Such representation makes it easier to calculate the stress-energy tensor, but is not very convenient for calculations of the correlation functions.

The action (4.34) may appear odd, because the presence of $i = \sqrt{-1}$ in the action raises doubts whether the theory is unitary. We shall prove, however, that for certain values of $Q$ the unitarity is preserved. It may also appear that by adding nonlinear terms to the action we make the theory untractable. It is not the case, however, because multiparticle correlation functions of $V_{\alpha_\pm}$-operators vanish and thus the perturbation expansion in $V_{\alpha_\pm}$ contains only a finite number of terms. Later in this section we shall consider how to apply this procedure for calculations of correlation functions in more detail. Now let us discuss some more elementary problems. We know that the model (4.34) supports only those correlation functions where the 'charges' satisfy the 'electroneutrality' condition (4.33). As we have already seen, this condition is easily satisfied for two operators and we have concluded that exponents with the charges $-\alpha$ and $\alpha - Q$ have the same conformal dimensions. Now we want to have nontrivial multi-point correlation functions of such operators. Without screening charges such functions will always vanish. However, if $\alpha$ belongs to the set

$$\alpha_{n,m} = -\frac{1}{2}(n-1)\alpha_- - \frac{1}{2}(m-1)\alpha_+ \qquad (4.36)$$

the four point correlation function is not zero:

$$\langle V_{\alpha_{n,m}}(1) V_{\alpha_{n,m}}(2) V_{\alpha_{n,m}}(3) V_{Q-\alpha_{n,m}}(4) \rangle_Q$$

$$= \lim_{R\to\infty} R^{2Q^2} \mu^{(n+m-2)} \int d^2\xi_1...d^2\xi_{n+m-2} \langle V_{\alpha_{n,m}}(1) V_{\alpha_{n,m}}(2) V_{\alpha_{n,m}}(3)$$
$$V_{Q-\alpha_{n,m}}(4) V_{\alpha_+}(\xi_1)...V_{\alpha_+}(\xi_{n-1}) V_{\alpha_-}(\xi_n)...V_{\alpha_-}(\xi_{n+m-2}) V_Q(R) \rangle \qquad (4.37)$$

Notice, that the perturbation expansion in powers of $\mu$ contributes only a single term unless $\alpha_+ p + \alpha_- q = 0$ for some integer $p, q$. According to Eq. (4.36) the latter would amount to an existence of a primary field with zero conformal dimension (this occurs, for instance, in the theory with $C = 0$ mentioned in the previous Section). If there are no such operators, however, the model (4.34) remains tractable despite being nonlinear. Quantization condition (4.36) determines the spectrum of conformal dimensions of the theory (4.34). Substituting (4.36) into Eq. (4.32) we get the formula for the permitted conformal dimensions:

$$\Delta_{n,m} = \frac{1}{2}\left[(\alpha_- n - \alpha_+ m)^2 - (\alpha_- + \alpha_+)^2\right] \qquad (4.38)$$

Now we have to calculate the central charge. The stress-energy tensor can be calculated straightforwardly differentiating the action (4.34) with respect to the metric. The answer is

$$T(z) = -2\pi : \partial\Phi\partial\Phi : +i\sqrt{4\pi}Q\partial^2\Phi \qquad (4.39)$$

Calculating the correlation function $\langle T(z)T(0) \rangle = C/2z^4$ we find the value of the central charge:

$$C = 1 - 3Q^2 \qquad (4.40)$$

The family of models described by the action (4.34) contains a special subset of unitary models with

$$C = 1 - \frac{6}{p(p+1)}; \quad \Delta_{n,m} = \frac{[pn - (p+1)m]^2 - 1}{4p(p+1)} \quad (4.41)$$

where $p = 3, 4....$ These models are called *minimal*. The number of primary fields in these models is finite. The minimal model with $p = 3$ has $C = 1/2$ and is equivalent to the critical Ising model, the model with $p = 4$ has $C = 7/10$ and is equivalent to the Ising model at the tricritical point, and the model with $p = 5$ has $C = 4/5$ and equvalent to the $Z_3$ Potts model at criticality. In order to get a better insight in to how the DF scheme works, let us consider a four-point correlation function of $O_{1,2}$ fields. This field can be represented by the exponents $V_{-\alpha_+/2}$ and $V_{Q+\alpha_+/2}$. The nonvanishing four-point correlation function is obtained if one adds one screening operator:

$$\lim_{R \to \infty} R^{2Q^2} \langle V_{-\alpha_+/2}(1) V_{-\alpha_+/2}(2) V_{-\alpha_+/2}(3) V_{Q+\alpha_+/2}(4) V_{\alpha_+}(\xi) V_{-Q}(R) \rangle$$

Again, one can check that there are no contributions from other screening operators. As a result we have (from now on we shall drop the subscript $Q$ in notations of correlation functions)

$$\langle \Phi(1)...\Phi(4) \rangle \equiv \langle O_{1,2}(1) O_{1,2}(2) O_{1,2}(3) O_{1,2}(4) \rangle$$
$$= \mu \frac{|z_{12} z_{13} z_{23}|^{\alpha_+^2/2}}{|z_{14} z_{24} z_{34}|^{\alpha_+(\alpha_++2Q)/2}} \int d^2 \xi \frac{|z_4 - \xi|^{\alpha_+(\alpha_++2Q)}}{|(z_1 - \xi)(z_2 - \xi)(z_3 - \xi)|^{\alpha_+^2}} \quad (4.42)$$

Let us make sure that the obtained expression is conformally invariant; that is, it can be written in the canonical form (4.16). To achieve this we perform a transformation of the variable in the integral:

$$\xi = \frac{z_{13} z_4 \eta + z_{34} z_1}{z_{13} \eta + z_{34}} \quad (4.43)$$

This transformation maps the points $\xi = z_1, z_2, z_3, z_4$ onto $0, x, 1$ and $\infty$ respectively. Substituting this expression into (4.42) and taking into account the fact that $\alpha_+(\alpha_+ - Q) = 2$ we obtain

$$\langle \Phi(1)...\Phi(4) \rangle = \frac{1}{|z_{13} z_{24}|^{4\Delta}} |x(1-x)|^{\alpha_+^2/2} \int d^2 \eta \, [|\eta(1-\eta)(x-\eta)|]^{-\alpha_+^2/2} \quad (4.44)$$

where $\Delta = \alpha_+(\alpha_+ + 2Q)/4$. The latter expression coincides with Eq. (4.16) with

$$G(x,\bar{x}) = |x(1-x)|^{\alpha_+^2/2+4\Delta} \int d^2\eta \, [|\eta(1-\eta)(x-\eta)|]^{-\alpha_+^2/2} \quad (4.45)$$

The integral over $\eta$ may diverge at certain $\alpha$. In this case it should be treated as analytic continuation from the area of $\alpha$ where it is convergent. The problem of calculation of such integrals was solved by DF who have obtained the following expression for a general four-point correlation function:

$$\langle O_{n_1,m_1}(1) O_{n_2,m_2}(2) O_{n_3,m_3}(3) O_{n_4,m_4}(4) \rangle$$
$$= \frac{|z_{13}|^{2[\Delta(\alpha_1+\alpha_3+\alpha_+)-\Delta_1-\Delta_3+\alpha_+\alpha_2]} |z_{24}|^{2[\Delta(\alpha_2+\alpha_4)-\Delta_2-\Delta_4+\alpha_+\alpha_2]}}{|z_{12}|^{2[\Delta_1+\Delta_2-\Delta(\alpha_1+\alpha_2)]} |z_{23}|^{2[\Delta_2+\Delta_3-\Delta(\alpha_2+\alpha_3)]}}$$
$$\times |z_{34}|^{-2[\Delta(\alpha_3+\alpha_4+\alpha_+)-\Delta_3-\Delta_4]} |z_{14}|^{-2[\Delta(\alpha_1+\alpha_4+\alpha_+)-\Delta_1-\Delta_4]}$$
$$\times \{ \sin[\pi\alpha_-(\alpha_1+\alpha_2+\alpha_3)] \sin[\pi\alpha_-\alpha_2] |I_1(x)|^2$$
$$+ \sin[\pi\alpha_-\alpha_1] \sin[\pi\alpha_-\alpha_3] |I_2(x)|^2 \} \quad (4.46)$$

where $\Delta(\alpha)$ is determined by the formula (4.32) and

$$I_1(x) = \int_1^\infty dt \, t^{\alpha-\alpha_1}(t-1)^{\alpha-\alpha_2}(t-x)^{\alpha-\alpha_3}$$
$$= \frac{\Gamma[-1-\alpha_-(\alpha_1+\alpha_2+\alpha_3)]\Gamma(\alpha_-\alpha_2)}{\Gamma[-\alpha_-(\alpha_1+\alpha_3)]} F[-\alpha_-\alpha_3, -1-\alpha_-(\alpha_1+\alpha_2+\alpha_3), -\alpha_-(\alpha_1+\alpha_3); x]$$

$$I_2(x) = \int_0^x dt \, t^{\alpha-\alpha_1}(t-1)^{\alpha-\alpha_2}(t-x)^{\alpha-\alpha_3}$$
$$= x^{1+\alpha_-(\alpha_1+\alpha_3)} \frac{\Gamma(1+\alpha_-\alpha_1)\Gamma(1+\alpha_-\alpha_3)}{\Gamma[2+\alpha_-(\alpha_1+\alpha_3)]} F[-\alpha_-\alpha_2, 1+\alpha_-\alpha_1, 2+\alpha_-(\alpha_1+\alpha_3); x]$$

### References

- J.-S. Caux, I. I. Kogan and A. M. Tsvelik, *Nucl. Phys.* **B466**, 444 (1996).
- J.-S. Caux, A. Lewis, I. I. Kogan and A. M. Tsvelik, *Nucl. Phys.* **B489**, 469 (1997).
- V. S. Dotsenko and V. A. Fateev, *Nucl. Phys.* **B240**, 312 (1984).
- L. Rozansky and H. Saleur, *Nucl. Phys.* **B376**, 461 (1992).

## 5. Current (Kac–Moody) algebras; the first assault

In this Lecture we shall discuss fermionic current operators. We shall also give an example of nontrivial conformal theory related to these operators.

Let $R^+_{\alpha,n}, R_{\alpha,n}$ ($L^+_{\alpha,n}, L_{\alpha,n}$) be the right- (left-) moving components of *free* massless fermions, with $\alpha$ and $n$ ($n = 1, 2, ...k$) being the spin and 'flavour' indices, respectively. It is assumed that the fermionic fields transform according to some Lie groups G and F, operating in the spin and flavour spaces. The reason for introducing the second index $n$ will become clear later. The current operators on group G are defined as follows:

$$J^a(z) = \sum_{n=1}^{k} L^+_{\alpha,n}(z) \tau^a_{\alpha\beta} L_{\beta,n}(z)$$

$$\bar{J}^a(\bar{z}) = \sum_{n=1}^{k} R^+_{\alpha,n}(\bar{z}) \tau^a_{\alpha\beta} R_{\beta,n}(\bar{z}) \tag{5.1}$$

where $\tau^a$ are matrices – generators of the Lie algebra $\mathcal{G}$. They satisfy the following relations:

$$[\tau^a, \tau^b] = i f^{abc} \tau^c$$

$$\text{Tr}\,\tau^a \tau^b = \frac{1}{2} \delta^{ab} \tag{5.2}$$

For the case of the SU(2) group $\tau^a = S^a$ are the matrices of spin $S = 1/2$ and $f^{abc} = \epsilon^{abc}$. In a similar way one can define the current operators on the group F, using for this purpose its generators $T^j$. Since the fermionic operators with different chiralities commute, so do the currents with different chiralities; currents with the same chirality satisfy the following Wilson operator expansion:

$$J^a(z) J^b(z') = \frac{k}{8\pi^2 (z-z')^2} \delta^{ab} + i f^{abc} \frac{J^c(z')}{2\pi(z-z')} + ... \tag{5.3}$$

To check the validity of this very important identity, one should consider two diagrams: one for the current-current correlation function depicted on Fig. 2.2 and another for the correlation function of three currents (see Fig. 7.1).

The term with a derivative of the delta function is anomalous; it appears only in infinite systems. The second term in the right-hand side of Eq. (5.4) being absent in the spinless case is an ordinary commutator and can be obtained by a straightforward calculation.

Often the Kac–Moody algebra is written for the Fourier components of current operators. In this case we assume that the system is placed on the strip $0 < x < L$ with periodic boundary conditions and expand the current operators into the series

$$J^a(x) = \frac{1}{L} \sum_n e^{-2i\pi nx/L} J_n^a$$

Substituting this expansion into Eq. (5.4) we get the Kac–Moody algebra for the Fourier components:

$$[J_n^a, J_m^b] = \frac{nk}{2}\delta^{ab}\delta_{n+m,0} + if^{abc} J_{n+m}^c \qquad (5.4)$$

The Kac–Moody algebra includes the original algebra G as its subalgebra; this is the algebra of the zeroth components of the currents:

$$[J_0^a, J_0^b] = if^{abc} J_0^c \qquad (5.5)$$

Therefore one can think of these zeroth components as about matrices – generators of the group G.

Let us now follow the example of spinless fermions and consider the operators defined as G-invariant quadratic forms of chiral currents.

$$T(z) = \frac{1}{k+c_v} : J^a(z)J^a(z) :$$
$$\bar{T}(\bar{z}) = \frac{1}{k+c_v} : \bar{J}^a(\bar{z})\bar{J}^a(\bar{z}) : \qquad (5.6)$$

The numerical value of the coefficient $1/(k+c_v)$, where $c_v$ is the Casimir operator in the adjoint representation, i.e., is defined by the identity

$$f_{abc} f_{\bar{a}bc} = c_v \delta_{a\bar{a}}$$

It is possible to show that the operators such defined are indeed chiral components of a stress energy tensor, i.e. they satisfy the Virasoro algebra (3.19) with the central charge

$$C = \frac{\dim G}{k+c_v} \qquad (5.7)$$

The easiest way to prove it would be to calculate the two- and three-point correlation functions of $T$ making use of the free fermion representation of current operators (5.1).

## 5.1 Sugawara Hamiltonian for Wess–Zumino–Novikov–Witten model

According to the general property of conformal theories a sum of the zeroth components of traceless stress-energy tensors $T_0$, $\bar{T}_0$ gives a Hamiltonian. From Eq. (5.6) we extract the corresponding Hamiltonian which traditionally carries the name of Sugawara:

$$\hat{H}(k,G) = \frac{2\pi v}{L}(L_0 + \bar{L}_0)$$

$$= \frac{2\pi v}{(k+c_v)L}\left[J_0^a J_0^a + 2\sum_{n>0} J_{-n}^a J_n^a + (J \to \bar{J})\right] \quad (5.8)$$

The Sugawara Hamiltonian defines the Hamiltonian of the Wess–Zumino–Novikov–Witten (WZNW) model whose action will be discussed later. Since we have introduced the Hamiltonian (5.8) axiomatically, the reader may remain perplexed about its physical meaning. It is clear that the WZNW model represents some conformal theory, but we have not explained relations between this theory and more conventional models. These relations are established by the following identity which we give only for the case G = SU($N$):

$$\int dx[-iR^+_{\alpha,n}\partial_x R_{\alpha,n} + iL^+_{\alpha,n}\partial_x L_{\alpha,n}] = H[U(1)] + H[k, SU(N)] + H[N, SU(k)] \quad (5.9)$$

$$H[U(1)] = 2\pi \int dx[:J(x)J(x): + :\bar{J}(x)\bar{J}(x):] \quad (5.10)$$

$$H[k, SU(N)] = \frac{2\pi}{(N+k)}\sum_{a=1}^{D_N}\int dx[:J^a(x)J^a(x): + :\bar{J}^a(x)\bar{J}^a(x):] \quad (5.11)$$

$$H[N, SU(k)] = \frac{2\pi}{(N+k)}\sum_{\lambda=1}^{D_k}\int dx[:J^\lambda(x)J^\lambda(x): + :\bar{J}^\lambda(x)\bar{J}^\lambda(x):] \quad (5.12)$$

where $J^\lambda$ are currents of the SU($k$) group. This representation corresponds to the decomposition of the nonsimple symmetry group of the free fermions into the product of simple groups U(1)× SU($N$)× SU($k$). It has been used here that for the SU($N$) group $c_v = N$ and dim SU($N$) = $N^2 - 1$. The decomposition (5.9) is a generalization of the decomposition considered later in Chapter 15 for a case of fermions with U(1)×SU(2) symmetry. It would be too complicated to give a complete rigorous proof of the validity of this identity. A simple argument in favour of this identity is equality of the central charges. The central charge of the theory free Dirac fermions is equal to the number of fermion species, so it is $Nk$. According to (5.7) the central charge of the SU($N$)-invariant WZNW model is given by

$$C = \frac{k(N^2 - 1)}{k + N} \tag{5.13}$$

where the number of generators is $D_N = N^2 - 1$. Thus we have

$$Nk = 1 + \frac{k(N^2 - 1)}{k + N} + \frac{N(k^2 - 1)}{k + N}$$

which is identically satisfied.

From Eq. (5.9) we see that the WZNW model on group G can be understood as a G-invariant subsector of the model of free fermions. So, take free fermions and project some states out. If you are careful and do this projection intellegently preserving the conformal symmetry, you will obtain a model with interesting properties. As we shall see later in Part II, many interactions do this job perfectly thus leaving the WZNW model to be an effective theory for low-lying excitations. Now we have to discuss the WZNW theory in more detail. It is instructive to construct the full basis of its eigenstates. One should do this using the current operators *only*, because only in this way can one be sure that all eigenstates are G-invariant. To fulfil this task, we need to know how the current operators commute with the Hamiltonian. The corresponding commutation relations follow from the definition of the stress-energy tensor (5.6):

$$[L_n, J_m^a] = -m J_{n+m}^a \tag{5.14}$$

Since $H = 2\pi v(L_0 + \bar{L}_0)/L$, we get

$$[H, J_m^a] = -\frac{2\pi v m}{L} J_m^a \tag{5.15}$$

Now we have everything we need to construct the eigenstates. Let us define the vacuum vectors $|h\rangle$ as states annihilated by positive Fourier components of the currents:

$$J_m^a |h\rangle = 0; \quad \bar{J}_m^a |h\rangle = 0 \tag{5.16}$$

*Lectures on ConformalField Theory*

The Hamiltonian (5.8) acting on these states is reduced to

$$H_{\text{reduced}} = \frac{2\pi v}{(k+c_v)L}(J_0^a J_0^a + \bar{J}_0^a \bar{J}_0^a) \tag{5.17}$$

Therefore $|h\rangle$ are eigenstates of the Hamiltonian if they realize irreducible representations of the left and right G groups:

$$J_0^a J_0^a |h\rangle = C_{\text{rep}}|h\rangle; \quad \bar{J}_0^a \bar{J}_0^a |h\rangle = \bar{C}_{\text{rep}}|h\rangle \tag{5.18}$$

Now we can use the negative components of the currents as creation operators. According to Eq. (5.15), the following vectors

$$\bar{J}^{a_1}_{-m_1}...\bar{J}^{a_p}_{-m_p} J^{a_1}_{-n_1}...J^{a_q}_{-n_q}|h\rangle \tag{5.19}$$

are eigenvectors of the Hamiltonian (5.8) with the energies

$$E_{pq}(h) = \frac{2\pi v}{L}\left(\frac{C_{\text{rep}}}{k+c_v} + \sum_{i=1}^{q} n_i + \frac{\bar{C}_{\text{rep}}}{k+c_v} + \sum_{i=1}^{p} m_i\right) \tag{5.20}$$

As we have seen in Chapter 4, each eigenstate in conformal field theories is associated with some conformal field, whose conformal dimensions are related to the eigenvalues of energy and momentum via Eqs. (2.9) and (2.10). In the given case the eigenstates of the Hamiltonian are eigenstates of the Casimir operators of the group G. The conformal dimensions are

$$\Delta = \frac{C_{\text{rep}}}{k+c_v} + \sum_{i=1}^{q} n_i$$

$$\bar{\Delta} = \frac{\bar{C}_{\text{rep}}}{k+c_v} + \sum_{i=1}^{p} m_i \tag{5.21}$$

It turns out that the basis of states (5.19) is overcomplete. To make it complete one has to restrict the number of vacuum states $|h\rangle$ choosing a finite number of irreducible representations of G. For example, in the case G = SU(2) where the irreducible representations are representations of spin operators with $C_{\text{rep}} = S(S+1), S = 1/2, 1, ...$, the basis is composed by $S = 1/2, 1, ...k/2$ (Fateev and Zamolodchikov (1986)). Each vacuum vector $|h\rangle$ can be considered as a state created from the lowest vacuum $|0\rangle$ by a corresponding primary field $g_h(\tau, x)$. Indeed, according to Eq. (2.7) we have

$$|h\rangle = \lim_{\tau \to +\infty} e^{\tau(E_h - E_0)}(L/2\pi)^{(2\Delta_h + 2\bar{\Delta}_h)} g_h(\tau, x=0)|0\rangle \tag{5.22}$$

Thus the primary fields are matrices $g_{ab}$ – tensors realizing irreducible representations of the group G. There is one more restriction on the vacuum states, namely a requirement that physical fields must have integer or half-integer conformal spins. If these fields are built exclusively from the WZNW fields, this means that

$$\frac{C_{\text{rep}}}{k+c_v} - \frac{\bar{C}_{\text{rep}}}{k+c_v} = n/2 \quad (5.23)$$

where $n$ is an integer. In general, this requirement is difficult to satisfy except for $n = 0$. In this latter case $C_{\text{rep}} = \bar{C}_{\text{rep}}$ and the primary fields are dim $C_{\text{rep}} \times$ dim $C_{\text{rep}}$. This is the case considered in the original paper by Knizhnik and Zamolodchikov (1984). There are other cases, however, where physical fields are products of fields of several WZNW models (the corresponding examples will be given in the chapters where we discuss spin–charge separation). Then the restriction on right and left representations is different.

## 5.2   Knizhnik–Zamolodchikov (KZ) equations

Now we shall derive differential equations for multi-point correlation functions of the WZNW primary fields. The derivation is based on the fact that the WZNW stress energy tensor is quadratic in currents (5.6). Writing this expression in components, we get

$$L_n = \frac{1}{c_v + k} \sum_m : J_m^a J_{n-m}^a : \quad (5.24)$$

where the normal ordering assumes that the operators $J_m^a$ with a positive subscript (annihilation operators) stay on the right. Among the Virasoro generators there is one whose action on operators is particularly simple – it is $L_{-1} \equiv \partial_z$. For $n = -1$ we have

$$\partial_z \equiv L_{-1} = \frac{2}{c_v + k}[J_{-1}^a J_0^a + \sum_{m=1}^{\infty} J_{-m-1}^a J_m^a] \quad (5.25)$$

Since primary fields are vacuum states, they are annihilated by positive components of current operators (5.16). Therefore from (5.25) it follows that any primary field satisfies the identity

$$\left(\partial_z - \frac{2}{c_v + k} J_{-1}^a J_0^a\right) \phi_h(z) = 0 \quad (5.26)$$

# Lectures on ConformalField Theory

We already know that $J_0^a$ acts simply as the generator of the group (see Eq. (5.18)). To proceed further we need to know the action of $J_{-1}$. This can be extracted from the Ward identity

$$\langle J^a(z)\phi(1)...\phi(N)\rangle = \sum_j \frac{\tau_j^a}{z-z_j}\langle \phi(1)...\phi(N)\rangle \qquad (5.27)$$

Thus we have

$$\langle J_{-1}^a\phi(1)...\phi(N)\rangle = \frac{1}{2\pi i}\int_C dz(z_1 - z)^{-1}\langle J^a(z)\phi(z_1)...\phi(N)\rangle$$

$$= \sum_{j\neq 1}\frac{\tau_j^a}{z_1 - z_j}\langle \phi(1)...\phi(N)\rangle \qquad (5.28)$$

where the contour $C$ encircles $z$.

Combining these results we conclude that the $N$-point function of primary fields satisfy the following system of equations:

$$\left[\frac{1}{2}(c_v + k)\partial_{z_i} - \sum_{j\neq i}\frac{\tau_i^a \tau_j^a}{z_i - z_j}\right]\langle \phi(1)...\phi(N)\rangle = 0 \qquad (5.29)$$

where a matrix $\tau_i^a$ acts on the indices of the $i$-th operator.

Let us consider this equation for the case $N = 2$. Then we have

$$\left[\frac{1}{2}(c_v + k)\partial_{z_1} - \frac{\tau_1^a \tau_2^a}{z_1 - z_2}\right]\langle \phi(1)\phi(2)\rangle = 0$$

$$\left[\frac{1}{2}(c_v + k)\partial_{z_2} + \frac{\tau_1^a \tau_2^a}{z_1 - z_2}\right]\langle \phi(1)\phi(2)\rangle = 0 \qquad (5.30)$$

From these two equations it follows that $\langle \phi(1)\phi(2)\rangle = G(z_{12})$ and the function $G(z)$ satisfy the following equation:

$$\partial_z G_{\alpha,\alpha'}^{\beta,\beta'}(z) - \frac{2\tau_{\alpha,\gamma}^a \tau_{\beta,\delta}^a}{(c_v + k)z}G_{\gamma,\alpha'}^{\delta,\beta'}(z) = 0 \qquad (5.31)$$

It is natural to suggest that a nonzero solution exists only if the second field is the Hermitian conjugate of the first one: $\phi(2) = \phi^+(1)$. It is essential that $\tau_2^a$ acting on this operator gives minus sign. Then $\tau_1^a \tau_2^a$ becomes a Casimir operator:

$$\tau_1^a \tau_2^a = -(\tau^a)^2 = -C_{\rm rep}$$

and we get

$$\langle \phi(1)\phi^+(2)\rangle = z_{12}^{-2\Delta_h} \qquad (5.32)$$

with the correct conformal dimension (5.21).

Solutions of KZ equations for four-point correlation functions of primary fields in the fundamental representation of the SU(N) group were found by Knizhnik and Zamolodchikov (1984). Solutions for four-point functions of all primary fields of the $SU_k(2)$ WZNW model were found by Fateev and Zamolodchikov (1986). There is a regular procedure for finding multi-point correlation functions, called the Wakimoto construction, which is based on the representation of the WZWN operators in terms of free bosonic fields (Wakimoto (1986)). The details of this procedure can be found in (Dotsenko (1990)).

**References**

- I. Affleck and A. W. W. Ludwig, *Nucl. Phys.* **B352**, 849 (1991); *Phys. Rev. Lett.* **67**, 3160 (1991).

- Vl. S. Dotsenko, *Nucl. Phys.* **B338**, 747 (1990); **358**, 547 (1990).

- V. A. Fateev and A. B. Zamolodchikov, *Yad. Fiz. (Sov. Nucl. Phys.)* **43**, 657 (1986).

- V. G. Knizhnik and A. B. Zamolodchikov, *Nucl. Phys.* **B247**, 83 (1984).

- A. M. Tsvelik, *Sov. Phys. JETP*, **66**, 221 (1987).

- M. Wakimoto, *Commun. Math. Phys.* **104**, 605 (1986).

- E. Witten, *Commun. Math. Phys.* **92**, 455 (1984).

# 6. Non-Abelian bosonization

## 6.1 WZNW model in the Lagrangian formulation

Despite the fact that we have already learned a lot about WZNW model, we still lack a procedure which would be as powerful for non-Abelian models, as Abelian bosonization for more simple ones. Such a procedure exists; to formulate it we need to have the WZNW model in its Lagrangian form. The Lagrangian form of the WZNW model was derived by Polyakov and Wiegmann (1983). We shall reproduce their

derivation in the next Section.
**Theorem.** The Euclidean action for the Sugawara Hamiltonian

$$\hat{H} = \frac{2\pi}{k+c_v} \sum_{a=1}^{D} \int dx [: J^a(x)J^a(x) : + : \bar{J}^a(x)\bar{J}^a(x) :]$$

where the currents $J^a$, $\bar{J}^a$ satisfy the Kac–Moody algebras for the group G with the central extension $k$ defined by Eq. (5.4), is given by $S = kW(U)$, where $U$ is a matrix from the fundamental representation of the group G and $W(U)$ is

$$W(U) = \frac{1}{16\pi} \int d^2x \mathrm{Tr}(\partial_\mu U^{-1}\partial_\mu U) + \Gamma[U] \quad (6.1)$$

$$\Gamma[U] = -\frac{i}{24\pi} \int_0^\infty d\xi \int d^2x \epsilon^{\alpha\beta\gamma} \mathrm{Tr}(U^{-1}\partial_\alpha U U^{-1}\partial_\beta U U^{-1}\partial_\gamma U) \quad (6.2)$$

It is supposed that the field $U(\xi, \mathbf{x})$ in Eq. (6.2) is defined on a three-dimensional hemisphere whose boundary coincides with the two-dimensional plane where the original theory is defined, so that $U(\xi = 0, \mathbf{x}) = \mathbf{U}(\mathbf{x})$. The appearance of the third coordinate may look mysterious, but close inspection reveals that the integral is, in fact, $\xi$-independent (or rather *almost* independent). The reason for this is that the integrand in (6.2) is a total derivative – a Jacobian of transformation from three dimensional plane coordinates to the coordinates of the group. The first remarkable fact about Lagrangian (6.1) is that it has local equations of motion:

$$\partial \bar{J} = 0, \; \bar{J} = \frac{k}{2\pi} U^{-1}\bar{\partial}U,$$
$$\bar{\partial} J = 0, \; J = -\frac{k}{2\pi} U\partial U^{-1} \quad (6.3)$$

In fact, as we shall show later, the currents $J$, $\bar{J}$ are the WZNW currents introduced in Chapter 7, i.e. they have the same correlation functions as the currents obeying the Sugawara Hamiltonian of the Theorem.

The easiest way to obtain the equations of motion (6.3) is to use the following identity:

$$W(gU) = W(U) + W(g) + \frac{1}{2\pi} \int d^2x \mathrm{Tr}(g^{-1}\bar{\partial}g U\partial U^{-1}) \quad (6.4)$$

The meaning of Eq. (6.4) is easy to grasp considering its Abelian limit. Let $U = \exp(i\phi)$, $g = \exp(i\eta)$, where $\phi$, $\eta$ are scalar fields. In this case

the daunting Γ-term vanishes identically and the WZNW action (6.1) becomes
$$W[\exp(\mathrm{i}\phi)] = \frac{1}{16\pi}\int \mathrm{d}^2x (\partial_\mu \phi \partial_\mu \phi)$$

Substituting this into the identity (6.4) we get
$$\frac{1}{16\pi}\int \mathrm{d}^2x [\partial_\mu(\phi+\eta)\partial_\mu(\phi+\eta)] =$$
$$\frac{1}{16\pi}\int \mathrm{d}^2x (\partial_\mu \phi \partial_\mu \phi) + \frac{1}{16\pi}\int \mathrm{d}^2x (\partial_\mu \eta \partial_\mu \eta) + \frac{1}{8\pi}\int \mathrm{d}^2x (\partial_\mu \eta \partial_\mu \phi)$$

In order to appreciate the necessity of introduction of the auxilary variable $\xi$ in Eq. (6.2), let us consider the case when $G = SU(2)$. There is a simple parametrization for the fundamental representation of the SU(2) which allows to express the action explicitly in terms of the group coordinates. We mean the Euler parametrization:

$$U = \exp\left(\frac{\mathrm{i}}{2}\phi\sigma^z\right)\exp\left(\frac{\mathrm{i}}{2}\theta\sigma^x\right)\exp\left(\frac{\mathrm{i}}{2}\psi\sigma^z\right)$$
$$= \begin{pmatrix} \cos(\theta/2)\exp[\mathrm{i}(\phi+\psi)/2] & \mathrm{i}\sin(\theta/2)\exp[\mathrm{i}(\phi-\psi)/2] \\ \mathrm{i}\sin(\theta/2)\exp[\mathrm{i}(\psi-\phi)/2] & \cos(\theta/2)\exp[-\mathrm{i}(\phi+\psi)/2] \end{pmatrix} \quad (6.5)$$

where $\sigma^a$ are the Pauli matrices. The group coordinates $\psi, \theta, \phi$ belong to the manifold which is a sphere of radius $2\pi$ whose boundary is equivalent to its central point. The matrix $U$ is invariant with respect to the transformations of coordinates $\phi \to \phi + 2\pi$, $\psi \to \psi + 2\pi$, $\theta \to \theta$.

We advise the reader to prove the following valuable identities:
$$\Omega^3_\mu \equiv -\mathrm{i}\mathrm{Tr}[\sigma^z U^{-1}\partial_\mu U] = \partial_\mu \phi + \cos\theta \partial_\mu \psi$$
$$\Omega^\pm_\mu \equiv -\mathrm{i}\mathrm{Tr}[\sigma^\pm U^{-1}\partial_\mu U] = \mathrm{e}^{\pm\mathrm{i}\phi}[\partial_\mu \theta \pm \mathrm{i}\sin\theta \partial_\mu \psi] \quad (6.6)$$

With these identities we can express $\mathrm{i}U^{-1}\partial_\mu U$ as follows:
$$-\mathrm{i}U^{-1}\partial_\mu U = \frac{1}{2}\sigma_a \Omega^a_\mu \quad (6.7)$$

Substituting the latter expression into the Eq. (6.2) we find that the integrand is equal to the Jacobian of transformation from $\psi, \theta, \phi$ to $\xi, x$:

$$\Gamma_{\mathrm{su}(2)} = -\frac{\mathrm{i}}{96\pi}\int_0^\infty \mathrm{d}\xi \int \mathrm{d}^2 x \epsilon^{\alpha\beta\gamma}\epsilon_{abc}\Omega^a_\alpha \Omega^b_\beta \Omega^c_\gamma$$
$$= \frac{\mathrm{i}}{4\pi}\int \mathrm{d}\xi \mathrm{d}^2x \frac{\partial(\phi,\theta,\psi)}{\partial(\xi,x_1,x_2)} = \frac{\mathrm{i}}{4\pi}\int \mathrm{d}^2x \epsilon_{\mu\nu}\phi\sin\theta \partial_\mu \theta \partial_\nu \psi \quad (6.8)$$

As we see, the result of integration cannot be expressed unambiguously in terms of $U(x)$; $U$ is periodic in $\phi$, but $\Gamma_{\mathrm{su}(2)}$ is not: when $\phi$ changes by $2\pi q$ it changes by

$$\delta\Gamma_{\mathrm{su}(2)} = \frac{iq}{2}\int d^2x\, \epsilon_{\mu\nu}\sin\theta\, \partial_\mu\theta\, \partial_\nu\psi \qquad (6.9)$$

The latter integral is a topological invariant; it is an integer counting the number of times the vector field $\mathbf{n} = (\cos\theta, \sin\theta\cos\psi, \sin\theta\sin\psi)$ covers the sphere.

Thus the WZNW action (6.1) is what is called *multivalued* functional. One may wonder how we can use such ambiguosly defined thing as an action of a field theory. However, as it often happens in quantum theory, we will turn this ambiguity to our advantage. Indeed, the fact that the action (6.1) is defined modulo $2\pi i$ does not prevent the equations of motion (6.3) from being local. Neither does it prevent the partition function from being well defined, *provided k is an integer*. This means that the $\Gamma$-functional can become a part of an action of a physical theory only if it has an integer coefficient $k$. In other words, the multivaluedness of the action yeilds the quantization condition for $k$! Nedless to say, $k$ does not renormalize.

In the Lagrangian formulation we identify the primary fields with the following matrix fields.

(i) $U$-field – a primary field transforming according to the fundamental representation of the group G. If G = SU($N$) ($c_v = N$) its scaling dimension is equal to (see Eq. (5.21))

$$\Delta_U = \bar\Delta_U = \frac{N^2 - 1}{2N(N+k)} \qquad (6.10)$$

(ii) $\phi_1^{ab} = \mathrm{Tr}(U^{-1}t^a U t^b)$ where $t^a$ are generators of $G$. This field is also a primary one and belongs to the adjoint representation; its scaling dimensions are equal to

$$\Delta_1 = \bar\Delta_1 = \frac{c_v}{c_v + k} \qquad (6.11)$$

(iii) The 'wrong currents' $K^a \sim \mathrm{Tr}(t^a U^{-1}\partial U)$, $\bar K^a \sim \mathrm{Tr}(t^a \bar\partial U U^{-1})$ whose scaling dimensions are equal to $(\Delta_1 + 1, \Delta_1)$ and $(\Delta_1, \Delta_1 + 1)$, respectively. These fields are not primary.

(iv) The Lagrangian density $\mathrm{Tr}(\partial_\mu U^{-1}\partial_\mu U)$ is also not a primary field and has scaling dimensions equal to $(\Delta_1+1, \Delta_1+1)$. As we see, the Lagrangian density is an irrelevant operator. Therefore the general theory with the action

$$W(\lambda; U) = \frac{1}{2\lambda} \int d^2 x \mathrm{Tr}(\partial_\mu U^{-1} \partial_\mu U) + k\Gamma[U] \qquad (6.12)$$

scales to the critical point $\lambda = 8\pi/k$. The crossover was described exactly by Polyakov and Wiegmann who solved this model by the Bethe ansatz (1984). In fact, it would be proper to call this model the Wess–Zumino–Novikov–Witten model; then the theory described by the action (6.1) is 'the WZNW model at critical point' or *critical* WZNW model. Using the above results we can compile the following table.

Table

| Critical U($N$) WZNW-model with k = 1 | Massless Dirac Fermions |
|---|---|
| Action | Action |
| $\frac{1}{2} \int d^2 x (\partial_\mu \Phi)^2 + W[g]; g \in$ SU(N) | $2 \int d^2 x (R_\alpha^+ \partial_{\bar z} R_\alpha + L_\alpha^+ \partial_z L_\alpha)$ |
| Operators | Operators |
| $1/(2\pi a) \exp\left[i(N/4\pi)^{1/2}\Phi\right] g_{\alpha\beta}$<br>$i(N/4\pi)^{1/2}\partial\Phi$<br>$-i(N/4\pi)^{1/2}\bar\partial\Phi$<br>$1/2\pi \mathrm{Tr}(\tau^a g^{-1}\bar\partial g)$<br>$-1/2\pi \mathrm{Tr}[\tau^a g \partial g^{-1}]$ | $R_\alpha^+ L_\beta$<br>$\frac{1}{2} : L_\alpha^+ L_\alpha :$<br>$\frac{1}{2} : R_\alpha^+ R_\alpha :$<br>$: R_\alpha^+ \tau_{\alpha\beta}^a R_\beta :$<br>$: L_\alpha^+ \tau_{\alpha\beta}^a L_\beta :$ |
| Critical O($N$) WZNW-model with k = 1 | Massless Real Fermions |
| Action | Action |
| $W[g]; g \in O(N)$ | $2 \int d^2 x \sum_{\alpha=1}^{N} (\chi_{R,\alpha} \partial_{\bar z} \chi_{R,\alpha} + \chi_{L,\alpha} \partial_z \chi_{L,\alpha})$ |
| Operators | Operators |
| $1/(2\pi a) g_{\alpha\beta}$<br>$1/2\pi \left(g \partial g^{-1}\right)_{\alpha\beta}$<br>$-1/2\pi \left(g \bar\partial g^{-1}\right)_{\alpha\beta}$ | $\chi_{R,\alpha} \chi_{L,\beta}$<br>$: \chi_{R,\alpha} \chi_{R,\beta} :$<br>$: \chi_{L,\alpha} \chi_{L,\beta} :$ |

*Lectures on ConformalField Theory* 93

The presence of the Wess-Zumino term in the action of WZNW model (6.12) is strictly necessary for existence of the critical point. Ones this term is removed, the model becomes an asymptotically free theory.

## 6.2 Derivation of the Lagrangian

Let us consider relativistic fermions described by creation and annihilation operators $\psi^+_{n\alpha}$, $\psi_{n\alpha}$ ($n = 1, ..., k$; $\alpha = 1, ..., N$) interacting with a non-Abelian gauge field

$$A^{\alpha\beta}_\mu = A^a_\mu \tau^{\alpha\beta}_a$$

where $\tau_a$ are matrices-generators of some Lie group G. These matrices act only on the Greek indices; the interaction is diagonal with respect to the English indices. The fermionic part of the action has the standard form:

$$S_f = \int d^2x \sum_{n=1}^{k} \bar\psi_{n\alpha} \gamma_\mu (\partial_\mu \delta_{\alpha\beta} + iA^{\alpha\beta}_\mu) \psi_{n\beta} \qquad (6.13)$$

Let us integrate over the fermions and consider the effective action for the gauge field:

$$\exp\{-S_{\text{eff}}[A]\} = Z^{-1} \int D\bar\psi D\psi\, e^{-S_f[\bar\psi,\psi,A]} \equiv e^{kW[A]} \qquad (6.14)$$

$$W[A] = \ln\det[\gamma_\mu(\partial_\mu\delta_{\alpha\beta} + iA^{\alpha\beta}_\mu)] - \ln\det(\gamma_\mu\partial_\mu) \qquad (6.15)$$

The field interacts only with the SU(N) currents; using the results of the previous chapter on the non-Abelian bosonization, we can separate the relevant degrees of freedom and rewrite the above effective action as an average over the WZNW action:

$$e^{-S_{\text{eff}}[A]} = \langle \exp\left(-i \int d^2x A^a_\mu J^{a,\mu}\right) \rangle_W \qquad (6.16)$$

At the present stage we are not yet in a position to write down an explicit expression for this average in path integral form. The reason is that we do not know the expression for the WZNW-Lagrangian. In order to find it, we need to do some preparatory work.

Our first step is to calculate $W[A]$. Let us define the current:

$$j_\mu(x) = ik \frac{\delta W[A]}{\delta A_\mu(x)}$$

The functional $W[A]$ is completely determined by the two equations satisfied by its derivatives $j_\mu(x)$:

$$\partial_\mu j_\mu + \mathrm{i}[A_\mu, j_\mu] = 0 \qquad (6.17)$$

$$\epsilon_{\mu\nu}(\partial_\mu j_\nu + \mathrm{i}[A_\mu, j_\nu]) = -\frac{k}{2\pi}\epsilon_{\mu\nu} F_{\mu\nu} \qquad (6.18)$$

where

$$F_{\mu\nu} = \partial_\mu A_\nu - \partial_\nu A_\mu + \mathrm{i}[A_\mu, A_\nu]$$

In the operator language these equations should be understood as equations for motion for the current operators. In the path integral approach we understand them as identities for the correlation functions. Eqs. (6.17) and (6.18) were derived by Johnson (1963) and solved by Polyakov and Wiegmann (1983). The first one, Eq. (6.17), follows from the gauge invariance and expresses the fact that the non-Abelian charge

$$Q^a = \int \mathrm{d}x_1 J_0^a(x)$$

is a conserved quantity. The total charge is the sum of charges created by left- and right-moving particles. Naively one can expect that for a massless theory, where there are no transitions between fermionic states with different chirality, right and left charges conserve independently. However, as we have already seen in Chapter 2, this expectation is false because the measure is invariant only with respect to simultaneous gauge transformations of right and left particles. The phenomenon of nonconservation of chiral components of the charge is called *chiral anomaly*. As it was shown by Johnson, the difference between right and left charges is proportional to the total flux of the gauge field and Eq. (6.18) expresses this fact. Thus we have two equations (6.17) and (6.18) for two unknown functions $j_\mu$ and thus can attempt to solve this system (see also Polyakov (1988)). In order to do it we rewrite these equations in a more convenient form, namely, in the light cone coordinates. The corresponding notations are:

$$\bar{j} = j_0 + \mathrm{i}j_x, \; j = j_0 - \mathrm{i}j_x$$

$$\bar{A} = A_0 + \mathrm{i}A_x, \; A = A_0 - \mathrm{i}A_x;$$

$$\partial_\tau = \partial + \bar{\partial}, \partial_x = \mathrm{i}(\partial - \bar{\partial})$$

The equations (6.17) and (6.18) are gauge invariant; we shall solve them in a particular gauge:

$$\bar{A} = 0, \; A = -2\mathrm{i}g\partial g^{-1} \qquad (6.19)$$

where $g$ is a matrix from the G group. In these notations Eqs. (6.17) and (6.18) acquire the following form:

$$\partial \bar{j} + \frac{i}{2}[A, \bar{j}] + \bar{\partial} j = 0 \tag{6.20}$$

$$\partial \bar{j} + \frac{i}{2}[A, \bar{j}] - \bar{\partial} j = -\frac{ik}{\pi}\bar{\partial} A \tag{6.21}$$

Substituting $\partial \bar{j} + \frac{i}{2}[A, \bar{j}]$ from the first equation into the second one we find

$$j = \frac{ik}{2\pi}A = \frac{ik}{\pi}g\partial g^{-1} \tag{6.22}$$

Then the first equation transforms into

$$\partial \bar{j} + [g^{-1}\partial g, \bar{j}] + \frac{ik}{\pi}\bar{\partial}(g\partial g^{-1}) = 0 \tag{6.23}$$

whose solution is given by

$$\bar{j} = -\frac{ik}{\pi}g^{-1}\bar{\partial} g \tag{6.24}$$

In order to find the effective action $W[A]$ one has to integrate the equation

$$k\delta W = \frac{i}{2}\int d^2x \text{Tr}(\bar{j}\delta A) = -\frac{k}{2\pi}\int d^2x \text{Tr}[\delta g g^{-1} \bar{\partial}(g\partial g^{-1})] \tag{6.25}$$

(in the derivation of this expression we have used the invariance of trace under cyclic permutations and the identity $\delta g^{-1} = -g^{-1}\delta g g^{-1}$).

To prove this theorem we use identity (6.4). Let us show that the integral (6.16) can be written as the path integral with the WZNW-action:

$$e^{-S_{\text{eff}}[A]} = \langle \exp(-i\int d^2x A_\mu^a J^{a,\mu})\rangle_W =$$

$$\frac{\int [U^{-1}DU]\exp[-kW(U) - (k/2\pi)\int d^2x \text{Tr}(U^{-1}\bar{\partial} U g\partial g^{-1})]}{\int [U^{-1}DU]\exp[-kW(U)]} \tag{6.26}$$

where $[U^{-1}DU]$ means the measure of integration on the group G. Now using the identity (6.4) and invariance of the measure with respect to the transformations

$$U \to Ug^{-1}$$

we can rewrite the integral as follows:

$$e^{-S_{\text{eff}}[A]} = \frac{\int [U^{-1}DU]\exp[kW(g) - kW(Ug)]}{\int [U^{-1}DU]\exp[-kW(U)]} = \exp[kW(g)] \quad (6.27)$$

and get the answer obtained before. The theorem is proven.

**References**

- K. Johnson, *Phys. Lett.* **5**, 253 (1963).
- V. G. Knizhnik and A. B. Zamolodchikov, *Nucl. Phys.* **B247**, 83 (1984).
- S. Novikov, *Usp. Math. Nauk* **37**, 3 (1982).
- A. M. Polyakov in *Les Houches, Session XLIX, Fields, Strings and Critical Phenomena, 1988*, ed. by E. Brezin and J. Zinn-Justin, Elsevier Science Publ. 1989.
- A. M. Polyakov and P. B. Wiegmann, *Phys. Lett.* **B131**, 121 (1983); **141**, 223 (1983).
- A. M. Tsvelik, *Phys. Rev. Lett.* **72**, 1048 (1994).
- E. Witten, *Commun. Math. Phys.* **92**, 455 (1984).

## 7. Semiclassical Approach to WZNW models

The WZNW action may be written in the form

$$S = \int d^2x \left[ \sqrt{g} g^{\mu\nu} G_{ab}[X] \partial_\mu X^a \partial_\nu X^b + \epsilon^{\mu\nu} B_{ab}[X] \partial_\mu X^a \partial_\nu X^b \right] \quad (7.1)$$

where $X^a$ are fields representing the coordinates on some group (or coset) manifold, $G_{ab}$ is the metric tensor on this manifold, and $B_{ab}$ is an antisymmetric tensor. We define the action on a curved (world sheet) surface with a metric $g_{\mu\nu}$ ($\mu, \nu = 1, 2$).

A very important feature of the action (7.1) is that the second term does not contain the world sheet metric. Consequently, the classical stress-energy tensor, $T_{\mu\nu} = \delta S / \delta g^{\mu\nu}$, is determined solely by the first term. In particular, the most important components for the critical model are given by,

$$T_{zz} = G_{ab}[X]\partial_z X^a \partial_z X^b, \quad T_{\bar{z}\bar{z}} = G_{ab}[X]\partial_{\bar{z}} X^a \partial_{\bar{z}} X^b \quad (7.2)$$

where $z = x_0 + ix_1, \bar{z} = x_0 - ix_1$. Here, the reader should not get the false impression that the Wess–Zumino term is not important. Unlike the *topological term* in action it does contribute to the equations of motion. Since the model (7.1) is supposed to be *critical*, these equations (to be understood as identities for correlation functions in the quantum theory) are

$$T_{z\bar{z}} = 0, \quad \partial_{\bar{z}} T_{zz} = 0, \quad \partial_z T_{\bar{z}\bar{z}} = 0. \tag{7.3}$$

Their fulfillment depends on the Wess–Zumino term through the dynamics of the underlying fields $X^a$.

The smallness of $1/k$ means that there are many fields in the theory with conformal dimensions much smaller than unity. This gives weight to the idea that the critical point occurs in the region where the coupling constant of the sigma model is relatively small. Thus, one may attempt to describe the critical point using the semiclassical approximation.

In the infinite plane, parametrized by coordinates $(z, \bar{z})$, conformal invariance restricts the two-point function of primary fields of conformal dimension $(h, \bar{h})$ to be of the form

$$\langle \phi(z_1, \bar{z}_1) \phi(z_2, \bar{z}_2) \rangle = (z_1 - z_2)^{-2h} (\bar{z}_1 - \bar{z}_2)^{-2\bar{h}}. \tag{7.4}$$

Under a conformal transformation of the plane, $w = w(z)$, this correlation function transforms like a tensor of rank $(h, \bar{h})$:

$$\langle \phi(w_1, \bar{w}_1) \phi(w_2, \bar{w}_2) \rangle = \prod_{i=1}^{2} \left( \frac{dz}{dw} \right)_{w_i}^{h} \left( \frac{d\bar{z}}{d\bar{w}} \right)_{\bar{w}_i}^{\bar{h}} \langle \phi(z_1, \bar{z}_1) \phi(z_2, \bar{z}_2) \rangle. \tag{7.5}$$

One may pass from the infinite plane to a strip of width $2\pi R$ by means of the conformal transformation $w = R \ln z$. Combining this transformation with (7.4) and (7.5), one obtains the two-point function in the strip geometry:

$$\langle \phi(w_1, \bar{w}_1) \phi(w_2, \bar{w}_2) \rangle = [2R \sinh(w_{12}/2R)]^{-2h} [2R \sinh(\bar{w}_{12}/2R)]^{-2\bar{h}}. \tag{7.6}$$

Introducing coordinates $(\tau, \sigma)$ along and across the strip respectively ($w = \tau + i\sigma$, $\bar{w} = \tau - i\sigma$; $-\infty < \tau < \infty$, $0 < \sigma < 2\pi R$), one may expand (7.6) in the following manner,

$$\langle \phi(\tau_1, \sigma_1) \phi(\tau_2, \sigma_2) \rangle = \sum_{n=0}^{\infty} \sum_{m=-\infty}^{\infty} C_{nm} e^{-(h+\bar{h}+n)|\tau_{12}|/R} e^{-i(h-\bar{h}+m)|\sigma_{12}|/R} \tag{7.7}$$

One may also obtain the two-point function in the operator formalism, leading to the Lehmann expansion:

$$\langle \phi(\tau_1, \sigma_1) \phi(\tau_2, \sigma_2) \rangle = \sum_{\alpha} |\langle 0|\hat{\phi}|\alpha \rangle|^2 e^{-E_\alpha |\tau_{12}| - i P_\alpha |\sigma_{12}|} \tag{7.8}$$

where $E_\alpha$ and $P_\alpha$ are the eigenvalues of the Hamiltonian and the momentum operator respectively, in the state $|\alpha\rangle$. Comparing (7.7) and (7.8) one obtains a relationship between the eigenvalues of the Hamiltonian and the momentum operator, and the scaling dimensions in the corresponding CFT:

$$E_\alpha = \frac{h + \bar{h} + n}{R}, \quad P_\alpha = \frac{h - \bar{h} + m}{R}. \tag{7.9}$$

Restricting our attention to fields with $h = \bar{h} = d/2$, one may rewrite (7.6) in the form

$$\langle \phi(\tau_1, \sigma_1) \phi(\tau_2, \sigma_2) \rangle = R^{-2d} \left[ 2\cosh(\tau_{12}/R) - 2\cos(\sigma_{12}/R) \right]^{-d}. \tag{7.10}$$

One observes that for $\tau_{12} \gg R$ the asymptotic form of the correlation function is *independent* of $\sigma$,

$$\langle \phi(\tau_1, \sigma_1) \phi(\tau_2, \sigma_2) \rangle \sim R^{-2d} \exp(-d\tau_{12}/R) \quad (\tau_{12} \gg R) \tag{7.11}$$

and in this limit one should set $n = m = 0$ in equations (7.7) and (7.9).

Let us now place the model (7.1) on a thin strip of width $2\pi R$, and neglect any $\sigma$-dependence of the fields $X_a$. From our considerations in the previous paragraph, such a procedure preserves the (large $\tau$) asymptotics of the correlation functions. The action (7.1) becomes

$$S = 2\pi R \int d\tau \, G_{ab}[X] \partial_\tau X^a \partial_\tau X^b, \tag{7.12}$$

which may be recognised as the action for a free, non-relativistic particle, of mass $m = 4\pi R$. The corresponding Hamiltonian is the Laplace–Beltrami operator (multiplied by $-1/2m$),

$$\hat{H} = \frac{-1}{8\pi R \sqrt{G}} \frac{\partial}{\partial X^a} \left( \sqrt{G} G^{ab} \frac{\partial}{\partial X^b} \right) \tag{7.13}$$

As we have already established, the eigenvalues of this Hamiltonian are related to the spectrum of scaling dimensions in our CFT by equation (7.9). Moreover, solution of this Schrödinger equation allows one to obtain explicit expressions for the eigenstates, $|\alpha\rangle$, appearing in (7.8).

Chapter 3

# MAGNETIZATION AND ORBITAL PROPERTIES OF THE TWO-DIMENSIONAL ELECTRON GAS IN THE QUANTUM LIMIT

S. Wiegers[1]*, E. Bibow[1], L. P. Lévy[1,2]†, V. Bayot[1,3], M. Simmons[4], M. Shayegan[5]

[1] *Grenoble High Magnetic Field Laboratory, CNRS and MPI-FKF, BP 166, F-38042 Grenoble, France*
[2] *Institut Universitaire de France and Université Joseph Fourier, BP 53, F-38400 St Martin d'Hères, France*
[3] *Université Catholique de Louvain, 1348 Louvain la Neuve, Belgium*
[4] *Cavendish Laboratory, Madingley Road, Cambridge CB3 OHE, UK*
[5] *Department of Electrical Engineering, Princeton University, Princeton N.J. 08544*

**Abstract**    In a two-dimensional electron gas, incompressible states at specific filling factors induce discontinuous jumps $\Delta M$ in the magnetization. The amplitudes of the jumps are related to the energy gap $\Delta_\mu$ between compressible states. When probing $\Delta M$, it is possible to learn how physical parameters, such as disorder, spin-polarization and Coulomb interactions affect $\Delta_\mu$, the incompressibility gap. In these lectures, the relation between the particular orbital structures found at integer and fractional filling factors, the thermodynamic density of states, the corresponding current distributions and the orbital magnetization are emphasized. The non-equilibrium states very often encountered at incompressible filling-factors due to the absence of dissipation are also discussed.

**Keywords:**    73.40.Hm Quantum Hall Effect
73.20.Dx Magneto-transport
71.10.Pm Thermodynamics

---

*Present address:* Research Institute for Materials, University of Nijmegen, P.O. Box 9010, 6500 GL Nijmegen, The Netherlands.
†*Corresponding author:* levy@grenet.fr

## 1. INTRODUCTION

In high magnetic fields, a two-dimensional electron gas (2DEG) with sufficiently low disorder exhibits quantized Hall (QH) plateaux at integer and fractional filling factors[1, 2]. While the microscopic origin of this quantization is different in integer and fractional regimes, the macroscopic transport and thermodynamic properties are dominated by the presence of a charged gap in the excitation spectrum: for example, the longitudinal resistivity $\rho_{xx}$ minima are thermally activated over the energy gaps[4]. With a charge gap, the two-dimensional electron fluid is "incompressible"[3], a fundamental property for most physical observable. The relation between energy gaps, incompressibility and the quantum Hall effect is so fundamental that it is worth being restated explicitly. The pressure of the two-dimensional charged fluid, ($P = -\partial E/\partial \mathcal{A}$, where $E$ is the total energy and $\mathcal{A}$ the area occupied by the electron fluid) can be related to the chemical potential using

$$-P = \frac{\partial E}{\partial \mathcal{A}} = n\frac{\partial E}{\partial N} = n\mu \qquad (1.1)$$

where $n = \frac{\partial N}{\partial \mathcal{A}} = \frac{N}{\mathcal{A}}$ is the aerial electronic density and $N$ the number of electrons. According to its definition, the inverse compressibility is proportional to the thermodynamic derivative of the chemical potential with respect to the aerial density:

$$\frac{1}{\kappa} = -\mathcal{A}\frac{\partial P}{\partial \mathcal{A}} = \mathcal{A}\frac{\partial^2 E}{\partial \mathcal{A}^2} = n^2 \frac{d\mu}{dn}. \qquad (1.2)$$

This quantity is infinite when the chemical potential jumps discontinuously across a charged gap at specific filling factors. Hence the presence of energy gaps implies incompressibility. Since the electron liquid is not only incompressible but also irrotational, its only low energy excitation can only be surface waves: in the context of the QH effect, they have been identified with the edge excitations of the QH fluid. If there is a true gap $\Delta$ in the bulk, there cannot be any change in the current density anywhere but at the edge as long as the chemical potential changes by an amount $\delta\mu$ small compared to $\Delta$. Since this current change takes place at the edges, the variation in orbital magnetization is simply $\delta M = \delta I \mathcal{A}$. As shown in Sec. II and IV, thermodynamics requires the change in $\delta M$ to be proportional to $\delta\mu$. Hence

$$\delta I = \frac{\nu}{\Phi_0}\delta\mu. \qquad (1.3)$$

This implies Hall quantization when the change in chemical potential $\delta\mu = e\delta V$ is driven by a potential difference $\delta V$ between different edges.

In practice, disorder can also introduce some gapless excitation in the bulk. They do no affect the quantization of $\sigma_{xy} = \nu\frac{e}{\Phi_0}$ provided that the corresponding electronic states are localized on a length scale small compared to sample size. On the other hand, the current distribution is affected in the vicinity of compressible regions localized in the bulk. It is therefore necessary to distinguish between the mobility gap $\Delta_{tr}$ probed in transport and the incompressibility gap $\Delta_\mu$ revealed in thermodynamic (i.e. magnetization) measurements. According to the above argument,

$$\Delta_\mu \leq \Delta_{tr} \qquad (1.4)$$

the equality being reached in the clean limit.

The purpose of these lectures is to illustrate some of the thermodynamic phenomena related to the presence of charge gaps in the QH effect with experiments carried out in the integer and fractional regime. The paper is organized as follows. Section II makes the connection between the incompressibility of QH states and the thermodynamics of the two-dimensional electron gas for "clean" and "dirty" samples. An experimentally observable consequence of incompressibility and gapless edge excitation are large non-equilibrium dissipationless currents which are illustrated in Sec. 3. The observed magnetic oscillation in the integer regime are discussed in Sec. 4 at high and low filling factors. The role of spin degrees of freedom and of electron-electron interaction are pursued further in Sec. 4.4. Finally the last section is devoted to the fractional regime where the role of disorder is more relevant. In particular, the presence of bulk current are quite manifest when comparing thermodynamic and transport gaps.

## 2. DE HAAS-VAN ALPHEN EFFECT AND THE THERMODYNAMIC OF CLEAN AND DIRTY 2DEG

It is convenient at this stage to ignore spin degrees of freedoms and electron-electron interactions. Disorder and other confining potentials will be introduced later. In a sufficiently large magnetic fields, electron occupy only a few Landau levels with energies $\epsilon_n = \hbar\omega_c(n + \frac{1}{2})$. The number of Landau level (LL) occupied depends only on the number of particles $N$ and on the magnetic field, through the LL degeneracy $D = \frac{\Phi}{\Phi_0} = \frac{AB}{\Phi_0}$ and is measured by the filling factor

$$\nu = \frac{N}{D} = N\frac{\Phi_0}{\Phi}, \quad \nu_i = \text{int}(\nu). \tag{1.5}$$

$\nu_i$ is here the number of fully occupied LL, while the $\nu_i + 1$ level has a partial occupation of $\delta\nu = \nu - \nu_i$. The total energy of the electron gas can be computed by summing the energies of all occupied electronic states. At sufficiently high magnetic field, only the lowest LL is occupied and the total energy is simply

$$\mathcal{E}_0 = \frac{\hbar\omega_c}{2}N. \tag{1.6}$$

When $\nu_i = 1$, only $N - \frac{\Phi}{\Phi_0} = \frac{\Phi}{\Phi_0}(\nu - 1) = \frac{\Phi}{\Phi_0}\delta\nu$ electrons occupy the $2^{nd}$ LL, while the lowest LL has its maximum occupation $D = \frac{\Phi}{\Phi_0}$. Consequently the total energy is

$$\mathcal{E}_1 = \frac{\hbar\omega_c}{2} \times \frac{\Phi}{\Phi_0} + \frac{3\hbar\omega_c}{2}\left(N - \frac{\Phi}{\Phi_0}\right). \tag{1.7}$$

The same argument can be repeated for higher LL: when the $\nu_i + 1$ level is partially occupied, the energy of the $\frac{\Phi}{\Phi_0}\delta\nu$ electrons occupying this LL has to be added to the energy of the $\nu_i$ fully occupied levels

$$\mathcal{E}_{\nu_i} = \frac{\Phi}{\Phi_0}\hbar\omega_c\left(\sum_0^{\nu_i-1}\left[j + \frac{1}{2}\right] + \delta\nu\left[\nu_i + \frac{1}{2}\right]\right) \tag{1.8}$$

$$M_{\nu_i} = -\frac{\partial \mathcal{E}_{\nu_i}}{\partial B} = 2\mu_* N\left(\frac{\nu_i}{\nu}[\nu_i + 1] - \left[\nu_i + \frac{1}{2}\right]\right), \tag{1.9}$$

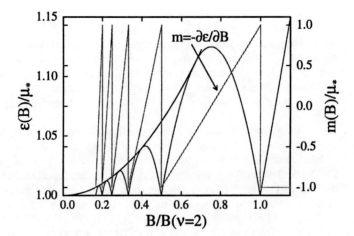

*Figure 1.1* Total energy and orbital magnetization of a 2DEG as a function of magnetic field. $\mathcal{E}$ is a piecewise parabolic function of B combining to the linear dependence of the LL energy and degeneracy with field.

where the magnetization is obtained as the partial derivative of the total energy with respect to the field. The energy of the 2DEG is a piecewise parabolic function of the magnetic field as shown in figure 1.1, while the magnetization is a sawtooth function periodic in $\nu$ ($1/B$) with amplitude $\mu_* N$ where $\mu_* = \frac{e\hbar}{2m_*}$ is the orbital magnetic moment. This unusual structure for a thermodynamic quantity was first derived by Peierls[7] and revisited by Vagner and co-workers in the context of the quantum Hall effect[8].

Each LL can be viewed as an angular momentum shell, since corresponding states in different LL can be labeled by their angular momentum. When the flux through the sample is increased by one quantum, there are $\nu_i$ electrons in the partially filled $\nu_i + 1$ LL that are transferred into the lower LL: the overall angular momentum decreases by

$$\Delta L = (\sum_{i=0}^{\nu_i} i - \nu_i(\nu_i + 1))\hbar = -\frac{\nu_i(\nu_i + 1)}{2}\hbar \qquad (1.10)$$

and the magnetization increases by $\Delta M = -2\frac{|e|}{m_*}\Delta L$, in agreement with Eq. 1.9. This accounts for the linear increase of the magnetization between LL.

The discontinuous jumps of the orbital magnetization can be traced to the depopulation of the edge states which carry a large orbital current. The physics at edges is best described by considering explicitly the potential $V(z)$ ($z = x + iy$) confining the 2DEG. In very high magnetic field, electronic states are localized around a center $z_{n,m} = x_{n,m} + iy_{n,m}$ with a spatial extend governed by the magnetic length $\ell_B = \sqrt{\Phi_0/(2\pi B)}$. If $V(z)$ varies slowly on the scale of $\ell_B$, a semiclassical description of LL states in the confining potential is straightforward: the LL energy levels are pushed upward at the edges by $V(z)$ according to

$$\epsilon_{n,m} = \hbar\omega_c\left(n + \frac{1}{2}\right) + V(z_{n,m}). \qquad (1.11)$$

# Magnetization of the 2DEG

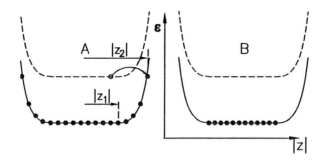

*Figure 1.2* Spatial dependence of LL energies illustrating the depopulation of edge states at integer filling factors.

The second term partially lifts the LL degeneracy, the degeneracy along equipotential lines being preserved. Fig. 1.2 shows the spatial dependence of the two lowest LL energies in terms of the center coordinates $z_{n,m}$, pictured as solid circles. The depopulation of edge states occurs over a small magnetic field interval around integer filling: for the lowest LL, the depopulation occurs between the configuration $A$ and $B$ represented graphically in Fig. 1.2. The change in total energy between $A$ and $B$ is dominated by the energy of edge states which, for a cylindrically symmetric confining potential, is

$$\Delta\mathcal{E} = \mathcal{E}_A - \mathcal{E}_B \approx \sum_{|z_{0,m}|\in[z_1,z_2]} V(z_{0,m}). \tag{1.12}$$

The corresponding change in magnetization is

$$\begin{aligned}
\delta M &= \frac{\partial \Delta\mathcal{E}}{\partial B} \approx \sum_{|z_{0,m}|\in[z_1,z_2]} \frac{\partial V}{\partial |z|} \frac{\partial |z_{0,m}|}{\partial B} \\
&= \frac{N}{B} \int_{z_1}^{z_2} \frac{\partial V}{\partial |z|} d|z| = \frac{N}{B}(V(z_2) - V(z_1))
\end{aligned} \tag{1.13}$$

since the density of state is constant in two-dimension $\frac{\partial |z_{0,m}|}{\partial B} = \frac{|z_{0,m}|}{B} = \frac{N}{B}(\delta|z_{0,m}|)$ (N is the total number of particles, $\delta z_{0,m}$ is the distance between neighboring center coordinates). $V(z_2) - V(z_1)$ is the energy gap $\Delta$ between LL, equal to $\hbar\omega_c$ in the ideal, disorder free and non-interacting limit. The magnetization "jump" at integer filling factors can therefore be related to the energy gap between LL according to

$$\boxed{\frac{\delta M}{N} = \frac{\Delta}{B}, \quad \delta M = \nu \frac{A}{\Phi_0} \Delta.} \tag{1.14}$$

This fundamental relation is in fact thermodynamic, and also holds in the fractional regime as will be shown in Sec. V. In this ideal limit, the magnetization jump per particle $\delta M/N$ is $\Delta/B = 2\mu_*$ in agreement with Eq. 1.9. Hence the overall current carried by a LL at the edges generate a magnetic moment which compensates exactly the magnetization of all the bulk LL states. This is in fact a leftover of the classical van Leuwen theorem stating that the magnetization of

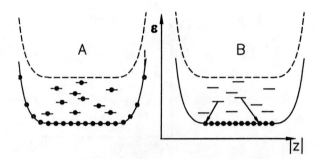

*Figure 1.3* In presence of short range disorder, the depopulation of edge states (between A and B) is accompanied by the transfer of localized states in the gap into the LL states.

skipping orbits at the edges compensate exactly the magnetic moment of all the cyclotron orbits in the bulk. It is because the energies at edges are higher than bulk states (a consequence of incompressibility) that the compensation takes the form of the Peierls sawtooth oscillation in a quantum description.

The role of disorder is now addressed. In 2D heterostructures or quantum wells, disorder can arise from defects or impurities at (or close) to the $AlGaAs - GaAs$ interface (short range disorder). Some electronic states can be localized on these defects and are usually spread over a broad energy range ($> \Delta$) as shown in Fig. 1.3. During the depopulation of edge states, electrons localized in the gap are transferred to the bulk lower LL states, giving rise to a correction $\delta M = n_i A \Delta (\mu_* - \mu_\ell)$ to the magnetization jump. Here $n_i$ is the density of localized impurity states and $\mu_\ell$ the orbital moment of the localized states (usually smaller than $\mu_*$). In high mobility sample $\delta M$ remains small. However, between A and B, the change in magnetic flux through the sample is determined by the number of additional states $n_i A \Delta$ necessary to accommodate the additional electrons. Hence, the width of the magnetization jump as a function of magnetic field

$$\delta B = n_i \Delta \Phi_0 \tag{1.15}$$

is controlled by $n_i$.

A more fundamental source of disorder arises from the ionized $Si$ donors usually placed at some setback distance $d$ (typically between 300 Å and 1500 Å) from the interface. Since the ionized donors are (nearly) randomly distributed, they scatter electrons via a slowly (on a scale larger than $d$) varying random potential, which can be added to the confining potential $V(z)$ as illustrated in Fig. 1.4. The amplitude of potential fluctuations induced by a mean density $C$ of charged $Si$ donors, are of the order of

$$W = \frac{e^2 \sqrt{C}}{\kappa} \tag{1.16}$$

for randomly distributed donors ($\kappa$ is here the average dielectric constant of the structure). For realistic samples, values for $W$ can range between 40 and 200 K, which are comparable or larger than the cyclotron ($\approx 20 - 200$ K) and fractional gaps ($< 10$ K). It is possible to reduce further $W$ after illumination of the sample by red or infrared light. Light smoothens the charge distribution

Magnetization of the 2DEG    105

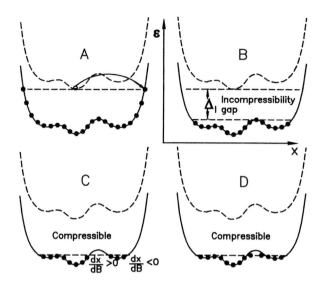

*Figure 1.4* The long range disorder is described by a slow variation of $V(z)$. Between configuration A and B, there are no compressible states in the bulk and the depopulation of edge states is rapid (incompressibility gap). In configuration C, the localized states in the bulk carry currents which oppose the edge current and reduce the magnetization jumps. Electron-electron interactions spread the charges in the classically forbidden region (D) to satisfy electro-neutrality but do not change qualitatively the current distribution in the compressible regions.

of ionized donors improving the sample mobility by a factor ranging from 20 to 60%. In a first step, we consider the effect of $W$ as a static disorder potential and we ignore screening, which considerably reduces the effect of the potential fluctuations in compressible regions. Screening is considered in a second step at the end of this section.

The evolution of the chemical potential and the spatial distribution of the electron density are sketched in Fig.1.4 in the vicinity of integer filling factors $\nu_i$ as a function of field. Between the configuration A and B, the only gapless excitation are at the sample edge: the depopulation of edge states yields to the same magnetization change as in the ideal case. The gap inferred from the magnetization swing $M_B - M_A$ using Eq. 1.14 defines the incompressibility gap $\Delta_\mu$. Beyond B, compressible states appear in the bulk (configuration C or D), localized on a lengthscale larger than the setback distance $d$. Since the area enclosed by these states localized on equipotential lines increases as the magnetic field increases, the magnetic moment induced by the localized currents opposes the magnetic moment of edge currents. In real samples, electro-neutrality which is violated in regions forbidden by the fluctuations of $V(z)$ is restored by electron-electron interactions, spreading localized states (configuration D in Fig. 1.4), without changing their compressible nature. The overall picture is the same as in the disorder-free case: the rapid magnetization swing can be attributed to incompressibility gaps always smaller than the transport and cyclotron gaps.

When compressible bulk states appear, they contribute to a reduction of the magnetization swing at integer filling factors.

The incompressible gap $\Delta_\mu$ is always smaller than the mobility gap $\Delta_{tr}$. Indeed, a mobility edge is reached when there is an extended state connecting sample edges through the bulk. This occurs when an incompressible region percolates through the sample, i.e. when the chemical potential is close to $\epsilon_n + \langle V(z) \rangle$, which is always smaller than $\epsilon_n + \max[V(z)]$, hence $\Delta_\mu < \Delta_{tr}$.

We now address the role of screening, which has dramatically different effects in compressible and incompressible regions as emphasized by Efros and coworkers[9, 10, 11]. In compressible regions, the chemical potential is nearly constant and the electron fluid adjusts its density to screen long range potential fluctuations induced by the donors. In these regions, the 2DEG can be considered as a good metal and the disorder potential is screened on long lengthscales. In incompressible regions, the constraint $\nu = \nu_i$ fixes the density and screening is limited to a readjustment of electron correlations. In this sense, the incompressible regions resemble a dielectric since an electric field $\mathcal{E}$ exists perpendicularly to the equipotential lines. This electric field "tears apart" the incompressible regions into narrow strips of width $w$ separating compressible regions with $\nu_- < \nu_i$ and $\nu_+ > \nu_i$. This width is specified by the condition that the jump in the electrochemical potential $\delta\mu \approx \mathcal{E}w$ equals the incompressibility gap $\Delta$. This point is more thoroughly discussed in Sec. 5.1. This condition yields an estimate[11, 10] for the width $w$ of the incompressible strips

$$w^2 = \frac{4\kappa\Delta}{\pi^2 e^2} \frac{1}{|\vec{\nabla} n(\vec{r})|} \qquad (1.17)$$

where $\vec{\nabla} n(\vec{r})$ is the density gradient in absence of magnetic gaps. For randomly distributed donors, the typical value of the density gradient is[10]

$$\langle |\vec{\nabla} n(\vec{r})|^2 \rangle^{1/2} = \Gamma\left(\frac{5}{4}\right)^2 \left(\frac{3C}{16\pi}\right)^{1/2} \frac{1}{d^2}. \qquad (1.18)$$

For realistic situations, this yields a width of the incompressible strips $w \approx 1.4\sqrt{(\Delta/W)}d$ ranging from 0.3 to 1.4 $d$. Hence, close to incompressible quantum Hall states, the difference of screening in compressible and incompressible regions (nonlinear screening) induces a highly divided structure where the incompressible fluids forms a rather dense network of long and narrow filaments covering only a partial area of the sample. How such a complex structure affect the relation (Eq. 1.14) between the magnetization and the incompressibility gaps is a subtle issue which is postponed to discussion of fractional gaps (Sec. 5) for which the ratio $\Delta/W$ is significantly less than one. Since the discontinuity in the chemical potential is associated to the incompressible fluid (IF), a knowledge of the sample fraction $f$ covered by the IF will be necessary to carry a quantitative analysis. In the limit where $\Delta/W$ is significantly less than unity, Efros and coworkers[12, 10] give an estimate for $f$

$$f = 0.57\sqrt{\frac{\Delta}{W}}, \qquad (1.19)$$

which will be used in conjunction with other experimental estimates. In the integer regime, $\Delta/W$ is much larger, and a large portion of the sample is in an

*Magnetization of the 2DEG* 107

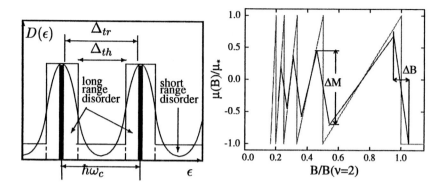

*Figure 1.5* Left: the thermodynamic density of states is separated into three pieces: the extended states (in black) are found close to the center of LL, states localized by long range potential fluctuations are modeled by rectangular blocks around the extended states, while short range disorder is represented by a small constant density of states ($n_i$) in between LL. Right: in this model, the jump width $\delta B$ is set by $n_i$, while long range disorder reduce $\Delta M \propto \Delta_\mu$ the incompressibility gap $\Delta_\mu$.

incompressible state. In this limit, a qualitative discussion of the overall density of states is sufficient to understand the experimental data.

In the thermodynamics density of states (DOS), states localized by long range potential fluctuations are found close to the LL center while short range disorder spreads states everywhere between LL as illustrated in Fig. 1.5. To summarize the role of short and long range disorders, it is instructive to separate their respective contributions as two adjacent rectangular blocks in the DOS (Fig. 1.5). The width $\delta B/\Delta \propto n_i$ of the magnetization jumps is in this model controlled by short range disorder, while the jump amplitude $\Delta M$ which still measures the incompressibility gap $\Delta_\mu$ is reduced by the long range disorder.

In the configuration C or D of Fig. 1.4, the sample topology is similar to a Corbino geometry where there is an "inner edge" (corresponding to states localized in the bulk) and an outer edge. In such multiply connected structures, any flux variation in the quantum Hall regime induces *dissipationless* non-equilibrium currents which are now described.

## 3. NON-EQUILIBRIUM CURRENTS

In the quantum Hall regime, the longitudinal conductivity $\sigma_{xx}$ is exponentially small at low temperature and there is no mechanism to relax a non-equilibrium charge distribution to equilibrium. Non-equilibrium charges drive chemical potentials gradients which in turn drive nonequilibrium current distributions. The intimate relation between non-equilibrium charges and currents distributions can be clearly illustrated in a Corbino geometry.

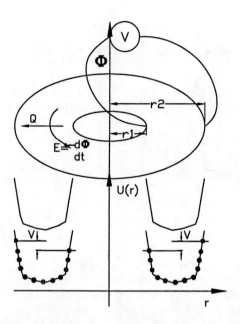

*Figure 1.6* Corbino geometry: a magnetic flux threaded through the central area transfers one electron from the inner to the outer edge. With this charge transfer, an induced voltage can be measured between edges.

## 3.1  CHARGING OF EDGES BY FIELD SWEEPS

When a magnetic field increases at a rate $dB/dt$, there is an azimuthal electromotive force $\int \mathcal{E} d\ell = 2\pi \mathcal{E}(\vec{r}) = -\frac{d\Phi}{dt} = -\pi r^2 \frac{dB}{dt}$ which drives a radial current

$$j(\vec{r}) = \sigma_{xy}\mathcal{E}(\vec{r}) = -i\frac{e^2}{h}\frac{r}{2}\frac{dB}{dt}, \quad (1.20)$$

where $i$ is the Landau Level index. This current flow drives a change in the local charge $\rho(r)$ density via the continuity equation,

$$\frac{\partial \rho(r))}{\partial t} = -\vec{\nabla} \cdot \vec{J} = i\frac{e^2}{h}\frac{dB}{dt} \quad (1.21)$$

Therefore, the change in charge density

$$\Delta n = \frac{\Delta \rho}{e} = i\frac{\Delta B}{\Phi_0}. \quad (1.22)$$

is entirely controlled by the magnetic field variation from the beginning to the end of the sweep. Since the system is incompressible, the charge density remains uniform in the bulk. The total charge being constant, the increase in charge density is fed from the edges and implies a charge transfer $Q_T$ from the inner to the outer edge of the Corbino structure. This charge transfer drives a change chemical potential $\Delta \mu = eV$, which can be measured capacitively. In the simplest configuration, the actual electrical capacitance $C_i$ and $C_o$ of the inner and outer edge with the reference gate electrode completes the electrical circuit of this Corbino structure. The charge transfer $Q_T$ is in this case specified by the overall voltage drop between edges which is the sum of the voltage across each capacitance

$$\frac{Q_T}{C} = \frac{q_i}{C_i} + \frac{q_o}{C_o} \qquad (1.23)$$

where $C$ is the series capacitance ($C^{-1} = C_i^{-1} + C_o^{-1}$). The charges $q_i$ and $q_o$ being controlled by the flux changes through the inner and outer disks, the charge transfer $Q_T$ depends only on the magnetic field change $\Delta B$

$$Q_T = i\frac{e^2}{h}\pi \frac{C_i r_o^2 + C_o r_i^2}{C_o + C_i} \Delta B \qquad (1.24)$$

and not on the rate $dB/dt$ at which $B$ changes. This quantized charge transfer has been checked experimentally in silicon MOSFET's and GaAs heterostructures, in an experiment[13, 14], schematically represented by the top panel in Fig. 1.7. The charge transfer is inferred from voltage across the large (compared to $C_o$ and $C_i$) external capacitance $C_0$[1] through

$$Q_T = C_0 V \approx i\frac{e^2}{h} \frac{\pi}{2}(r_o^2 + r_i^2)\Delta B. \qquad (1.25)$$

When the longitudinal conductivity vanishes, the charge transfer varies at the same rate as the magnetic field (Fig. 1.7, center panel), and the slope of $Q_T$ as a function of $B$ is quantized by $\sigma_{xy}$ (Fig. 1.7, bottom panel). This quantization is destroyed when equilibrium between edges can be restored, i.e. when the current backflow driven by the voltage difference $V = \Delta\mu/e$ flows through the sample: this occurs when $\sigma_{xx}$ is nonzero, or when the electric built up between edges exceed a breakdown value of order of $E_c = \frac{\alpha\Delta}{e_*\ell_B}$[17]. This dissipative regions delimits the characteristic hysterisis loops (Fig. 1.7, center panel), with quantized linear dependence between charge and field variation from the edge of on each Hall plateaux.

## 3.2 NONEQUILIBRIUM DISTRIBUTIONS OF CHARGES AND CURRENTS

In quantum Hall systems, nonequilibrium charge and current distributions are intimately related. In metals, the current response is usually determined by the local electric field $\vec{j} = \underline{\sigma}\mathcal{E}$ where $\underline{\sigma}$ is the conductivity tensor and the local electric field $\mathcal{E} = \vec{\nabla}\mu$ is set by the spatial variations of the local electrochemical potential. In high magnetic fields, there is a second contribution to the local current: each quantum state being a "local current loop", any gradient in charge density implies a net current. In a quantum Hall fluid, these two contributions are (nearly) spatially separated, the charge density being uniform in incompressible regions, while screening pins the electrochemical potential in compressible regions[15, 16]. In the incompressible regions, the charge density is constant and the local chemical potential follows the electrochemical potential. Since there is

---

[1]With a large capacitance $C_0$ across the electrometer, the charge transferred $Q_T$ is mostly on $C_0$, and the measured voltage $V$ no longer depends on the poorly known capacitance $C_i$ and $C_o$. An absolute measurement of $Q_T$ is thus achieved.

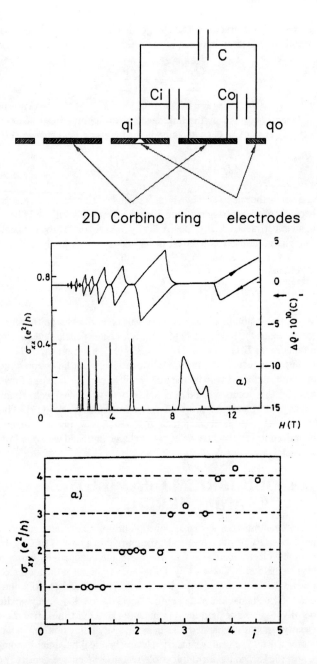

*Figure 1.7* Top: equivalent circuit used to measure the charge transfer between inner and outer edges. A known capacitance $C_0 \gg C_o, C_i$ is placed between the inner and the outer electrode, so the measured voltage difference across $C_0$ no longer depends on the poorly known capacitances $C_i$ and $C_o$. Middle: dependence of the charge transfer and the longitudinal conductance when the magnetic field is swept at a constant rate in the up and down direction. Off plateaux, the finite longitudinal conductivity bleeds the nonequilibrium to zero. Bottom: slopes of $Q_T(B)$ for each quantum Hall plateaux, displaying explicitly the quantization of the charge transfer.

# Magnetization of the 2DEG

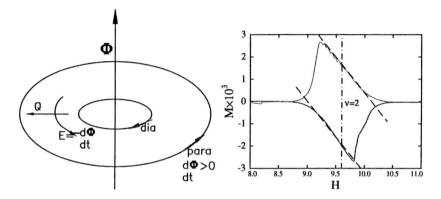

*Figure 1.8* Left: Electric fields and currents in the Corbino geometry. The current direction is set by the sign of $dV/dr$. Right: example of the non-equilibrium magnetization observed in a rectangular sample when the magnetic field is swept up and down. The similarity of the parallelogram shape hysteresis loop with the non-equilibrium charge detected at the edge of a Corbino structure (Fig. 1.7) makes the connection between the physics of charging at edges and nonequilibrium explicit.

no screening, nonequilibrium "surface" charges at the edge of the incompressible fluid forces the chemical potentials gradient which drive the local current density

$$\vec{j}_{QH}(\vec{r}) \simeq -\frac{e^2}{h}\nu(\vec{r})\vec{\nabla}\mu(\vec{r}) \times \hat{z}. \quad (1.26)$$

Here $\nu(\vec{r}) = n(\vec{r})\Phi_0/B$ is the local filling factor, equals to $\nu_i$ in incompressible regions and $\mu(\vec{r})$ is the local electrochemical potential. Since $\nu$ is a constant $\nu_i$ in regions occupied by a QH fluid, Eq. 1.26 can be recognized as the usual quantum Hall current. In compressible regions, local equilibrium is reached and chemical potential is pinned. On the hand, the local charge density which screens local potential variations fluctuates and drives a second contribution to the current density

$$\vec{j}_D(\vec{r}) \approx -\gamma_i \frac{e\omega_c}{4\pi}\vec{\nabla}\nu(\vec{r}) \times \hat{z} \quad (1.27)$$

where $\gamma_i = 2\nu_i + 1$ in the integer regime. $j_D$ is sometimes referred to as a transverse (to the density gradient) diffusion current.

To illustrate the effect of non-equilibrium charges on currents, the Corbino geometry represented in Fig. 1.8 is considered in an incompressible QH state. There are two sources for the gradients in electrochemical potential driving the currents (1.26). First, the confining potential drives the usual edge currents, $j_{QH} = \nu_i e^2/h\vec{\nabla}V \times \hat{z}$ flowing in opposite direction at the inner and outer edges: as long as the states remain occupied, this is an equilibrium contribution to the current. Second, the charge transfer $Q_T$ between edges driven by the flux change induces a voltage difference $V$ between them. Assuming the chemical potential at the inner and outer edge change by $\delta\mu = \pm eV/2$, the nonequilibrium edge currents are

$$\delta I_{i,o} = \pm \frac{\nu_i}{\Phi_0} \frac{eV}{2}, \quad (1.28)$$

according to Eq. 1.3. The corresponding magnetization change is

$$\delta M = i\frac{e^2}{h}\frac{AV}{2} \tag{1.29}$$

where $\mathcal{A} = \pi(r_o^2 - r_i^2)$ is the area of the Corbino ring. The actual voltage difference $V$ depends on the overall electrostatics of the circuit and has been evaluated in the last paragraph (Eq. 1.25) for a specific experimental situation. Nevertheless, $V$ is always proportional to $\Delta B$ when $\sigma_{xx} = 0$ and hence $V$ changes sign when the field sweep direction is changed. On the other hand, this is *not* an eddy current since $\delta I$ only depends on the overall change $\Delta B$ and not on the sweep rate $dB/dt$. An example of magnetic hysteresis observed for the $\nu = 2$ plateau in a $GaAs/GaAlAs$ heterostructure is shown in the right panel of Fig. 1.8: a parallelogram hysteretic loop is measured which shape is nearly identical to the charge transfer loop observed by [13] in the Corbino geometry (Fig. 1.7). This makes the connection between non-equilibrium currents and charges (cf. Eq. 1.29) explicit. This may appear surprising considering that this sample is not etched in a Corbino structure: on the other hand, it was argued in Sec. II, that long range disorder introduces compressible areas in the bulk: these regions are "holes" in the incompressible fluid, effectively making the sample multiply connected and topologically equivalent to a Corbino sample. Recently, Wiegers et al.[18] have investigated in realistic samples the actual current distribution in QH states. They found that although Eq. 1.29 is always satisfied, disorder induces a significant current redistribution at integer filling factors. In Sec. V, the contributions of compressible regions (Eq. 1.27) explain consistently this observation and imply an significant difference between transport and incompressibility (thermodynamic) gaps.

At first sight, these nonequilibrium currents appear unavoidable making the true thermodynamic ground state inaccessible whenever the electron fluid is incompressible. Fortunately, at finite temperature ($k_B T \ll \Delta_{tr}$) dissipation is weak but non-zero. Furthermore, the prefactor for the longitudinal conductivity is large in very clean sample. Using high mobility heterostructures, it is possible to reduce the potential fluctuations which pins the chemical potential by illuminating the sample with an infrared diode. Following this procedure there is a sufficient amount of dissipation in the sample to quench nonequilibrium current and measure meaningful thermodynamic quantities.

## 4. MAGNETIC OSCILLATIONS IN THE INTEGER REGIME

The measurement of "incompressible" gaps is based on Eq. 1.14. This fundamental thermodynamics relation is a consequence of the Maxwell relation associated to the grand potential ($d\Omega = -MdB - Nd\mu - SdT$)

$$\left(\frac{\partial M}{\partial \mu}\right)_B = \left(\frac{\partial N}{\partial B}\right)_\mu \tag{1.30}$$

At the edge of a magnetization jump, the constraint on the chemical potential ($\mu = cst$) is equivalent to maintaining the filling factor at a fixed value as illustrated on Fig. 1.9. Hence,

*Magnetization of the 2DEG* 113

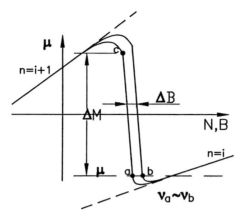

*Figure 1.9* When the field is changed by $\Delta B$, the entire magnetization curve as a function of $N$ is shifted. The points $a$ and $b$, at the same chemical potential, are also at the same filling factor since the magnetization jump occurs at a fixed value of $\nu = \nu_i$.

$$\left(\frac{\partial N}{\partial B}\right)_\mu = \left(\frac{\partial N}{\partial B}\right)_\nu = \frac{N}{B} \quad (1.31)$$

Integrating the left hand side of Eq. 1.30 over the chemical potential jump (between $a$ and $c$) yields the same relation thermodynamic Eq. 1.14 between magnetization jump and thermodynamic gap. This thermodynamic derivation is quite general and can also be used in the fractional regime. The only constraint is that the magnetization jump must be sufficiently sharp so the width $\Delta B$ between points $a$ and $c$ is small.

## 4.1 TORQUE MAGNETOMETRY

Although other techniques have been used[19], torque magnetometry is probably the most sensitive technique[20] in high magnetic field. Since it only detects the anisotropic part of the magnetization, it is almost exclusively sensitive to the orbital magnetism of the 2D gas, spurious signals arising only from spin anisotropies from impurities present in the sample and the magnetometer. The two-dimensional sample is place on a lightweight hysol rotor suspended by a thin ($25\mu m$) phosphor-bronze torsion wire (see Fig. 1.10). The position of the rotor is detected capacitively, using three arrays of electrodes (one on the rotor (c) and two on the stator (a and b)). The angular rotation $\delta\theta$ of the rotor is proportional to the capacitance difference $\delta C_{ca} - \delta C_{cb}$. A small current loop is also placed on the rotor for an absolute calibration of the torque (with a 10% accuracy) free of all the geometrical factors entering in the measurement. Also, an in-situ background subtraction can be achieved by injecting a small current in this loop proportional to the applied field.

The normal of the 2D sample is tilted by an angle $\theta \approx 30°$ with respect to the field axis. On very general grounds the torque measured is

$$\tau = M_\perp B = MB\sin\theta \quad (1.32)$$

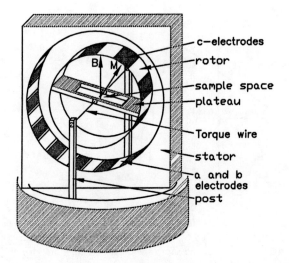

*Figure 1.10* Schematic drawing of a torque magnetometer used in the measurement magnetic oscillations. The normal of the 2D sample makes an angle $\theta \approx 25 - 30°$ with respect to the magnetic field. The rotor is designed so that the center of mass is as close the torque wire as possible, in order to decouple translation and rotation motions. The position of the rotor is detected capacitively as explained in the text.

where $M_\perp$ is the component of the magnetization perpendicular to the magnetic field. Landau Level energies and degeneracies are determined by the magnetic field component perpendicular to the sample, while the Zeeman energies depends on the total field $|\vec{B}|$. The magnetometer sensitivity increases with field (cf. Eq. 1.32) and is of the order of $10^{-13}$ J/T.

## 4.2 MAGNETIC OSCILLATIONS IN HIGH LANDAU LEVELS

A variety of GaAs/AlGaAs structures have been used in these studies. Some are quantum wells (GaAs sandwiched between two AlGaAs layers), some are heterostructures (AlGaAs/GaAs interface). Doping is realized with $Si$ donors which can be placed in an atomic layer set back at a distance $d$ from the AlGaAs/GaAs interface(s) ($\delta$-doping). In other samples, the donors are distributed in a layer of thickness $\delta$ set back at a distance $d$ from the interface (selective doping). These differences are sufficient to introduce some variation in the long-range potential fluctuation and are exploited to understand the role of disorder on thermodynamic properties[2].

---

[2]Some of the physical differences between $\delta$-doped and volume doped samples:
- For a given setback, ionized donors are more uniformly distributed in volume doped samples reducing the disorder potential.
- Spin-flip scattering could be larger in volume-doped samples, due to the random direction of the local electric fields.

*Table 1.1* Structure, dopants, density and mobility of the samples studied in this work. (QW ≡ Quantum well), (HS ≡ heterostructure). Except for the multiquantum wells sample (MQW), all the electronic densities are measured after illumination by an infrared diode.

| Sample | Type | doping | $n$ (cm$^{-2}$) | $\mu$ (cm$^2$/Vs) |
|---|---|---|---|---|
| sqw1 | QW | $d = 400$Å, $\delta = 2000$Å | $14 \times 10^{11}$ | $1.2 \times 10^6$ |
| 1649 | HS | $d = 300$Å, $\delta = 400$Å | $16.2 \times 10^{11}$ | $1.45 \times 10^5$ |
| T19 | HS | $d = 800$Å, $\delta = 1000$Å | $8.8 \times 10^{11}$ | $4.4 \times 10^5$ |
| T35 | HS | $d = 500$Å, $\delta = 800$Å | $2.6 \times 10^{11}$ | $3.7 \times 10^6$ |
| T55 | HS | $d = 500$Å, $\delta = 1000$Å | $3.6 \times 10^{11}$ | $1.4 \times 10^6$ |
| T95 | HS | $d = 500$Å, $\delta = 800$Å | $4.9 \times 10^{11}$ | $2.3 \times 10^6$ |
| T139 | HS | $d = 500$Å, $\delta = 1000$Å | $8 \times 10^{10}$ | $3 \times 10^6$ |
| M280 | MQW | $d = 1250$Å | $8.5 \times 10^{10}$ | $7 \times 10^5$ |

Table 1.1 summarizes the main parameters of the samples studied. As explained in the preceding section, non-equilibrium currents are usually quenched after illuminating the sample by an infrared diode. The rearrangement of charged donors driven by their photoinization reduce sufficiently the disorder potential, so that the value of $\sigma_{xx}$ in the middle of the Hall plateau remain finite. All thermodynamic measurements were done in this way except for the multilayer samples.

In Fig. 1.11, de-Haas van-Alphen oscillations are observed in a variety of samples up to filling factors $\nu \geq 50$. These curves are obtained dividing the measured torque by $B \sin \theta$ and subtracting an overall background. This subtraction is done by fitting the measured magnetization to a third or fourth order polynomial which is subtracted from the signal. As long as many oscillations are present in the magnetic field interval studied, this procedure is quite precise. In all the panels, the oscillations are damped at higher filling factors (lower magnetic fields). Disorder is one of the sources of the damping for de-Haas van-Alphen oscillations. In this section, a semiclassical description of the scattering by *long range* potential fluctuations is shown to account for this behavior

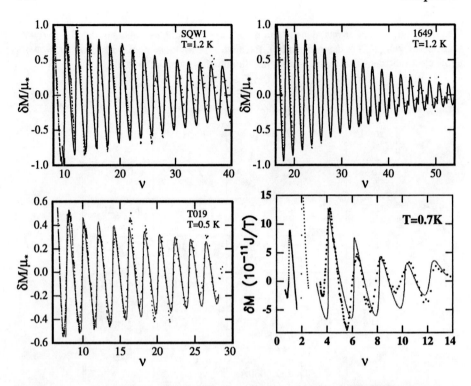

*Figure 1.11*  de-Haas van-Alphen oscillation observed in two heterostructures (1649 and T019) and two quantum wells (SQW1 and M280) described in Table 1.1. The slight discrepancies observed in the multiquantum-well M280 can mostly be attributed to an imperfect background subtraction: for lower filling factors, a smaller number of oscillations are available and the subtraction procedure is less precise.

using the mean potential fluctuation $W = \sqrt{(\delta V)^2}$ as *sole* parameter for the disorder. At high $\nu$, a quantum state can be represented semiclassically as a ring or radius $r_c$, the cyclotron radius and width $\lambda_F$, the Fermi wavelength. The energy shift induced by long-range potential fluctuations on each quantum state is the average of $\delta V(\vec{r})$ over this "ring-wavefunction". If $\xi$ ($> \lambda_F$) is the spatial scale over which $\delta V$ fluctuates (of the order of the setback distance $d$), there are $2\pi r_c/\xi$ independent blocks where $V$ assumes distinct random values over the spatial extent of $\psi(\vec{r})$. Hence the mean fluctuation of Landau level energies is of the order of

$$\Gamma^2(B) \equiv (\delta\epsilon)^2 = \frac{\xi}{2\pi r_c}W^2 = \frac{\xi\omega_c}{2\pi v_F}W^2. \qquad (1.33)$$

The broadening $\Gamma(B)$ of each Landau level, increases as the square root of the magnetic field. How are the de-Haas van-Alphen oscillations affected by this field dependent LL broadening? A generalization of the concept of Dingle temperature [22] is here straightforward. Since the scattering by potential fluctuations is physically similar to scattering by phonons[3], it is reasonable to incorporate the field dependent broadening $\Gamma(B)$) as an effective temperature $k_B T^*(B) = k_B T + \Gamma(B)$. With this substitution, it is possible to fit the de-Haas van-Alphen oscillations using the finite temperature expression for a clean system[21].

In Fig.1.11, the de-Haas van-Alphen oscillations observed in heterostructures (samples 1649, T019) and quantum wells (samples SQW1 and M280) are fitted to Eq. 3 of reference [21] using $k_B T^*$ as a phenomenological broadening (see Eq. 1.33). The amplitude dependence, the phase and the asymmetric shape of the oscillations are all very well reproduced. Considering that $W$ and the overall scale factor are the only fitting parameters (the electronic densities are known from transport), the agreement is excellent. This gives confidence that long-range disorder fluctuations dominates the damping of de-Haas van-Alphen oscillations at high filling factors.

## 4.3 LOW FILLING FACTORS

At low filling factors, the shape of the oscillation is much closer to the sawtooth pattern expected for a clean sample as illustrated in Fig. 1.12. Since the amplitude of the oscillations is close to the theoretical value $2\mu_* N$ and depends weakly on magnetic field (or filling factor), most of the relevant thermodynamic information is contained in the sharp magnetization steps. They occur precisely in the middle of Hall plateaux, i.e. precisely at integer filling factors as expected for samples with a fixed number of electrons. Two quantities are used to parameterize each magnetization steps: the magnitude $\Delta M$ of the overall magnetization swing is a direct measure of the incompressibility gap as argued in Sec. II and IV. No other thermodynamic measurement give such a direct access to this physical parameter. The second parameter, the jump width $\Delta B$ is a measure of the residual density of states in the gap. While short range disorder contribute directly to $\Delta B$, long range potential fluctuations have also an effect on the broadening of the step.

In Fig. 1.12, the jump width $\delta B_2$ at $\nu = 2$ (corresponding to 12T, for sample T95) is 0.4T. Hence, only 3% ($\Delta B/B$) of the states lie in the incompressibility gap. Using Eq. 1.14, the energy gaps are found to be close to the theoretical value $\hbar\omega_c$. This gives an upperbound for the LL broadening induced by long-range disorder: referring to the physical picture proposed in Fig. 1.4, the incompressibility gap $\Delta_\mu$ equals to $\hbar\omega_c - \Gamma$, where $\Gamma$ is the overall broadening of LL induced by long-range potential fluctuations. In this picture, the magnitude $\Delta M \approx (\hbar\omega_c - \Gamma)/B$ is within 20% of the theoretical value $2\mu_* N$, indicating that LL width is at most 20% of $\hbar\omega_c$ at low filling factors. If the difference $\hbar\omega_c - B\Delta M$ is attributed to compressible regions in the disordered potential, similar portion of localized states in the sample are found.

Disorder is not the sole parameter affecting the de-Haas van-Alphen amplitudes. Coulomb repulsion increases the overall energy of the 2DEG. How does this Coulomb contribution depends on LL index? The relative wavefunction for two electrons in LL $n$ is strongly peaked at a separation $2\ell_B\sqrt{n}$ where $\ell_B$ is the magnetic length. Hence, the Coulomb energy decreases as $e^2/(2\ell_B\sqrt{n})$ with LL index. As a result, the incompressibility gap, which involves the energy

---

[3]$GaAs$ being piezoelectric, the long wavelength potential fluctuations induced by phonons scatter electrons as a weak disorder potential.

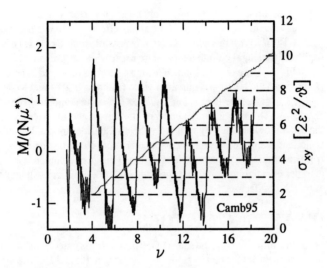

*Figure 1.12* Left scale: sawtooth magnetization of sample T95 as a function of $\nu$; rightscale: Hall conductance measured in a transport experiment on the same wafer under similar experimental conditions (after illumination).

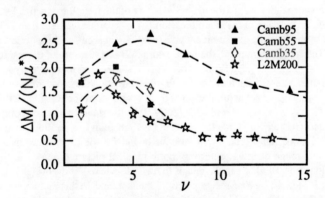

*Figure 1.13* Magnitude of the jumps $\Delta M$ as a function of filling factors for a variety of samples. The decrease observed at low filling factors can be attributed to electron-electron interactions or to subband mixing by the in plane magnetic field.

difference between a filled LL and a filled LL plus a few particles, is *reduced* by Coulomb interactions, the reduction being the largest at $\nu = 2$ (see Fig. 1.14).

In Fig. 1.13, the jump magnitudes observed in a variety of samples are plotted as a function of filling factor. For $\nu > 6$, there is a monotonous decrease of this amplitude with $\nu$: this behavior is consistent with the normal damping of de-Haas van-Alphen amplitudes induced by disorder. On the other hand, a decrease of the jump magnitudes is also observed at low filling factors after a maximum around $\nu = 6$. This may be attributed to the larger Coulomb interaction in lower

## Magnetization of the 2DEG

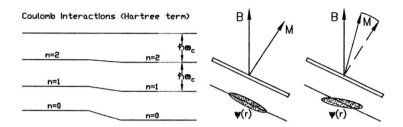

*Figure 1.14* Left: the 2DEG energy is increased by Coulomb interactions. This energy shift represented in the left column of the top panel decreases as a function of filling factor since the average distance between electrons is larger in higher LL. As a result the energy gaps are reduced between lower LL. Right: a component of the magnetic field in the plane of the 2DEG mixes electric subbands. The resulting wavefunctions are "rotated" with respect to 2D plane reducing the effecting torque at low filling factors.

LL as argued in the preceding paragraph. We have illustrated the shifts in the LL spectrum by Coulomb interactions on the left panel in Fig. 1.14. There is however an alternative explanation. In a torque experiment, there is an in-plane magnetic field, $B_\| = B \sin \theta$ which mixes adjacent electric subbands and consecutive Landau levels. This mixing can be described by the perturbing Hamiltonian[24]

$$\mathcal{H}_2 = z p_x \omega_c^\|. \quad (1.34)$$

The admixture of the lowest Landau level of the second electric subband in the second Landau level of the lowest subband can be estimated using perturbation theory,

$$\eta = \frac{B \sin \theta \langle 1|z|2 \rangle}{(B_c - B \cos \theta)\ell_B}. \quad (1.35)$$

where $\ell_B$ is the magnetic length. The critical field $B_c = (\epsilon_2 - \epsilon_1)/\mu_*$ is reached when the cyclotron energy $\hbar \omega_c^\perp$ equals the electric subband separation $\epsilon_2 - \epsilon_1$. In GaAs heterostructures, this energy difference depends sensitively on the electronic density and the aluminum concentration in the $GaAlAs/GaAs$ structure[25]. For realistic structures, $B_c$ is of the order of 170 T, and

$$\eta \approx 0.25 \frac{B^{3/2}}{B_c} \sin \theta \quad (1.36)$$

where $B$ is expressed in Tesla. The largest admixture $\eta \approx 0.016$ occurs in the highest density sample (T95), where $\nu = 2$ is reached when $B \approx 8$T. This value of $\eta$ is too small to account for the observed decrease of the magnetic torque at low filling factors[26].

In conclusion, the observed magnetic oscillations at low filling factors follow closely the Peierls sawtooth pattern, with a decrease of the sawtooth amplitude at low filling factors. This decrease, most pronounced in lower density sample is attributed to Coulomb interactions.

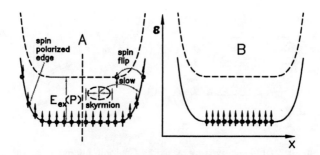

*Figure 1.15*  As $\nu \to 1$ there are spin-flip of single particle excitations in the $\nu = 2$ LL or of skyrmion collective objects into the available states at the sample edges. Since the overlap of the states are weak and may involve many-body processes, the relaxation processes are slow. At $\nu \approx 1$ the system is fully polarized after the depopulation of the spin-polarized edge.

## 4.4 ODD FILLING FACTORS: QUANTUM HALL FERROMAGNETS

In $GaAs/AlGaAs$ heterostructures, spin-orbit coupling reduce significantly the effective g-factor of the 2D electrons. Hence the Zeeman term is a smaller energy than all other spin-dependent exchange energies arising from Coulomb interactions. More precisely the relative values of cyclotron, Hartree, exchange and Zeeman energies are of the order of

$$\hbar\omega_c > E_h = \alpha_n E_c > E_{ex} = 0.5\beta_n E_c \mathcal{P} > g\mu_B B$$
$$200K > 50K > 25K > 2.7K \qquad (1.37)$$

where $E_c = \frac{e^2}{4\pi\epsilon\epsilon_0 \ell_B}$ is the Coulomb energy scale, $\mathcal{P} = \frac{n_\uparrow - n_\downarrow}{n_\uparrow + n_\downarrow}$ is the polarization of the $n^{th}$ Landau level, while the coefficients $\alpha_n$ and $\beta_n$ are of the order unity for the lowest Landau level and decrease as $\sqrt{n}$ in higher LL. Finally, in disordered samples, a last energy scale, the LL broadening $\Gamma(B)$ controls most of the experimentally observed magnetic oscillations at odd filling factors. The exchange energy is proportional to the polarization of the 2DEG, or equivalently to the occupation difference of the ↑ and ↓ subbands. These spin-subbands occupation factors are determined by the magnetic field component $B_\perp = B\cos\theta$ perpendicular to the 2D plane. Hence, the exchange energy unlike the Zeeman term is anisotropic and can be detected in a torque experiment.

In a quantum Hall ferromagnet, the polarization of the 2D gas is not a simple function of the filling factor: it is determined self-consistently by the relative energies of the spin subbands which are a function of their occupations through the spin-exchange energy, and by their occupations which depends on their energies relative to the chemical potential. In a realistic model, it is also necessary to include the screened long range potential fluctuations $V(z)$ which broaden each spin subband thus reducing the exchange interaction. In other words, it is no longer possible to optimize simultaneously the exchange and the Hartree energies in presence of $V(z)$, which requires the local electronic density to stick as closely as possible to the solution of the local electrostatics (Poisson equation). Fogler and Shklovskii[27] have proposed a mean-field Stoner model appropriate to

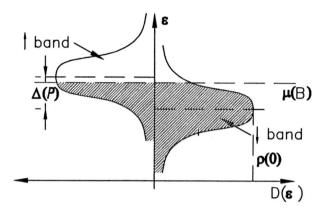

*Figure 1.16* Density of states of the spin ↑ and spin ↓ subbands around $\nu = 1$. Each subband is broadened by the local disorder. The overall splitting $\Delta$ between spin subbands is a self consistent solution with each subband occupation

weakly disordered Quantum Hall ferromagnets which is summarized here. Very generally, the local spin occupations around the point of coordinate $z = x + iy$ may be written as

$$n_\uparrow(z) = f(V(z) + Z/2 + \epsilon_\uparrow(z)) \quad (1.38)$$
$$n_\downarrow(z) = f(V(z) - Z/2 + \epsilon_\downarrow(z)) \quad (1.39)$$

where $f$ is the Fermi function and $Z = g\mu_B B$ is the Zeeman energy. Here $V(z) + Z/2 + \epsilon_\uparrow(z)$ and $V(z) - Z/2 + \epsilon_\downarrow(z)$ are the local energies for each spin orientation including exchange and Hartree energies. Depending on the relative strength of the disorder and the exchange energy, the density of states of each subband represented in figure 1.16 can overlap thus reducing the exchange energy. The splitting

$$\Delta_S = Z + \overline{\epsilon_\uparrow(z)} - \overline{\epsilon_\downarrow(z)} \quad (1.40)$$

between the subbands centers must therefore be determined self-consistently.

Since electronic wavefunction have an extension of the order of the magnetic length, a local Hartree-Fock approximation is possible: the energy of each subband becomes

$$\epsilon_\uparrow(z) = \epsilon_0 + \frac{1}{2}\beta E_c \frac{n_\uparrow(z)}{n_\uparrow(z) + n_\downarrow(z)} \quad (1.41)$$

$$\epsilon_\uparrow(z) = \epsilon_0 - \frac{1}{2}\beta E_c \frac{n_\downarrow(z)}{n_\uparrow(z) + n_\downarrow(z)} \quad (1.42)$$

where $\epsilon_0$ is the lowest Landau level energy where the average Hartree correction has been added to the zero point energy $\hbar\omega_c/2$. Potential fluctuations also induce local fluctuation of the filling factor: hence Hartree and exchange energies also fluctuate. In this context, phase separation between coexisting spin-polarized and unpolarized phases is physically possible. On the other hand, if disorder is sufficiently weak, it is reasonable to assume that the electronic fluid remains homogenous, even for a partial spin-polarization. If this is so, it is legitimate to average over the whole sample (mean field Stoner model). Then, the effect of disorder may be represented by a broadened density of states $\rho(\epsilon)$ (per spin-subband) of the lowest Landau level. The chemical potential $\mu$ is set by the number of particles available,

$$N = \int_{-\infty}^{\mu} (\rho(\epsilon - \Delta_S) + \rho(\epsilon + \Delta_S)) d\epsilon, \qquad (1.43)$$

while the net polarization is defined as

$$\mathcal{P} = \frac{1}{N}\left(\int_{\mu - \frac{\Delta_S}{2}}^{\mu + \frac{\Delta_S}{2}} \rho(\epsilon) d\epsilon\right). \qquad (1.44)$$

Within the Hartree-Fock approximation (Eq. 1.42), the spin-splitting $\Delta_S$ is

$$\Delta_S = Z + \frac{1}{2}\beta E_c \mathcal{P}. \qquad (1.45)$$

It is instructive to compare the mean-field equations (1.44) and (1.45), which specify the polarization of the 2DEG, to the Stoner model[29] of itinerant ferromagnetism. The spin-splitting has exactly the same form, if we identify $\beta E_c$ with the exchange integral $I$. On the other hand, Eq. 1.44 differs from the Stoner mean field equation

$$\mathcal{P} = \tanh \frac{\Delta_S}{2k_B T} \qquad (1.46)$$

by its analytical form and the difference in the physical parameters: in Eq. 1.44, disorder broadening plays the same role as the thermal broadening in Eq. 1.46. Since there is a critical temperature in the Stoner model below which a spontaneous polarization $\mathcal{P}$ appears, it is natural here to seek a critical value for the disorder below which a spontaneous spin polarization can develop.[4] To determine this value, the splitting $\Delta_S$ is assumed to be small compared to $\Gamma$, the LL width. Eq. 1.44 may be linearized

$$\mathcal{P} = \frac{\Delta_S \rho(\mu)}{N}, \qquad (1.47)$$

where $\Delta_S, \rho(\mu)$ and $\mathcal{P}$ are represented graphically in Fig. 1.16. Using Eq. 1.45, a phase transition occurs at the critical density of states $\rho_c$,

$$\rho_c = \frac{2N}{\beta E_c} \qquad (1.48)$$

at zero Zeeman energy. In practice, experiments are carried out varying the filling factor $\nu = n\Phi_0/B$ or equivalently the chemical potential $\mu(B)$. In terms of theses physically relevant variables, the transition takes place at a critical filling factor $\nu_*$ such that the chemical potential $\mu_*(B)$, which follows the lowest LL energy below the transition, satisfies the condition $\rho(\mu(B_*)) = \rho_c$. This is illustrated in figure 1.17, where the chemical potential is plotted as a function of

---

[4] Just as in the Stoner model, the Zeeman energy broadens the ferromagnetic transition with disorder.

*Magnetization of the 2DEG*

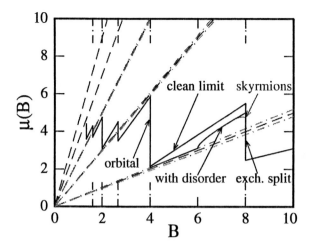

*Figure 1.17* Magnetization at odd filling factors. A spontaneous spin polarization occurs at a magnetic field $B_*$ such that $\rho(\mu(B_*)) = \rho_c$ which lies between filling factors $\nu = 2$ and $\nu = 1$. Close to $\nu = 1$, the contribution of topological excitations (skyrmions) reduce further the magnetization jump at $\nu = 1$.

magnetic field. When $\rho(\mu(B)) < \rho_c$, there is a field induced spin-splitting and a finite spin-polarization

$$\Delta_S = \frac{Z\rho_c}{\rho_c - \rho(\mu)} \text{ and } \mathcal{P} = \frac{Z}{N}\frac{\rho_c\rho(\mu)}{\rho_c - \rho(\mu)}. \qquad (1.49)$$

This mean field discussion excludes inhomogenous ground states. In addition to the phase separated electron-fluids mentioned earlier, other inhomogenous states arise as spin textures: they can be viewed as topological excitations of the local spin density. These objects known as skyrmions[30, 31, 32, 33] have been shown to be stable in absence of disorder at $\nu = 1$. They have a slightly lower energy than single spin-flip excitations and they may reduce slightly the chemical potential increase as $\nu$ is decreased towards $\nu = 1$. At present, the effect of disorder on skyrmions is not known, although spin-textures at the edge of $\nu = 1$ regions have been shown[28] to be stable for sufficiently slowly varying disorder potentials. On the other hand, since the spin polarization decrease rapidly with the skyrmions density, the ferromagnetic ground state should be quenched rapidly away from $\nu = 1$. From this point of view, skyrmions which are fluctuations from the mean field solution contribute to an increase of the critical density $\rho_c$ necessary to drive the transition, i.e. fluctuation corrections to the Stoner transition previously discussed are significant. For this reason, the interval $[\nu_*, 1]$ over which the quantum Hall ferromagnet (QHF) is thermodynamically stable is expected to be smaller than in the Stoner picture.

From an experimental point of view, quenching of non-equilibrium effects is more difficult at $\nu = 1$ and requires often elevated temperatures ($T \geq 0.7$ K). Spin relaxation between physically non-overlapping spin-polarized QH states is naturally slow and may require complex cooperative processes (see Fig. 1.15). The existence of a critical $\rho_c$ necessary to drive the QHF transition is seen

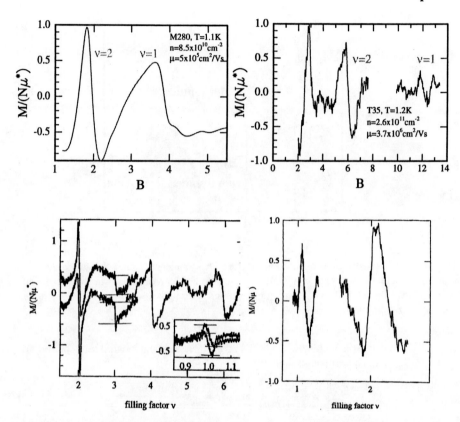

*Figure 1.18* de-Haas van-Alphen oscillation observed at odd filling factors in four different samples. From the top: sample M280, the oscillation at $\nu = 1$ is in $\delta$-doped QW sample closes to a sawtooth pattern; in sample T35 (selectively doped heterostructure), a structure in the magnetization is observed only very close to $\nu = 1$; for sample T19 (lower left panel), a similar structure is observed at $\nu = 1$ and $\nu = 3$; lower right panel, sample T139: in this low density heterostructure a sharp jump is also observed at $\nu = 1$.

experimentally using samples with different mobilities, but it is not easy to place the transition point in term of the physical parameters (mobility) characterizing the samples. Out of the eight samples studied, four of them did not show any clear magnetization jump at $\nu = 1$ although nice sawtooth pattern were observed at even $\nu$'s and transport mobilities were relatively large. We now analyze the magnetization curves of samples M280, T35, T19 and T139 (shown in Fig.1.18) for which a magnetization step is observed around $\nu = 1$.

The magnetization of the $\delta$-doped quantum-wells M280 is closest to a sawtooth shaped oscillation from $\nu = 2$ to $\nu = 1$: in other words, the critical value $\nu_*$ for this sample appear to be close to 2. The incompressibility gap at $\nu = 1$ can be estimated as $\Delta_\mu \approx 17 \pm 1.8$ K using Eq. 1.14, taking the overall magnetization swing at $\nu = 1$ for $\Delta M$. This is 25% below the transport gap $\Delta_{tr} \approx 22$ K (see table 1.2). This can easily be attributed to the presence of a compressible phase covering nearly 25% of the sample due the presence of the long-range disorder potential. This may be expressed in an equivalent way by noting that the reduction of the thermodynamic gap induced by disorder is here comparable to what is expected for even filling factors: using the the damping of de-Haas van-Alphen oscillations at higher $\nu$'s (see Fig. 1.11), we conclude that a 25% decrease of the de-Haas van-Alphen amplitude is expected when the magnetic field is halved (from $\nu = 1$ to $\nu = 2$).

For samples T35 and T19, the magnetization jumps are much smaller at $\nu = 1$ than at $\nu = 2$. Furthermore, the shape of the magnetic oscillations deviates significantly from the ideal Peierls sawtooth pattern: the oscillation is observed only relatively close to filling factor one. This may be attributed to a smaller critical filling factor $\nu_*$ where the QHF transition occurs, thus reducing the interval $[\nu_*, 1]$ where this phase is . Since the sample mobility of T35 is five times larger than that of M280, the effect of disorder on the QHF transition is probably more subtle than the Stoner model would suggest. The samples studied are anyway quite different, M280 being a $\delta$-doped quantum well while the T-samples are selectively-doped heterostructures (see Table 1.1 for details). In sample T19, similar magnetic jumps are observed at $\nu = 3$ and $\nu = 1$. Since it is believed that topological excitation are possible at $\nu = 1$ while unstable at $\nu = 3$, this data shows that the thermodynamics is sensitive to the energy of the excitations (which are similar for single spin-flip and skyrmions) and not to their nature. Finally, the $\nu = 1$ jump observed in the low density sample T139 (Fig. 1.18, last panel) has a relative amplitude to $\nu = 2$ comparable to what is observed in sample M280 (Fig. 1.18, first panel). This observation suggests larger magnetic oscillations at odd filling factors in lower density samples. Since the ratio between Coulomb and cyclotron energies $E_c/\hbar\omega_c$ varies as $1/\sqrt{B} \equiv 1/\sqrt{n\Phi_0}$ and increases at lower density, the QHF phase is naturally favored with a larger Coulomb exchange energy. How this energy is balanced by disorder or other effect is not experimentally clear since the QHF phase is quenched in some samples with good mobilities, . Experimentally, QHF phases are found to be fragile, even more so than fractional QH phases. Local probes (NMR, tunneling microscopy) would help sorting out the spatial structure of all possible QHF phases.

From a purely experimental point of view, it is the variety of behaviors observed in different samples at $\nu = 1$, which is difficult to place in a general picture of QH ferromagnetism. Schematically three generic behavior are observed:

- Quantum Hall ferromagnetism (QHF) is quenched at all filling factors (sample T95).
- QHF is present throughout the interval $[\nu = 2, \nu = 1]$ (samples M280 and L200).
- QHF is observed only close to $\nu = 1$ (samples T19, T35, T139).

While a Stoner model can reproduce these three behaviors, the region over which the QHF phase is stable should increase for smaller LL broadening. However, the LL width[5] $\Gamma$ of sample T95 is significantly smaller than for sample M280

---

[5]$\Gamma$ is measured (a) by the damping of de Haas-van Alphen oscillations at high filling factors, (b) from the width of magnetization steps at even $\nu$'s and (c) more indirectly by the sample mobility.

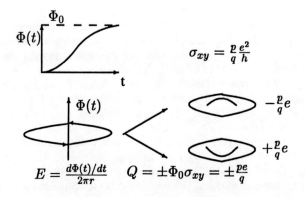

*Figure 1.19* Excitation at fractions: when a flux quantum is threaded through a fractional quantum Hall state a pair of fractionally charged excitations is generated.

while no QHF is observed. The electronic density seems to play a considerably larger role than in the Stoner model considered.

To conclude this section, we note that some physical aspects independent of skyrmion physics, particularly away from $\nu = 1$, are not properly understood. The most obvious way to account for all the behaviors observed is to assume that quantum Hall ferromagnets form very inhomogenous states competing with compressible fluids and/or unpolarized quantum Hall states. Images of incompressible regions by scanning probe microscopy would be very helpful to clarify this physics.

## 5. MAGNETIZATION AT FRACTIONS

When all the electrons are in the lowest Landau level, the kinetic energy is quenched: the dominant Coulomb interactions controls which of the available states are occupied. Since the Coulomb energy between two electrons is weaker when their relative angular momentum is large, it is energetically favorable to first fill states with large relative angular momentum. Once an "angular momentum shell" is full, there is a discontinuous jump[36] of the chemical potential $\Delta_\mu = E(N+1) - E(N)$ when an extra electron is added, because of the larger Coulomb interaction in the lower angular momentum states. At these specific fractional filling factors, the electrons form an incompressible correlated liquid as argued by Laughlin[3] and Haldane[6]. In contrast with the integer regime, there is a difference between the thermodynamic gap $\Delta_\mu$ defined as the difference in total energy of electron fluids differing by one particle and the transport gaps $\Delta_{tr}$ which probes the presence of quasi-particle/quasi-hole pairs. It is straightforward to show that these excitations are fractionally charged.

The azimutal electromotive force $\mathcal{E}_\phi$ induced by an arbitrary variation of magnetic flux $\Phi(t)$ through a disk of radius R is $\mathcal{E}_\phi = -\frac{1}{2\pi R}\frac{d\Phi(t)}{dt}$. If the system is in the $1/q$ FQH state, this electromotive force drive a purely radial current $j_r(t) = \sigma_{xy}\mathcal{E}(t) = \frac{e^2}{qh}\mathcal{E}(t)$. When the overall magnetic flux is increased or decreased by one flux quantum, a state is added or removed to the lowest LL. What is the corresponding excitation of the FQH fluid? The radial current flow reduces or increases the charge inside the disk of radius R by

$$Q = 2\pi R \int_0^\infty j_r(t) dt = \frac{1}{q} \frac{e^2}{h} \Phi_0 = \frac{1}{q} e, \qquad (1.50)$$

and an opposite charge $-Q$ appears outside the disk, since the system is electrically neutral. The resulting excitation is composite with a fractionally charged $+Q$ quasiparticle and a corresponding quasihole $-Q$. Its energy $\Delta_C = \epsilon_Q + \epsilon_{-Q}$, can easily be identified with the activation gap $\Delta_{tr}$ of the longitudinal conductivity, $\sigma_{xx} \propto \exp(-\frac{\Delta_{tr}}{2k_B T})$: in transport, the polarizing electric field couples directly to such composite excitations and the conductivity is determined by their density. Identifying the Boltzmann factors, we conclude

$$\Delta_C = \Delta_{tr}. \qquad (1.51)$$

Since a composite excitation is generated when adding or removing of a flux quantum, $\Delta_C$ is also the difference in total energy for the $N$ electrons in magnetic fluxes $qN\Phi_0$ and $qN\Phi_0 \pm \Phi_0$,

$$\begin{aligned} \Delta_C &= E(qN\Phi_0) - E(qN\Phi_0 - \Phi_0) \qquad (1.52) \\ &= \frac{1}{q}(E(qN\Phi_0) - E(q(N-1)\Phi_0)) = \frac{\Delta_\mu}{q}. \end{aligned}$$

Combining the arguments leading to Eq. 1.51 and 1.52, we conclude that transport and thermodynamic gaps in a disorder-free sample are related through the composite excitation energy $\Delta_C$ as

$$\Delta_C = \Delta_{tr} = \frac{\Delta_\mu}{q} \qquad (1.53)$$

at the fraction $1/q$. Assuming that the incompressible fluid at $\nu = 1/q$ occupies an area $\mathcal{A}$, the magnetization ($M = -\partial E(B)/\partial B$) change as the flux through this area is reduced by $q\Phi_0$ is

$$\Delta M_\nu = \frac{\mathcal{A}}{q\Phi_0}(E(qN\Phi_0) - E(q(N-1)\Phi_0)) = \nu \frac{\mathcal{A}}{\Phi_0} \Delta_\mu, \qquad (1.54)$$

which is precisely the same relation as Eq. 1.14 derived in Sec. 4 using purely thermodynamic arguments. To summarize, a comparison between transport and magnetization measurement is done in two steps: (a) the thermodynamic gap is inferred from the magnetization jump using Eq. 1.54 and (b) the result is compared to $q \times \Delta_{tr}$ (cf. Eq. 1.53) where $\Delta_{tr}$ is inferred from transport.

Fractional gaps are usually considerably smaller than the typical cyclotron energies. This means that other interactions such as disorder potentials can be at least as large as the many-body gaps occurring at fractions. Hence in realistic situations, the incompressible fluid rarely occupy the entire sample area. Physical states are very often heterogeneous with coexisting compressible and incompressible regions. Because ot the poor screening of incompressible states, the long-range disorder potential usually "confines" the incompressible QH fluid at filling factor $\nu_q$ to a narrow strip of width $w$ (see Eq. 1.17 in Sec. 2, for an estimate of the strip width) separating compressible regions $+$ and $-$ with filling factors $\nu_+ > \nu_q > \nu_-$. In transport, a FQH plateau is well defined when an incompressible region percolates through the sample driving the longitudinal

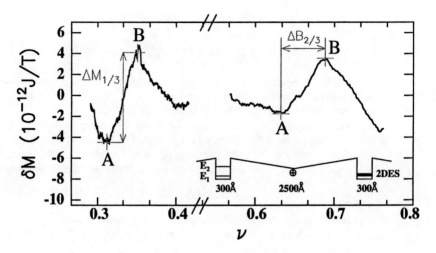

*Figure 1.20* Magnetization signal measured at fractional filling factor 1/3 and 2/3 by torque magnetometry. Inset: electronic bandstructure of the multi quantum-wells structures used in this work.

conductivity insulating: the finite conductivity of compressible regions has little effect and the corresponding areas are "inert". In thermodynamics, all regions contribute, but only the magnetization of the incompressible fluid has a singularity at fractional filling factors which can be related to its many-body gap using Eq. 1.54. It is not a-priori obvious that this relation can still be used even in presence of a moderate amount disorder. We argue below that the magnetization *of the incompressible area* $\mathcal{A}$ can still be related to the thermodynamic gap $\Delta_\mu$ using Eq. 1.54.

The M280 sample (see Table 1.1) used in magnetization measurements[40] at fractions is a 100 layers GaAs/AlGaAs multi-quantum wells 300Å wide, separated by a 2500Å AlGaAs barrier. The Si $\delta$-doping is placed in the middle of the barrier as illustrated in the inset of Fig. 1.20. The average electronic density in each layer is $n_{2D} = 8.5 \times 10^{10} cm^{-2}$ and varies by only a few percents (< 3%) from layer to layer. This sample which has a transport mobility $\mu_{tr} = 7 \times 10^5 cm^2/V \cdot s$ and shows well developed Hall plateaux at fractional filling factors, $\nu$=2/3, 3/5, 2/5 and 1/3.

The magnetization is measured by placing the sample on the magnetometer described in Sec. 4.1 inside the mixing chamber of a dilution refrigerator. The same background compensation by an in-situ current loop is used during the measurement. An additional background subtraction with a third order polynomial fit is applied to isolate the singular contributions of the 2DEG to the magnetization. The signal around fractional filling factors 1/3 and 2/3 is shown in Fig. 1.20 after background subtraction, respectively at temperature T=370 mK

*Table 1.2* Comparison between the thermodynamic gap $\Delta_\mu \propto \Delta M$ estimated from the magnetization jumps (cf. Eq. 1.14) and the corresponding gap $q \times \Delta_{tr}$ (cf. Eq. ) inferred from transport measurements on the same heterostructure (for $\nu \geq 1, q = 1$). Theoretical estimates $\Delta_{th}$[37, 38] (the lower numbers include corrections due to LL mixing and finite well width) are also included. Note the reduction of the thermodynamic gaps in the integer regime due to disorder.

| $\nu$ | 1/3 | 2/3 | 1 | 2 |
|---|---|---|---|---|
| $\Delta_\mu$(K) | $7 \pm 0.9$ | $2.1 \pm 0.3$ | $17 \pm 1.8$ | $20 \pm 1.8$ |
| $q \times \Delta_{tr}$(K) | $15.0 \pm 2$ | 4.5 | $22 \pm 1.1$ | - |
| $\Delta_{th}$(K) | 20...54 | 8...19 | 40 | $\frac{eB_\perp}{m_*} \approx 35$ |

and T= 144 mK[6]. Magnetization jumps, albeit broadened by disorder, can clearly be identified. From an experimental point of view, the peak to peak amplitude of the magnetization swing (between the point A and B shown in Fig. 1.20) is the most unambiguous determination of the jump magnitude $\Delta M_\nu$. Because of the finite jump width, the background subtraction introduces an unavoidable distortion in the wings of each jump which is a source of error in the determination of $\Delta M_\nu$. Using different extrapolation schemes, an upper bound of 30% for this error on $\Delta M_\nu$ could be set.

It is necessary to estimate the area $\mathcal{A}$ occupied by the incompressible fluid to estimate the many-body gap $\Delta_\mu$ with Eq. 1.54. There are different ways to estimate this area. From the deep minima observed in the longitudinal resistivity at $\nu = 1/3$ and 2/3, the incompressible fluid is known to percolate through the sample. A theoretical estimate for the sample fraction occupied by the incompressible fluid is Eq. 1.19 (Sec. 2) based on Pickus and Efros model appropriate to small gaps $\Delta$ compared to the disorder $W$. An estimate for the sample fraction occupied by the incompressible fluid of 30 to 45% is inferred, but large uncertainties exist considering that $W$ is not measured. This lower bound for $\mathcal{A}$ can be improved using the "imaging" of compressible and incompressible areas achieved in earlier studies of the recombination dynamics of electrons with holes bound to a $\delta$-doped layer[41]. For samples with comparable densities and mobilities, these authors estimate that 60-70% of the sample area is incompressible at $\nu = 1/3$. Using this rough estimate, the area occupied by the incompressible liquid is $\mathcal{A} \approx 0.15 \pm 0.04$ cm$^2$. Since there are 100 layers in the sample, the inferred thermodynamic gaps at 1/3 and 2/3, $\Delta_\mu = \frac{\Phi_0}{\nu \mathcal{A}} \frac{\Delta M_\nu}{100}$ are summarized in Table 1.2.

It is now instructive to compare the thermodynamic gaps with the activation energies $\Delta_{tr}$ necessary to excite a quasi-particle quasi-hole pair, a quantity

---

[6]Although non-equilibrium currents (see Sec. 3) induced by the flux variation through the sample were observed at $\nu = 1/3$ at the lowest temperature, they are not present in any of the data presented here.

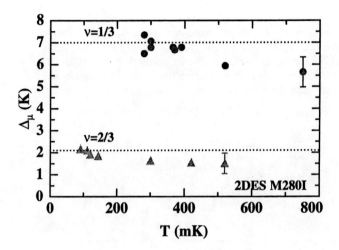

*Figure 1.21* Temperature dependence of the thermodynamic gap $\Delta_\mu$ at $\nu = 1/3$ and $2/3$. The dotted lines are the zero temperature value of the gaps.

accessible in a transport measurement on a Hall bar from the same wafer: reliable values for $\Delta_{tr}$ are obtained with an activation plot of the minima in $\rho_{xx}$. All gap values are collected in Table 1.2 and compared with theoretical estimates appropriate to disorder free samples. The activations gaps are only 20% smaller to those measured in the highest mobility single layer heterointerface [39]. On the other hand, *thermodynamic gaps $\Delta_\mu$ at fractions are at most one half of $q \times \Delta_{tr}$ while $\Delta_\mu$ and $\Delta_{tr}$ are comparable in the integer regime.*

To address this discrepancy, the first issue to consider is the reduction of incompressible gaps by a finite density of thermally excited quasiparticles. The magnetization swings $\Delta M$ at $\nu = 1/3$ and $2/3$ show a weak decrease with increasing temperature. This decrease is translated into a reduction of $\Delta_\mu \propto \Delta M$, in Fig. 1.21: for temperature small compared to the gap, the density thermally excited quasiparticles is small and has a weak effect on $\Delta M$ (or $\Delta_\mu$). It is therefore reasonable to assume that a small density of quasiparticles in the compressible regions has no dramatic effects on Eq. 1.54.

The subtle nature of electronic correlations and the smaller values of the gaps are some of the reasons why fractional states are more sensitive to disorder than integer states. Disorder can be characterized using several physical measurements. From the transport mobility, a mean free path $\ell = 2 \times 10^{-6}$m and a transport scattering rate $1/\tau_{tr} = 7 \times 10^{11} s^{-1}$ ($h/(k_B\tau_{tr}) = 3.4$K) may be derived and compared to the energy gaps. More relevant to thermodynamic quantities is the broadening of Landau levels, controlled by the average potential fluctuation over an electronic wavefunction. At high filling factors, this broadening increases with field as $\Gamma(B_\perp) = C \times \sqrt{B_\perp}$ and can be measured by fitting the damping of Haas van Alphen oscillations with increasing filling factor as explained in Sec. 4.3. The magnetization of our sample in the integer regime is shown in Fig. 1.11. Above $\nu = 4$, the overall decrease of the oscillations is reproduced with $C = 2.1$K (when $B$ is expressed in Tesla). Even if the $\sqrt{B_\perp}$ increase of the LL broadening saturates well below $B_\perp = 10.5$T ($\nu = 1/3$ in our

# Magnetization of the 2DEG

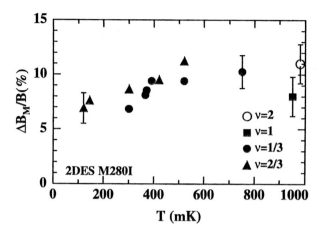

*Figure 1.22* Fractional widths of the magnetization jumps $\nu = 1/3, 2/3$ and $1$ at measured at different temperatures. Within experimental errors, the relative with of the jumps are the same.

sample), it is clear that $\Gamma(\nu = 1/3)$ is only marginally smaller than $\Delta_{tr} \approx 5K$. For integer states, the relative width $\Delta B_M/B$ of the magnetization jumps is set by the ratio $\Gamma(B)/\Delta$. As shown in Fig. 1.22, the relative width $\Delta B_M/B$ at $\nu = 1/3, 2/3, 1$ and $\nu = 2$ are comparable even though the transport gaps $\Delta_{th}$ differ by a factor of 4. Since the width $\Delta B_M$ remain small even for very small fractional gaps, it is clear that residual potential fluctuations are averaged out by electronic correlation over much larger distances than the magnetic length : it is only the very long range disorder fluctuation which broaden fractional states. This very subtle interplay between interactions and disorder in the fractional regime prompts us to reconsider the validity of Eq. 1.54.

## 5.1 ORBITAL MAGNETIZATION AND CURRENTS DISTRIBUTION

As soon as a deep minima in $\rho_{xx}$ is observed, there is at least one sample spanning incompressible region which separates the sample in at least two compressible parts (+ and -), with filling factors $\nu^+ > \nu > \nu^-$ as illustrated on the left hand side of Fig. 1.23. Over compressible regions (+ and -) screening pins the chemical potential while the density and the filling factor fluctuate accordingly. On the other hand, the electronic density and hence the filling factor $\nu$ are fixed in the incompressible fluid while the energy of a quasiparticle ($\mu_+$) or a quasihole $\mu_-$ follow the local self-consistent potential. In presence of disorder, the constancy of the local density can be achieved by balancing the variations in the disorder potential with opposite changes of interaction and correlation energies. A low density of quasiparticles with chemical potential $\mu(\nu^+)$ can be found in the + areas while a gas of quasiholes with chemical potential $\mu(\nu^-)$ partially occupies the - areas. Since the electronic density is constant in the incompressible region separating the + and - areas, the change in self consistent potential across this region must equal the incompressibility gap $\Delta_\mu$ as shown on the right hand side of Fig. 1.23.

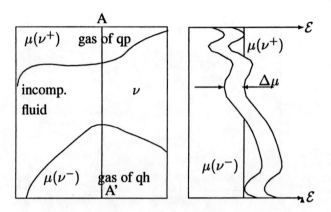

*Figure 1.23* Left: spatial distribution of compressible and incompressible region in presence of slowly varying disorder potential: the incompressible fluid at filling factor $\nu$, separates compressible regions occupied respectively by gases of quasiparticles and quasiholes. Right: energy diagram for the quasiparticle and quasihole excitations across the cut AA' through the samples. The energy gap of the incompressible fluid being the separation between these two energies, it is possible to relate the difference in local magnetization across the incompressible region to the energy gap.

In an effort to describe the current distributions and variations in the local magnetization for an heterogeneous QH fluid, Geller and Vignale[15, 16] have generalized the Kohn-Sham density functional theory to a current density functional formalism appropriate to 2DEG in high magnetic fields. This theory, being essentially exact in the limit of slowly varying potentials, is quite relevant to real experiments. As previously mentioned in Sec. 3.2, there are two distinct contributions to the local current density. The first contribution, a transverse "diffusion current" is mainly driven by the local gradient in electronic density (or filling factor $\nu(\vec{r}) = n(\vec{r})\Phi_0/B$),

$$\vec{j}_D(\vec{r}) = -\frac{e\omega_c}{4\pi}\vec{\nabla}\nu(\vec{r}) \times \hat{z} + \vec{\nabla} \times \overline{m}_{xc} \qquad (1.55)$$

which differ from the diffusion current previously discussed in the integer regime (cf. Eq. 1.27) by the exchange and correlation contribution of the orbital magnetization $\overline{m}_{xc}$, with the discontinuities removed. This diffusion current can only occur in regions where orbitals in the lowest LL are not equally occupied, namely in compressible regions. This contribution to local currents is unusual from the point of view of electrodynamics, since it is not controlled by local electric fields but by density gradients or equivalently by gradients in the occupations of orbitals. Physically, this is because all orbitals carry currents, which compensate each-other only in the homogenous limit. The second contribution to the current distribution is the more familiar quantum Hall current

$$\vec{j}_{QH}(\vec{r}) = -\frac{e^2}{h}\nu(\vec{r})\vec{\nabla}V_{eff}(\vec{r}) \times \hat{z}. \qquad (1.56)$$

where the self consistent disorder potential $V_{eff}(\vec{r}) = V(\vec{r}) + V_H(\vec{r}) + V_{xc}(\vec{r})$ is the sum of disorder, Hartree, exchange and correlation energy. In compressible regions, this contribution to the current distribution vanishes due to the "perfect

# Magnetization of the 2DEG

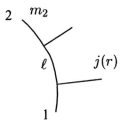

*Figure 1.24* The integrated current flowing across the line joining the points and equals the difference in local magnetization between points and . The unit vector is perpendicular to the integration path $\ell$.

screening theorem" (in the $T \to 0$ limit[16]), which may be expressed in a simplified manner in the regions with filling factors and as

$$\text{constant} = \mu(\nu_+) + \frac{\Delta_\mu}{2} + V_{eff}(\nu_+)$$
$$\mu(\nu_-)\quad \Delta_\mu \quad + V \quad (\nu\ ) \tag{1.57}$$

Since in equilibrium $\mu(\nu_+) = \mu(\nu_-) = \mu$, this theorem proves that the change in the self-consistent potential across the incompressible region equals the incompressible gap $\Delta_\mu$. To summarize, the contributions to the current density are physically separated in space: the transverse diffusion current (Eq. 1.55) only contribute to compressible regions while the quantum Hall current (Eq. 1.56) is non-zero when the electronic fluid is in an incompressible state. In the vicinity of a boundary separating compressible and incompressible regions, the spatial variations in self-consistent potential $V\ (\vec{r})$ and filling factors $\nu(\vec{r})$ have usually opposite sign since $\nu$ decreases as $V_{eff}$ increases. Hence "diffusion" and "quantum Hall" current flow in opposite direction close to the boundary. This is why compressible regions always contribute to a reduction of the magnetization jump $\Delta M$ for all QH filling factors.

This current distribution $\vec{j}(\vec{r}) = \vec{j}_D(\vec{r}) + \vec{j}_{QH}(\vec{r})$, can now be related in a straightforward way to the magnetization density $m(\vec{r})$ through,

$$\vec{j}(\vec{r}) = \vec{\nabla} \times m(\vec{r})\hat{z} \tag{1.58}$$

since in equilibrium $\vec{\nabla} \cdot \vec{j}(\vec{r}) = 0$. This makes it possible to express the current flux across an arbitrary path $\ell$ (see Fig. 1.24) joining any two points 1 and 2 of the 2DEG as the difference of the local magnetization at the end points

$$I_{1,2} = \int_1^2 \vec{j}(\vec{r}) \cdot \hat{n}(\vec{r}) d\ell = \int_1^2 \left[\vec{\nabla} \times m(\vec{r})\right] \cdot \hat{n}(\vec{r}) d\ell = m_1 - m_2 \tag{1.59}$$

When the end points lie just across an incompressible strip at filling factors $\nu_+$ and $\nu_-$, only the QH current (Eq. 1.56) contribute: hence the line integral of the gradient in self consistent potential across the strip yields

$$m(\nu_+) - m(\nu_-) = \frac{e\nu}{h}\Delta_\mu, \tag{1.60}$$

where we have used the "perfect screening theorem" (cf. Eq. 1.57). Hence in presence of disorder, the thermodynamic relation between incompressible gaps and magnetization steps still holds provided that the "jump" $\Delta M/\mathcal{A}$ coincide with the difference in local magnetization across the incompressible strip.

The *total* magnetization shown in Fig. 1.20 is measured while sweeping the magnetic field or equivalently the filling factor. When the average $\bar{\nu}$ (over the sample) is swept, the region + occupied by a low density gas of quasiparticles grows pushing gradually the incompressible region $\nu$ (and $\nu_-$) out of the sample. It is reasonable to assume that at the point $B$ (or $\bar{\nu} = \nu_B$) on the magnetization curve the region + occupies the whole sample. Similarly at point $A$ ($\bar{\nu} = \nu_A$) the region − covers the whole sample. Between $\nu$ and $\nu_A$ (resp. $\nu_B$), the incompressible regions are progressively replaced by a compressible gas of quasiholes (resp. quasiparticles). If in $\nu_A$ (resp. $\nu_B$) the magnetization of the gas of quasiholes (resp. quasiparticles) was uniform and equal to $m(\nu_+)$ (resp. $m(\nu_-)$), the analysis carried out in the last section would be accurate. On the other hand, the compressible regions are naturally inhomogenous and furthermore there is also a change in the magnetization between $\nu_+$ and $\nu_A$ which has not been considered. While these disorder specific contributions are at present difficult to quantify, they clearly contribute to a reduction of the magnetization jump since the orbital contribution of compressible and incompressible regions have opposite signs.

This detailed discussion shows that thermodynamic quantities are largely understood even when the QH fluid is inhomogenous. When the samples have a sizable amount of disorder (as in sample $M280$), not all contributions can be accurately quantified. On the other hand, as the fraction occupied by compressible regions decreases, the comparison between thermodynamic and transport gaps could be made as precise as in the integer regime.

## 6. OTHER THERMODYNAMIC QUANTITIES

In this section, a few other thermodynamic measurements are briefly mentioned. Most of them do not show the same sharp structures and cannot be analyzed in a model-independent fashion. In some instance, the broadening of thermodynamic oscillations can be attributed to extrinsic sources, such as disorder, backgrounds etc. This was the case in early magnetization measurements which were carried on low mobility samples[43, 44, 45] and in most magneto-capacitance studies[50, 51] of the 2DEG. For other measurements, such as the specific heat[46, 47, 48] thermodynamic oscillations are always smooth, even for disorder-free samples, and extrinsic disorder sources simply reduce the amplitude of the oscillations. Indeed, the electronic specific heat is exponentially small in incompressible states and reaches a maximum between filling factors, where the configuration entropy for quasiparticles is largest. In other words, the electronic specific heat is largest when the density of states is maximum. In this sense, one can state that the specific heat is sensitive to *the presence of states* (LL subbands) while the magnetization is most sensitive to the *absence of states* (energy gaps). The oscillations in the specific-heat experimentally observed in the integer regime are significantly smaller than theoretically expected. Some models for the density of states have been used to model the observed specific heat. Unfortunately, since several DOS models fit the specific heat, it has been difficult so far to extract precise informations from specific heat measurements.

Two other thermodynamic measurements have been able to determine integer and fractional gaps. In a first technique, the electronic compressibility $\kappa$ is measured in a double layer structure and the gaps are inferred by integrating $\kappa^{-1} = N^2 \partial \mu / \partial N$ as a function of filling factor [52]. In this experiment, one of the 2DEG is in an incompressible states while the other 2DEG of lower density is used to probe the electronic screening. Perhaps the most spectacular measurement of gaps in the integer and fractional regime come from from luminescence studies of the radiative recombination of 2D electrons with photoexcited holes bound to remote acceptors[54]. The first moment $M_1$ ("center of gravity") of the luminescence line shows flat steps as a function of magnetic field (or filling factors). Although $M_1$ is not a strictly speaking a thermodynamic quantity, it can be related[55] to the total energy of the 2DEG, with some assumptions. In this model, the magnetic field derivative of the first moment $\partial M_1 / \partial B$ is proportional to the magnetization of the 2DEG. This quantity is directly comparable to the magnetization curves presented in this paper. Most features and even the gap values inferred from these optical studies are all in excellent agreement with the direct thermodynamic measurements. Since these optical measurements could in principle be carried out on a local scale, the gaps could be measured locally.

Theoretical estimates of gaps at fractions have been considerably refined over the years[56] including LL mixing and the finite width of quantum wells. While they still exceed most experimental determination they now approach the values observed in the highest mobility samples.

## 7. CONCLUSIONS

Quantum Hall systems have current carrying (chiral) ground states. How the currents flow and how they redistribute themselves as a function of the filling factor is an issue which can be addressed by measuring its orbital magnetization. The incompressible nature of QH fluids pushes their quasiparticles excitations at the edges of the incompressible regions: this leads to the fundamental relation between the energy gaps and the singular contribution of QH phases to the magnetization (jumps). This relation, being thermodynamic, holds quite generally no matter how complex electronic correlations are.

Much of this paper has been devoted to understand thermodynamics properties in presence of disorder, an issue involving the interplay between disorder and interaction in QH systems. There are now beautiful pictures[57] taken by scanning capacitance spectroscopy showing that the QH fluids are inhomogenous for low even-integer filling factors. This inhomogenous nature is fully corroborated by the presence of non-equilibrium currents and other observations. While illumination by infrared diodes removes most of the inhomogeneities in high mobility samples in the integer regime, they persist at fractions. In this paper, we have attempted to demystified inhomogenous QH fluids. Because "diffusive" and quantum Hall currents are physically separated in space, it is relatively straightforward to isolate the quantum Hall contribution which are singular at integer or fractional filling factors. Diffusive contribution which are driven by the inhomogenous occupation of LL are a neat signature of the chiral nature of compressible ground states. In many ways, most experimental observations fit reasonably well within this general picture of orbital current at even integer and fractional filling factors, even if the gaps extracted from the magnetization jumps are smaller than expected. Unfortunately, the same cannot be said of odd-integer filling factors.

An interesting part of the physics left out of this paper is the energetics of interfaces between QH phases and the excitations at their edges. So far only edge magnetoplasmons have been probed experimentally. Interfacial energies and edge spin-waves have not been measured so far.

Further understanding of the magnetic properties of QH systems will requires other probes. NMR studies[58, 59, 60, 61] have already contributed considerably to the understanding of skyrmion physics at $\nu = 1$. This technique can easily discriminate between gapped and ungapped phases. Scanning tunneling and capacitance spectroscopy[34] have also been able to probe the relaxation dynamics at $\nu = 1$. Finally optical probes can also access local properties.

We wish to thank Peter van der Linden for his help in this work, S. Melinte for his measurements of the transport gaps, V. I. Falko, N. Freytag and B. Shklovskii for their suggestions and comments throughout this work. The GHMFL is "Laboratoire conventionné aux Université Grenoble I and INPG".

# References

[1] K. von Klitzing, G. Dorda, and M. Pepper, Phys. Rev. Lett **45**, 494 (1980).

[2] D.C. Tsui, H.L. Stormer and A.C. Gossard, Phys. Rev. Lett. **48**, 1559 (1982).

[3] R.B. Laughlin, Phys. Rev. Lett. **50**, 1395 (1983).

[4] R.R. Du, H.L. Stormer, D.C. Tsui, L.N. Pfeiffer and K.W. West, Phys. Rev. Lett. **70**, 2944 (1993).

[5] B.I. Halperin, Phys. Rev. B **25**, 2185 (1992).

[6] F.D. Hadane, Phys. Rev. Lett. **51**, 605 (1983).

[7] R. Peierls, Z. für Physik **81**, 186 (1933).

[8] I.D. Vagner, T. Maniv and E. Ehrenfreund, Phys. Rev. Lett. **51**, 1700 (1983).

[9] A.L. Efros, Solid State Comm. **65**, 1281 (1988); *ibid.* **67**, 1019 (1988); A.L. Efros, Phys. Rev. B **45**, 11354 (1992).

[10] F.G. Pickus and A.L. Efros, Phys. Rev. B **47**, 16395 (1993).

[11] D.B. Chklovskii, B.I. Shklovskii and L.I. Glazman, Phys. Rev. B **46**, 4026 (1992).

[12] A.L. Efros, Phys. Rev. B **60**, 13343 (1999).

[13] V.T. Dolgopolov, N.B. Zhitenev and A.A. Shashkin, Europhys. Lett. **14**, 255 (1991).

[14] V.T. Dolgopolov, A.A. Shashkin, N.B. Zhitenev and S.I. Dorozhkin, K. von Klitzing, Phys. Rev. B **46**, 12560 (1992).

[15] M.R. Geller and G. Vignale, Phys. Rev. B **50**, 11714 (1994).

[16] M.R. Geller and G. Vignale, Phys. Rev. B **52**, 14137 (1995). *ibid*, Physica B **212**, 283 (1995).

[17] J.P. Watts, A. Usher, A.J. Matthews, M. Zhu, M. Elliott, W.G. Herrenden-Harker, P.R. Morris, M.Y. Simmons, and D.A. Ritchie, Phys. Rev. Lett. **81**, 4220 (1998).
[18] S.A.J. Wiegers, J.G.S. Lok, M. Jeuken, U. Zeitler, J.C. Maan and M. Henini, Phys. Rev. B (1999).
[19] I. Meinel et al., Appl. Phys. Lett. **70**, 3305 (1997)
[20] S.A.J. Wiegers et al., Rev. Sci. Instr. **69**, 2369 (1998).
[21] K. Jauregui and I.D. Vagner, Phys. Rev. B **41**, 12922 (1990).
[22] R.B. Dingle, Proc. Roy. Soc. (London) **A211**, 500 (1952); *ibid* **A211**, 517 (1952).
[23] S.A.J. Wiegers, M. Specht, L.P. Lévy, M.Y. Simmons, D.A. Ritchie, A. Cavanna, B. Etienne, G. Martinez and P. Wyder, Phys. Rev. Lett. **79**, 3238 (1997).
[24] G. Bastard, *Wave mechanics applied to semiconductor heterostructures*, Les Editions de Physique (1996).
[25] F. Stern and S. das Sarma, Phys. Rev. B **30**, 840 (1984).
[26] V.E. Kirpichev, I.V. Kukushkin, V.B. Timofeev and V.I. Falko, JETP Lett. **51**, 436 (1990).
[27] M.M. Fogler and B.I. Shklovskii, Phys. Rev. B **52**, 17366 (1995).
[28] A. Karlhede, K. Lejnell and S.L. Sondhi, Phys. Rev. B **60**, 15948 (1999).
[29] L. P. Lévy, *Magnétisme et Supraconductivité* EDP-Sciences/Editions du CNRS (1997) [english translation, Spinger-Verlag (1999)].
[30] D.-H. Lee and C.L. Kane, Phys. Rev. Lett. **64**, 1313 (1990).
[31] S.L. Sondhi, A. Karlhede, S.A. Kivelson and E.H. Rezayi, Phys. Rev. B **47**, 16419 (1993).
[32] H.A. Fertig, L. Brey, R. Côté and A.H. MacDonald, Phys. Rev. B **50**, 11018 (1994).
[33] N. Read and S. Sachdev, Phys. Rev. Lett. **75**, 3509 (1995). (1993).
[34] H.B. Chan, R.C. Ashoori, L.N. Pfeiffer and K.W. West, Phys. Rev. Lett. (1999).
[35] A.H. MacDonald, Phys. Rev. Lett. (1999).
[36] D. Yoshioka, B.I. Halperin and P.A. Lee, Phys. Rev. Lett. **50**, 1219 (1983).
[37] For a review see "*The Quantum Hall effect*", R.E. Prange and S.M. Girvin Eds. (Springer-Verlag, N.Y. 1987).
[38] K. Park and J.K. Jain, Phys. Rev. Lett. **81**, 4200 (1998).
[39] R.L. Willett *et al.*, Phys. Rev. B **37**, 8476 (1988).
[40] S.A.J. Wiegers, E. Bibow, L.P. Lévy, V. Bayot, S. Melinte and M. Shayegan, sublitted to Phys. Rev. Lett. (1999).
[41] I.V. Kukushkin, V.I. Fal'ko, R.J. Haug, K. von Klitzing, K. Eberl and K. TÏtemayer, Phys. Rev. B **72**, 3594 (1994).

[42] I. Meinel, T. Hengstmann, D. Grundler and D. Heitmann, Phys. Rev. Lett. **82**, 819 (1999).
[43] J.P. Eisenstein et al., Phys. Rev. Lett. **55**, 875 (1985).
[44] I.M. Templeton, J. Appl. Phys. **64**, 3570 (1988).
[45] A. Potts et al., J. Phys. Cond. Matter **8** 5685 (1996).
[46] E. Gornik et al., Phys. Rev. Lett. **54**, 1820 (1985).
[47] J.K. Wang, J.H. Campbell, D.C. Tsui and A.Y. Cho, Phys. Rev. B **38**, 6174 (1988).
[48] J.K. Wang, D.C. Tsui, M. Santos and M. Shayegan, Phys. Rev. B **45**, 4384 (1992).
[49] V. Bayot, M.B. Santos and M. Shayegan, Phys. Rev. B **46**, 7240 (1992).
[50] T.P. Smith, B.B. Goldberg, P.J. Stiles and M. Heiblum, Phys. Rev. B **32**, 2696 (1985).
[51] R.C. Ashoori and R.H. Silsbee, Solid State Commun. **81**, 821 (1992).
[52] J.P. Eisenstein, L.N. Pfeiffer and K.W. West, Phys. Rev. B **50**, 1760 (1994).
[53] I.V. Kukushkin, R.J. Haug, K. van Klitzing and K. Ploog, Phys. Rev. Lett. **72**, 736 (1994).
[54] I.V. Kukushkin and V.B. Timofeev, Adv. in Physics **43**, 147 (1996).
[55] E.I. Rashba, *Optical Phenomena in Semiconductor Structures of Reduced Dimensions*, p. 63, D.J. Lockwood and A. Pinczuk Ed., Kluwer Scientific.
[56] R. Shankar, Phys. Rev. B **63**, 85322 (2001).
[57] S.H. Tessmer, P.I. Glicofridis, R.C. Ashoori, L.S. Levitov, M.R. Melldoch, Nature **392**, 51(2001).
[58] S.E. Barrett *et al.*, Phys. Rev. Lett. **72**, 1368 (1994).
[59] S.E. Barrett *et al.*, Phys. Rev. Lett. **74**, 5112 (1995); R. Tycko *et al.*, Science **268**, 1460 (1995).
[60] P. Khandelwal *et al.*, Phys. Rev. Lett. **81**, 673 (1998); N. N. Kuzma *et al.*, Science **281**, 686 (1998).
[61] S. Melinte, N. Freytag, M. Horvatić, C. Berthier, L. P. Lévy, V. Bayot and M. Shayegan, Phys. Rev. Lett. **84**, 354 (2000).

# Chapter 4

# SKYRMION EXCITATIONS OF THE QUANTUM HALL FERROMAGNET

Robin J. Nicholas
*Department of Physics,*
*Clarendon Laboratory,*
*Parks Road, Oxford, OX1 3PU, United Kingdom*

**Abstract**

The concept of a Skyrmion excitation of the quantum Hall ferromagnetic state is reviewed in terms of its theoretical development and the experimental evidence for the existence of Skyrmions. Evidence comes from measurements of the polarization of the electron system which is found to decrease rapidly away from a level occupancy $\nu = 1$ and from measurements of the excitation gaps for the creation of Skyrmion-antiSkyrmion pairs which are found to be strongly dependent on the ratio of the Zeeman Energy and the Coulomb interaction energy.

## 1. Introduction

The magnetic fields at which the integer quantum Hall effect (IQHE) [1] occurs correspond to the formation of a strongly interacting electron gas which can have a number of novel and interesting excitations. This is particularly true when an odd number of levels are occupied, which makes the system spin polarized. At $\nu = 1$ (where the filling factor $\nu = n_e h/eB$ measures how many Landau levels (LL) are filled) the ground state should be regarded as a ferromagnet since all the spin down states in the lowest LL are occupied while all the spin up states are empty and the system will be completely polarized by even an infinitesimally small Zeeman splitting due to the strong Coulomb interaction s. This has led several authors to suggest that novel charged excitations with non-trivial spin order, known as charged spin-texture excitations or Skyrmions, might occur [2, 3] when the Zeeman energy is small. Such excitations are charged, Soliton types of quasiparticles with a charge determined in magnitude and sign by the degree of their spin twist. A schematic representation of such excitations is shown in Fig 1.1. In GaAs the single particle Zeeman energy (ZE) $g\mu_B B$ is very small, 0.3 K/T, compared to the large value of the Coulomb energy $E_c = e^2/4\pi\epsilon l_B = 50.55\sqrt{B}$ K where $l_B = \sqrt{\hbar/eB}$ is the magnetic length, and there have been a series of experiments using heterostructures and quantum wells in the GaAs/Ga$_{1-x}$Al$_x$As system which show evidence for the existence of Skyrmion excitations.

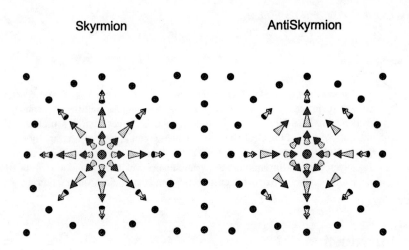

*Figure 1.1.* A schematic representation of a Skyrmion and and Anti-Skyrmion.

The $\nu = 1$ ferromagnetic ground state is significantly different from the more familiar Heisenberg ferromagnet since the spontaneous magnetization occurs in the presence of a quantising magnetic field, not at zero field, and the spins associated with the charge carriers are free to move, hence it is known as an 'itinerant' ferromagnet [3, 2]. Two types of charged excitations from the ferromagnetic ground state that produce a well separated spin up electron and spin down hole have been identified. The first is a spin wave (really a spin exciton) [5] whose energy depends on wavevector, usually given by the dimensionless quantity $kl_B$. At long wavelength, i.e. $kl_B = 0$, corresponding to zero spatial separation between the electron and the hole, the spin wave energy is just equal to the Zeeman Energy (ZE) $g\mu_B B$ as measured in spin resonance experiments [6]. When the electron and hole are well separated, i.e. at large $kl_B$, as might be found in an electrical transport measurement then the excitation requires a large addition contribution from the Coulomb exchange energy $E_c = e^2/4\pi\epsilon l_B$. The energy to create such an ionized spin wave is $E_{sw} = g\mu_B B + \kappa E_c$, where $\kappa$ is the spin stiffness calculated [5] to be $\sqrt{\pi/2}$ in the ideal case.

The second type of excitation is based on a spin texture that consists of a central reversed spin surrounded by rings of spin that gradually cant over until at the edge they are aligned with the external magnetic field. These are usually known as Skyrmions, although strictly this term is reserved for objects of infinite extent at zero ZE. The essential differences between this two-dimensional spin texture and a spin wave are that the net spin may be greater than one and on a path taken around the central spin there will be a change of spin orientation equivalent to a winding number of unity. In a system with zero ZE the Skyrmions should have infinite extent but for finite ZE they have a finite size that can be characterised by an additional number of reversed spins $\tilde{S}$ (or $\tilde{A}$) contained [7] in the Skyrmion (or anti-Skyrmion). When the filling factor moves away from $\nu = 1$ the ground state will contain a number of Skyrmions (quasi-holes) or anti-Skyrmions (quasi-electrons) and the first direct evidence for their existence was from measurements of the degree of spin polarisation in nuclear magnetic resonance [8] and absorption experiments [9]. Both of these measurements suggest that $\tilde{S} \sim 2 - 3$ for straight forward GaAs/Ga$_{1-x}$Al$_x$As quantum wells. At exactly $\nu = 1$ the excitations of the system will consist of Skyrmion–antiSkyrmion pairs which, when they are fully separated cost only half the exchange energy (for infinite Skyrmions) required for a simple spin spin exciton, but $s(= 1 + \tilde{S} + \tilde{A})$ times the ZE. One way to think of this is that the spin texture of the Skyrmions dresses the electrons. Eventually, at large enough ZE, $s = 1$ and the Skyrmion-antiSkyrmion pairs are indistinguishable from the undressed spin excitons.

The balance between the SP ZE and the Coulomb energy is determined by the parameter $\tilde{g} = g\mu_B B/E_c$ which determines whether Skyrmions with $S > 1$ (small $\tilde{g}$) or spin waves (large $\tilde{g}$) will be the lowest lying excitations. The crossover is calculated to be at $\tilde{g} = 0.018 - 0.054$ [3, 10, 11]. For GaAs $\tilde{g} = 0.006\sqrt{B}$ so Skyrmionic excitations are expected to be favoured at low magnetic fields and small $g$-factors.

## 2. Theory

The prediction that quantum Hall ferromagnets contain Skyrmions in their quasiparticle excitation spectra was first made by Sondhi et al [2] who used field theoretical techniques to show that the ionization energy at $\nu = 1$ for a Skyrmion-AntiSkyrmion pair was exactly half that of the spin wave in the limit of ZE $\to 0$. Their analysis was based on a long wavelength effective action for the spin dynamics, supplemented by a charge-density-topological density constraint that allowed an analytic computation of the properties of the Skyrmions in the limit of small ZE. They found that their calculated Skyrmion-AntiSkyrmion creation energies were quite close to those measured earlier by Usher et al [4] at $\nu = 1$. Similar ideas were also reported by Fertig et al [3] who used Hartree-Fock theory to study the effects of finite Zeeman Energy more quantitatively and to look at the size of the Skyrmions. They concluded that the Skyrmion excitations might be present but that they may be quite small for typical experimental situations.

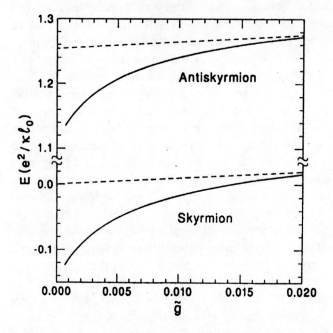

*Figure 1.2.* Solid lines show the energies of the Skyrmion type (charged spin texture) excitations as a function of $\tilde{g}$. Limiting values at $\tilde{g} \to 0$ are $-\frac{1}{4}\sqrt{\frac{\pi}{2}}$ for the Skyrmion and $\frac{3}{4}\sqrt{\frac{\pi}{2}}$ for the anti-Skyrmion. The dashed lines show the Zeeman energy of a single quasielectron or quasihole [3].

Figure 1.2 shows the energies for Skyrmion and antiSkyrmion excitations as a function of $\tilde{g}$ [3] from which it can be seen that for values of $\tilde{g} > 0.01$ the energies of the Skyrmion and single spin flip excitations are within $0.02E_c$ of each other. Kamilla, Wu and Jain have estimated the number of reversed spins in an anti-Skyrmion by minimising its energy [12] and find that for $\tilde{g} = 0.0017$ one excitation can contain of order 20 spin flips. In a subsequent publication Fertig *et al.* [10] have included finite thickness effects in calculating the Skyrmion size (measured in terms of the numbers of additional spin flips $\tilde{S}$ or $k$) as a function of $\tilde{g}$ as shown in Figure 1.3. A more recent extension of the effective-action studies has also allowed Lejnell et al [11] to use the effective

# Skyrmion excitations of the quantum Hall Ferromagnet

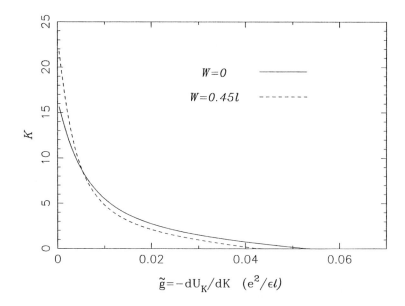

*Figure 1.3.* Number of reversed spins per hole (labelled as $K$ which is the same as $\tilde{A}$ used below) in a $\nu = 1$ quantum Hall ferromagnet as a function of $\tilde{g}/(e^2/\epsilon\ell)$. Results are shown for both strictly 2D and finite-thickness models. Note that for self-consistent solutions of the Hartree-Fock equations $\tilde{g} = -dU(K)/dK$ [10].

action to study the size of the Skyrmions. This theory works best in the low $\tilde{g}$ limit but predicts that the Skyrmions are the lowest energy charged excitations for $\tilde{g} < 0.018$ compared to the Hartree-Fock result of $\tilde{g} < 0.054$ [3] which is expected to be more accurate for small Skyrmions.

As the density (or magnetic field) is changed so that the occupancy moves away from $\nu = 1$ then the system is expected to generate Skyrmions (or Anti-Skyrmions). As a result of their large numbers of reversed spins the polarization of the system will fall rapidly. This is taken as one of the most unambiguous signs of the existence of Skyrmions. In the model proposed by Barrett et al. [8] the polarization is scaled by the parameter $S$ that gives the number of spin flips per unpaired flux quantum, however for ease of comparison with other data [7, 13] it is convenient to write this in terms of $\tilde{S} = S - 1$ ($\tilde{A} = A - 1$) which gives the number of additional spin flips per unpaired flux quantum above (below) $\nu = 1$. In this model the spin polarization is

$$\mathcal{P}(\nu) = \begin{cases} (\tilde{S}+1)\left(\frac{2-\nu}{\nu}\right) - \tilde{S} & \nu > 1 \\ \frac{1}{\nu} - (2\tilde{A}+1)\left(\frac{1-\nu}{\nu}\right) & \nu < 1 \end{cases} \quad (1.1)$$

which gives in the region around $\nu = 1$ with $\delta\nu = 1 - \nu \ll 1$

$$\mathcal{P}(\nu) = 1 - 2(\tilde{S}+1)\delta\nu, \tag{1.2}$$

Particle-hole symmetry suggests that the size of the Skyrmion will be the same as the Anti-Skyrmion, $S = A$, giving a rapid symmetric loss in polarization about $\nu = 1$ for $S > 1$. Once the population of these skyrmions becomes significant then they may begin to interact and there have been several predictions of the formation of a Skyrme crystal, similar to a Wigner lattice [14, 15, 16, 17]. These predictions remain to be verified, but the resulting phase diagram is quite rich and contains both hexagonal and square lattices [18] as shown in Figure 1.4.

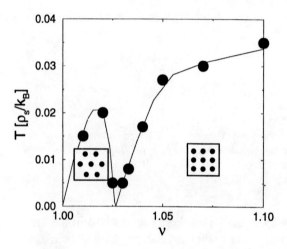

*Figure 1.4.* Thermodynamic phase diagram of a many-body Skyrmion system [18].

A more microscopic description appropriate to small Skyrmions has been introduced by MacDonald et al [19] who developed explicit wave functions describing Skyrmions with well defined spin quantum numbers based on a hard-core model Hamiltonian. This has been used to show that both Landau-level mixing and finite thickness effects (which generally reduce the strength of electron-electron interactions) reduce the size of the Skyrmions [20], although Hartree-Fock calculations have shown that finite thickness effects may favour the formation of Skyrmions [21].

In experiments attempting to study the Skyrmion formation at very low values of $\tilde{g}$ the Zeeman energy has been tuned by varying the contribution of the spin orbit term to the Zeeman energy (see below for further details) and this has prompted Fal'ko and Iordanskii [22] to consider how spin-orbit interactions might influence the quantum Hall ferromagnet state. They conclude that as the Zeeman energy approaches zero the system can form domains polarized along one of the easy-axis magnetic states, and finally for very small values of $\tilde{g}$ the spin polarization aquires a helically twisted spin texture with a mesoscopic scale which can be of order $500 nm$ for typical experimental situations. The presence of such domains may well be responsible for the collapse in the spin gap observed in several transport measurements discussed below.

The ground states at higher odd filling factors will also be ferromagnetic but only a fraction $1/\nu$ of the electrons are involved in the collective motion. The remainder in full LLs may act to screen any Coulomb interactions. Skyrmions also appear in the excitation spectrum at higher odd filling factors but for an ideal 2DEG they are calculated to have higher energies than the single spin exciton at vanishing ZE [23]. However, when the finite thickness $z$ of a real 2DEG is taken into account Skyrmions may become the lowest energy excitation at $\nu = 3$ [10] and also at all other odd filling factors for sufficiently extended wavefunctions [24]. The stability and size of the Skyrmions is predicted to increase with $z$ but to be reduced by finite ZE and LL mixing. For example, at $\tilde{g} = 0$ Skyrmions become the lowest excitation at $\nu = 3$ once $z > 0.1 l_B$, while for $z = l_B$ the transition occurs at $\tilde{g} = 0.0037$. The experimental picture at these higher odd filling factors is not clear yet.

A further factor which is now being considered is the role of disorder in influencing Skyrmion formation. It has been predicted [25] that the presence of large potential fluctuations on the scale of the Skyrmion size will give rise to Skyrmion-AntiSkyrmion pairs at $\nu = 1$, however in general the potential fluctuations in real samples are relatively small compared to the magnitude of the Coulomb energy. Sinova et al [26] have shown that for small disorder potentials $\gamma = E_c/\sigma > 1.5$ (where $\sigma$ is a measure of the disorder potential) the system retains total spin polarization as shown in Figure 1.5.

## 3. Experiment

Experimental work on Skyrmion properties falls generally into two main groups. Firstly there are the works such as the original pioneering experiments of Barrett *et al.* [8] which measure properties which are related more or less directly to the polarization of the spin system. These are generally in quite good agreement with the theoretical predictions. Secondly, there are measurements, such as the analysis of thermally activated magnetoresistance, which attempt to measure the energy gap associated with Skyrmion-antiSkyrmion pair formation. These often have significant differences from what is theoretically expected since they notably give values for the gaps which are substantially different from that expected from theory.

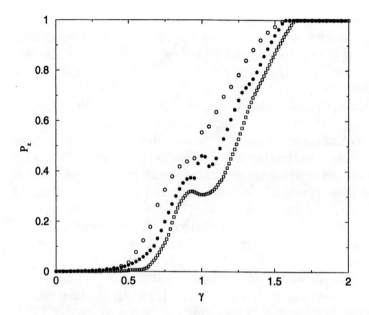

*Figure 1.5.* Global polarisation for two values of the ZE as a function of the relative order in the system $\gamma$ [26]. For disorder potentials only somewhat smaller than the Coulomb energy the system rapidly becomes 100% polarized

## 3.1 Polarization measurements

The works of Barrett *et al.* [8, 13] use the technique of optically pumped nuclear magnetic resonance (OPNMR). Optical pumping of interband transitions using polarized light is used to generate non-equilibrium spin populations of electrons and holes in doped GaAs/AlGaAs quantum wells, which then transfer some of their excess polarization to the the nuclei through the contact hyperfine coupling. This causes a large enhancement of the signal from nuclei in the quantum well region.

As a result of the averaged interaction with the equilibrium electron spin polarisation there is a Knight shift $K_S$ of the resonance from the $^{71}$Ga nuclei which is $K_s(\nu, T) = A_{zz} \langle S_z(\nu, T) \rangle$ for $\mathbf{B} \| \mathbf{z}$, where $A_{zz}$ is the hyperfine coupling constant for nuclei in in the well. The resonance remains sharp as it has been "motionally narrowed" through interacting with the average spin population of the electrons. After each measurement the nuclear polarization is then reset to zero with an RF saturation pulse. In order to measure the occupancy dependence of the polarization the sample is tilted with respect to the magnetic field direction so that the nuclear resonance occurs at different values of the perpendicular field component $\mathbf{B}_z$. An example of this type of measurement is shown in Figure 1.6 where the shifted resonance is seen to appear after a

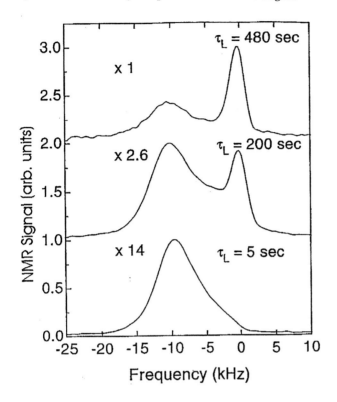

*Figure 1.6.* $^{71}$Ga NMR spectra for a GaAs/GaAlAs quantum well for various optical pumping times $\tau_L$. The two resonances arise in the wells and the barriers and their separation gives the Knight shift $K_s$ [8].

sufficiently long optical pumping time to cause the nuclear spin polarization. From the dependence of $K_s$ upon tilt angle the resulting spin polarisation was deduced as shown in Figure 1.7. This was fitted to Eq. (1.1) to give values of $\tilde{A} = \tilde{S} = 2.6 \pm 0.3$ in good agreement with the predictions of theory [2, 3]. Very similar numbers have been observed in a number of subsequent works [13, 48]. In contrast the temperature dependence of the polarization at $\nu = 1$ was found to be rather difficult to reconcile with simple models of the exchange enhanced spin gap since it persisted to quite high temperatures of order 4K but then collapsed rather rapidly. In fact this is to be expected [27] since thermal population of the upper spin state will cause a critical collapse of the exchange energy.

In a similarly direct measurement Aifer *et al.* [9] measured the optical absorption of polarized light around $\nu = 1$ in a single quantum well where the substrate had been thinned to $\sim 0.5\mu m$. Their data are shown in Figure 1.8 which shows that due to the selection rules transitions can be observed

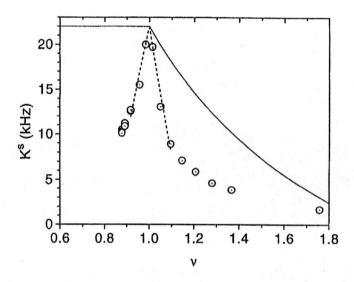

*Figure 1.7.* The dependence of the Knight shift $K_s$ as a function of filling factor. The solid line gives the fit to Eq. (1.1) with $\tilde{A} = 0$ while the dashed line has $\tilde{A} = 2.6$ [8].

into both electron spin states from one polarization of light, but due to the separation of the transitions in energy it is possible to deduce the total spin polarization of the system. They also concluded that $\tilde{A} = \tilde{S} = 1.5 - 2.5$ in remarkably good agreement with Barrett *et al.* [8]. In a subsequent work [28] they also studied the temperature dependence of the polarization and found that it could be fitted very well with a theory based on a continuum quantum ferromagnet [29].

Attempts to observe Skyrmion signatures in simple photoluminescence (PL) measurements have so far proved inconclusinve due to the complex excitonic interactions which take place around $\nu = 1$ [30, 31]. A measurement of the polarization of the electron system has been reported, however, by Kukushkin *et al.* [32] in which the polarization of time resolved radiative recombination of the electrons with photoexcited holes bound to acceptors is first measured. The polarization of the electrons is then deduced using the polarization of the hole system which is measured by observing recombination with higher Landau levels. The conclusion from this measurement was that no evidence for Skyrmion formation could be found. In a later publication [33], however, it has been suggested that this same measurement technique did find clear evidence of Skyrmion formation, but only under conditions of strong optical pumping with polarized light. This was attributed to a strong optically induced polarization of the nuclei giving a static effective nuclear magnetic field which then acts back on the electron system changing its Zeeman energy (the well

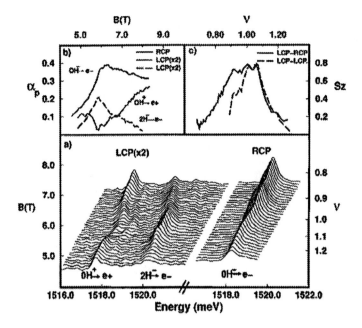

*Figure 1.8.* Absorption spectra for LCP and RCP light for a GaAs/AlGaAs quantum well in the region of $\nu = 1$. The large absorption for the $0H^+ - e^+$ transition for $\nu > 1$ is due to the presence of a large population of Skyrmions [9].

known Overhauser shift). This is known to be comparable to the electron Zeeman energy [34] and hence could compensate this for one direction of the nuclear field. For the highest pumping intensities values of $\tilde{S}$ up to 5 were reported [33]. It has been suggested by these authors that the reason for the relatively large Skyrmion effects seen by Barrett *et al.* [8] and Aifer *et al.* [9] could have been due to some uncontrolled optically induced nuclear spin polarization which therefore reduced the effective value of $\tilde{g}$, although for the absorption measurements at least this seems rather unlikely since similar conclusions were drawn from both right and left handed polarization when the nuclear fields would have been opposite.

In a remarkable series of experiments on heat capacity Bayot *et al.* [35] found that there was a giant enhancement of the specific heat capacity C at low temperatures around $\nu = 1$ with an increase of several orders of magnitude as C became dominated by the Schottky nuclear heat capacity. This observation implies a strong enhancement of the nuclear spin-lattice relaxation rate which was attributed to the presence of Skyrmions. Melinte *et al.* [36] used this phenomenon to examine the dependence of the Skyrmion formation on $\tilde{g}$. Figure 1.9 shows the heat capacity measured for a multi-quantum well as a function of filling factor for several different tilt angles of the sample with respect to the magnetic field direction. As the angle $\theta$ between the surface normal and the magnetic field is increased the ZE increases due to the increasing total field, while $E_c$ remains constant so that $\tilde{g} \propto \frac{1}{\cos\theta}$. This shows the remarkably large heat capacity that exists over the range of $0.8 < \nu < 1.2$ which is relatively unaffected by tilting up to an angle of $\theta \sim 80°$. Once the angle increases above this however C is found to collapse. This is shown in Figure 1.10 where the ratio of the heat capacity measured to that of the of the calculated nuclear heat capacity of the quantum well $C_{QW}$ is shown as a function of $\tilde{g}$. The striking thing about this data is that the heat capacity enhancement is completely destroyed at $\tilde{g} \sim 0.04$ in good agreement with the theoretical calculations for the upper limit of the ZE for Skyrmion formation [3, 2, 11]. At some intermediate angles it was also found that the heat capacity could collapse close to $\nu = 1$ and it has been suggested that this behaviour may be related to some of the more exotic Skyrmion phases [18, 14] although the evidence is very uncertain at present.

## 3.2 Excitation gaps

The next class of experiments involves temperature studies of the activation energy associated with the formation of Skyrmion-antiSkyrmion pairs. It has long been known that Coulomb effects cause a large enhancement of the spin splitting in Quantum Hall systems (see [4, 37, 39] and references therin) as has been elegantly shown by Leadley *et al.* [37]. They used the well know Lifshitz-Kosevich formula to analyse the depth of the resistivity minima [38]:

$$\Delta \rho_{xx} \propto \frac{X}{\sinh X} \exp\left(-\frac{\pi}{\omega_c \tau_s}\right) \cos(\pi(\nu + 1/2)), \qquad (1.3)$$

where the factor containing $X = 2\pi^2 kT/\Delta$ arises from the width of the Fermi function at finite temperature (when there is no spin splitting the gap $\Delta$ is the cyclotron energy $\hbar\omega_c$). Using this method they showed that at $\nu = 1$ the spin splitting was $\propto \sqrt{n}$ and was dominated by the Coulomb contribution $E_c$ where:

$$E_{ex} = CE_c, \quad \text{with} \quad C = 0.21. \qquad (1.4)$$

This represents the Coulomb contribution to the energy required to excite fully separated electron-hole pairs in the empty and full levels. The fact that this value is so small compared to the theoretical value [5] of $\sqrt{\pi/2}$ is what prompted Sondhi *et al.* [2] to propose the existence of Skyrmion excitations for the quantum Hall ferromagnet. Some reduction in this value can be explained by the finite thickness of the real 2DEG which softens the Coulomb interaction,

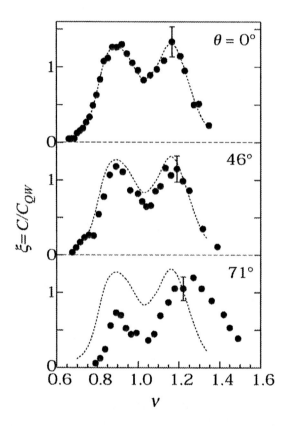

*Figure 1.9.* The heat capacity of a MQW GaAs/GaAlAs sample as a function of filling factor for three different tilt angles [36].

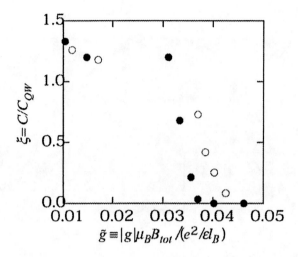

*Figure 1.10.* The heat capacity as a function of $\tilde{g}$ for $\nu > 1$ (•) and $\nu < 1$ (○) studied by varying the tilt angles [36].

but this usually only accounts for a factor of 2 at most [10, 40]. Including both the Zeeman and Coulomb terms we can write:

$$\Delta = |g_0|\mu_B B_T + E_{ex} \tag{1.5}$$

where $E_{ex}$ is determined only by the carrier density and hence only the perpendicular component of magnetic field, as was shown recently by Leadley et al. for high index Landau levels [41]. If the excitations, however, are Skyrmion-antiSkyrmion pairs with a total spin $s = 1 + \tilde{S} + \tilde{A}$ then the Zeeman Energy plays a greater role and Eq. (1.5) is modified to:

$$\Delta = s|g_0|\mu_B B_T + E_{ex}. \tag{1.6}$$

which gives the basis for a determination of the magnitude of s by tilting the sample with respect to the magnetic field and hence changing only the Zeeman component. An anomalous increase of the Ahhrenius activation gap on tilting (deduced from $\rho \sim \exp(\Delta/2kT)$) was first noticed by Nicholas et al. [39] but was not understood at the time. With the advent of the Skyrmion picture Schmeller et al [43] found that for several different samples with different degrees of disorder and confinement potentials that their data could be fitted well to Eq. (1.6) with a value of s=7 in good agreement with the Knight shift [8] and optical absorption [9] data. At this point one might conclude that the overall picture of agreement between experiment and theory was very good. Unfortunately, however, as for the spin wave the absolute values of the measured gaps are substantially smaller than theory predicts and their systematic dependence on Zeeman energy also suggests that there are some unexplained aspects of the data.

The idea behind a further series of experiments to test the Skyrmion picture was to reduce the g-factor of the electrons still further in order to move more strongly into the large skyrmion limit. This can be done in GaAs and GaAs/GaAlAs heterostructures in several different ways. In GaAs at ambient pressure $g = -0.44$, as a result of subtracting band structure effects driven by the spin-orbit interaction from the free electron value of 2. Simple **k.p** theory [42] says that if the band gap is increased then the band structure contribution reduces, and so the magnitude of $g$ passes through zero at a band gap of $\sim 1.67eV$. The $g$- factor can be calculated using **k.p** theory [42] and may be approximated by the following expression:

$$g = 2 - 20000 \left( \frac{1}{1500 + \hbar\omega_c + \Delta E_g} - \frac{1}{1850 + \hbar\omega_c + \Delta E_g} \right) + 0.062, \quad (1.7)$$

which gives the region where g tends to zero at around $\Delta E_g = 0.17eV$ which can be achieved in several different ways. Using $Ga_{1-x}Al_xAs$ with an Al content of 0.13, applying a hydrostatic pressure of $\sim 18$ kbar confining the carriers in a quantum well of thickness $\sim 50 \overset{\circ}{A}$ or even applying a magnetic field of $\sim 90T$ are all ways which have been used to study the limit as $\tilde{g} \to 0$.

Figure1.11 shows the activation energy measured as a function of $\tilde{g}$ in a $Ga_{0.87}Al_{0.13}As/Ga_{0.65}Al_{0.35}As$ quantum well where the g-factor is estimated to be $g = 0.043 \pm 0.010$ by Shukla *et al.* [44]. Also shown are the earlier results of Schmeller *et al.* [43]. As $\tilde{g} \to 0$ it is deduced that the value of $s \to 50$. The striking feature of this data however is that the skyrmion gap is tending towards 0 as $\tilde{g} \to 0$. The results of Hartree-Fock calculations [3, 24] are also shown in Figure1.11. For the infinitely thin 2-D limit the calculations are much larger than the measured values and the Skyrmion gap remains large and finite as $\tilde{g} \to 0$. Even when finite thickness corrections are taken into account [24] the predicted gaps only decrease by $\sim 30$. Some further reductions can also be expected due to landau level mixing effects [20, 45] but these are also relatively small. Agreement was found on making the arbitrary assumption that the calculated skyrmion gap should be reduced to 0 as $\tilde{g} \to 0$ as shown by the dotted line.

The picture of substantial disagreement with the absolute values is repeated in the studies as a function of pressure and of confinement. Figure 1.12 shows a study of the value of $\Delta/E_c$ as a function of $\tilde{g}$ as varied by pressure [37]. In this case the results show that the value of the spin gap is essentially independent of $\tilde{g}$ until $\tilde{g} < 0.005$ suggesting that the Skyrmions do not begin to form until this point. Beyond this the decrease in gap corresponds to a value of s=35 but the gap remains finite at $\sim 0.04E_c$. The value of s=35 is consistent with

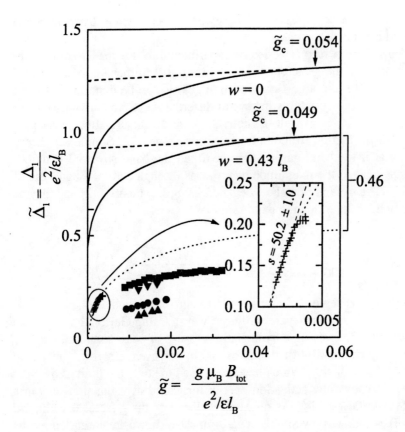

*Figure 1.11.* Tilted field studies of a $Ga_{0.87}Al_{0.13}As/Ga_{0.65}Al_{0.35}As$ quantum well. The normalized activation energy $\tilde{\Delta}_1 = \Delta_1/(e^2/\epsilon\ell_B)$ is plotted vs normalized Zeeman energy $\tilde{g}$ from experiments and calculations. The experimental data are from [44] (+) and [43] (closed symbols). In the inset, the asymptote (dashed line) fit to the lower range of sample A data reveals $s = 50.2 \pm 1.0$. The results of Hartree-Fock calculations for a 2DES with zero layer-thickness ($w = 0$) and for $w = 0.43\,\ell_B$ are also shown in the main figure. In the inset, the $w = 0.43\,\ell_B$ skyrmion excitation gap (dotted line) shifted down by $0.46\,e^2/\epsilon\ell_B$ matches the lower range of sample A data. [44].

the estimate of Kamilla, Wu and Jain [12] of s=36 at $\tilde{g} = 0.0017$ but again contradicts the large pair gap of $0.627\,E_c$ which is very much larger than that observed experimentally.

Rather better qualitative agreement with theory was found by Maude *et al.* [46] who used a combination of quantum well confinement and pressure to tune $\tilde{g}$ completely through 0. Their results are shown in Figure1.13 together with the theoretical values which have been reduced by a constant fitting factor of 0.17. In this case they did see the expected halving of the gap as $\tilde{g} \to 0$ but

# Skyrmion excitations of the quantum Hall Ferromagnet

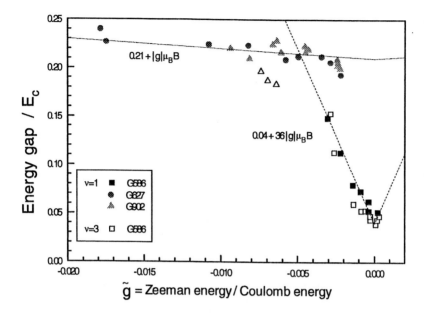

*Figure 1.12.* Pressure tuned values of $\Delta_1/(e^2/\epsilon\ell_B)$ vs normalized Zeeman energy $\tilde{g}$ for a series of GaAs/GaAlAs heterojunctions where the ZE is tuned with pressure. For $\tilde{g} < 0.005$ the slope gives s = 36. [37].

clear Skyrmion-like behaviour was only observed for $\tilde{g} < 0.002$ where a value was deduced of s=33.

## 3.3   Skyrmions at $\nu = 3$

We now turn to the data at $\nu = 3$. The initial work from tilted field measurements [43] showed no evidence for Skyrmion formation at higher occupancies. This picture was reinforced by Maude et al. [47] who again found no evidence for Skyrmion excitations even in the low $\tilde{g}$ limit, but these measurements used 8.2nm quantum wells which may be too narrow to allow the Skyrmion formation [24]. Evidence has been found however by two other measurements by Song et al. [48] and Leadley et al. [37] who used wider quantum wells and heterojunctions. Figure 1.12 also shows the data from [37] at $\nu = 3$ from which it can be seen that a similar reduction in energy gap occurs as $\tilde{g} \to 0$ as was seen for $\nu = 1$. This suggests that the same mechanism is responsible for the excitations at both of these filling factors and led to the conclusion that Skyrmionic excitations also occur at $\nu = 3$ with a value of s $\sim$22. This idea is consistent with theoretical predictions [24] where for $z = 2l_B$ Skyrmions will be excited at $\nu = 3$ provided $\tilde{g} < 0.0044$ and at $|\eta| = 0.002$, $s = 15$ and $E_g = 0.48E_c$. Once again the measured energy gaps are much

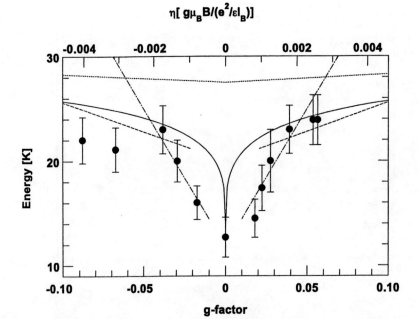

*Figure 1.13.* The measured energy gap as a function of g-factor and $\tilde{g}$ as tuned by pressure in a 6.8nm GaAs/AlGaAs quantum well. The solid line is the theoretical prediction [2] scaled to fit the data, while the short-dashed line gives the expected dependence for a bare ZE with s=1. The lines with slopes corresponding to s=7 and s=33 are shown for comparison [46].

smaller than predicted but the threshold for Skyrmion formation and the sizes are in agreement with theory. Song *et al.* [48] used OPNMR measurements to show further evidence for $\nu = 3$ Skyrmions at much larger values of $\tilde{g} \sim 0.01$ but the Skyrmion sizes which they deduce were correspondingly much smaller with values of $\tilde{S} = 1.2 \pm 0.4$ and $\tilde{A} = 0.4 \pm 0.3$. PL measurements [49] have been reported where anomalous reductions in polarization can be seen at $\nu = 3, 5, 7$ which have been attributed to Skyrmion formation, however it is not understood why this should occur.

## 3.4 Skyrmions at $\nu = 1/3$

The composite Fermion (CF) model of the fractional quantum Hall effect (FQHE) allows us to map the behaviour of strongly interacting electrons onto an integer quantum Hall effect (IQHE) of CFs [50, 51, 52]. Thus the physics of the state at filling factor $\nu = 1/3$, where there is one completely occupied CF Landau level (LL), is explained by analogy with $\nu = 1$. In GaAs the electronic ZE $g\mu_B B$ has similar magnitude to the gaps between CF LLs, which arise from electron-electron correlations and scale like $E_c$. CF LLs originate from

$\nu = 1/2$ where there is an offset of $g\mu_B B_{1/2}$ between fans of each spin, which provides an essential difference from the IQHE. This gives the possibility of level crossings as the ZE and magnetic field are varied and leads to the observed disappearance and re-emergence of fractions [42, 53, 54]. Although the ground state at $\nu = 1/3$ will be fully spin polarized, the states at $\nu = 2/3$ or 2/5 may be either fully polarized or unpolarized depending on the relative sizes of the ZE and the CF LL gaps. Similarly the excitations may involve either spin flips or inter CF LL transitions. At $\nu = 1/3$ and small ZE, we might expect a repeat of the picture where either a single spin flip of one CF or a Skyrmionic excitation of the Composite Fermions may occur which would be a Composite Skyrmion.

Evidence for the existence of Composite Skyrmions was first presented by Leadley *et al.* [55] who also used pressure to tune $\tilde{g}$ through zero. Figure 1.14 shows the scaled energy gaps measured for $\nu = 1/3$ and 2/5 as a function of $\tilde{g}$. For $\tilde{g} > 0.01$ the FQHE state at $\nu = 1/3$ has a constant Coulomb gap, which may correspond to either the spin-wave or more probably the CF gap. For $\tilde{g} < 0.01$ the data can be fitted using Eq. 1.6 with $s = 3$, suggesting that there is a small Composite Skyrmion being formed. In a rough estimate Sondhi *et al.* suggested that a Skyrmion formed at $\nu = 1/3$ occurring at 1 T should contain 'a couple of reversed spins' [2]. They also estimate the Skyrmion–antiSkyrmion pair gap as $0.024E_c$ at $\tilde{g} = 0$. The minimum gap seen in Figure 1.14 is $0.01E_c$, which compares better with theory than much of the data at $\nu = 1$. In a more detailed calculation the energy to create an antiSkyrmion at $\nu = 1/3$, i.e. the energy to remove one spin at fixed magnetic field, was found to be $E_{1/3}/E_c = 0.069 + 0.024 \exp\left(-0.38 S^{0.72}\right) + \tilde{g}S$ [12], and minimising this gives an estimate for $S$ of 1 for $\tilde{g} > 0.004$; $S = 3$ at $\tilde{g} = 0.002$ and $S = 6$ at $\tilde{g} = 0.001$, in reasonable agreement with the data. Also visible in the same picture is the data for $\nu = 2/5$ which shows a minimum gap for a finite value of $\tilde{g}$ indicating a phase transition from a polarized to an unpolarized ground state and an increasing Coulomb gap as $\tilde{g} \to 0$.

Evidence for Composite Skyrmions has also now been seen from measurements of OPNMR in which Khandelwal *et al.* [13] found that for unpressurized GaAs there was a small depolarization around $\nu = 1/3$ as shown in Figure 1.15 from which they were able to deduce that $\tilde{S} = 0.15 \pm 0.04$ and $\tilde{A} = 0.085 \pm 0.005$ at $\tilde{g} = 0.02$ consistent with being close to the onset of Composite Skyrmion formation.

## 4. Conclusions

Taking the broad body of evidence presented above there is obviously a well founded theoretical basis for the existence of Skyrmionic excitations of the quantum Hall ferromagnet. The experimental evidence is also generally quite convincing, but displays some considerable inconsistencies depending on the

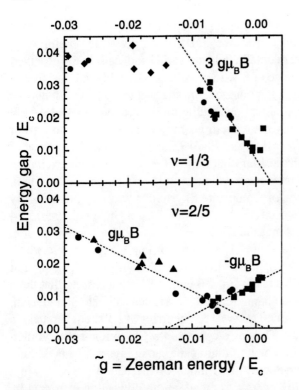

*Figure 1.14.* The measured energy gap for $\nu = 1/3$ and $\nu = 2/5$ as a function of $\tilde{g}$ as tuned by pressure in a GaAs/AlGaAs heterojunction. The dashed line gives the expected dependence for s=3 at $\nu = 1/3$ and for single spin flips at $\nu = 2/5$ [55].

techniques used to measure the sizes, energy gaps and thresholds for Skyrmion formation. Table 1.1 summarises the data described above for the Skyrmion sizes as a function of $\tilde{g}$. It in fact shows a remarkable degree of agreement between the different authors and techniques.

In order to demonstrate this more clearly we plot in Figure 1.16 the values for the spin s of the Skyrmion-antiSkyrmion pair as a function of $1/\tilde{g}$ for both experiment and theory [10]. This shows a remarkable consistency with the only exceptional values being those from Kukushkin *et al* [33], where there is a substantial amount of analysis in deducing the effective Zeeman energy and it has also been calculated that the role of the valence band hole involved in the PL will be to reduce the Skyrmion size [31]. Even more remarkable is that for a significant range of $1/\tilde{g}$ there is an almost linear dependence of the Skyrmion-antiSkyrmion pair size with (Zeeman energy)$^{-1}$ which can be written:

$$s = 1 + 0.1(\tilde{g}^{-1} - \tilde{g}_c^{-1}). \tag{1.8}$$

# Skyrmion excitations of the quantum Hall Ferromagnet

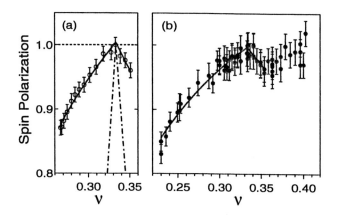

*Figure 1.15.* Dependence of $\mathcal{P}$ on filling factor at fixed temperature for two GaAs/AlGaAs MQW samples (a) and (b) at $\nu = \frac{1}{3}$. The dashed line is for: $\tilde{A}=\tilde{S}=0$; the solid line for $\tilde{A}=0.085$ and $\tilde{S}=0.15$ and the dash-dotted line has $\tilde{A}=\tilde{S}=1$ (dash-dotted line) [13].

*Table 1.1.* Summary of Skyrmion sizes as a function of Zeeman Energy

| Reference | Technique | $\tilde{g}$ | s=$(1 + \tilde{S} + \tilde{A})$ |
|---|---|---|---|
| Sondhi et al. [2] | Theory | 0.054 | >1 |
| Fertig et al. [3] | Theory | 0.014 | 7 |
|  |  | 0.0044 | 18 |
| Barrett et al. [8] | OPNMR | 0.017 | 6.6 |
| Aifer et al. [9] | Magnetoabsorption | 0.016 | 4-6.4 |
| Melinte et al. [36] | Nuc. Heat Capacity | 0.04 | >1 |
| Schmeller et al. [43] | Activ. Energy | 0.012 | 7 |
| Kukushkin et al. [33] | Polarized PL | 0.01 | 1 |
|  |  | 0.0033 | 13 |
| Maude et al. [46] | Activ. Energy | 0.0013 | 33 |
| Leadley et al. [37] | Activ. Energy | 0.0025 | 36 |
| Shukla et al. [44] | Activ. Energy | 0.002 | 50 |

where $\tilde{g}_c = 0.05$ is the critical value for Skyrmion formation. It should be rememebered that there are considerable errors in determining the values of $\tilde{g}$, particularly as $\tilde{g} \to 0$ because of the uncertainties in the exact band parameters and the values of $\tilde{g}$ quoted are estimates as to where the tangent to the activation energy curve as a function of $\tilde{g}$ best fits the data in the activation energy measurements. Nevertheless this suggests that there is both good agreement with theory and an essentially a constant energy associated with the pair formation of $\sim 0.1 E_c$ over a wide range of $\tilde{g}$.

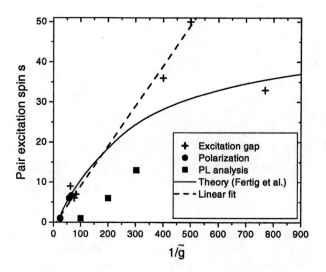

*Figure 1.16.* The Skyrmion-AntiSkyrmion total spin s as a function of $1/\tilde{g}$ for both experiment and theory [10]. The dashed line is empirical and is given by Equ. 1.8.

Finally it is necessary to consider the possible role of disorder in reducing the spin splitting. It has been suggested that disorder may play an important role in determining the spin stiffness of the system [27]. If the disorder potential is smaller than the ZE, it will play no significant role and spin waves will be created as normal [26]. However, once the disorder potential is comparable to the ZE reversed spins will already exist in the ground state [25]. This reduction in spin stiffness makes it is easier to perform additional spin flips so the spin waves become dressed, which in turn reduces the spin stiffness further. The excitations at very small ZE will thus contain many reversed spins. It is not clear whether this mechanism will lead to a Skyrmionic spin texture or merely a multiple spin exciton. If the transition from single reversed spin excitations to multiple reversed spin excitations is critically driven by disorder then the energy gap may decrease more rapidly as $g = 0$ is approached than was the case in the variable sized Skyrmion model discussed above, where there is a smooth change in size of the collective excitation. Is the disorder then responsible for many of the observations? In the majority of the small $\tilde{g}$ measurements [37, 44, 46] the methods used to reduce the g-factor also lead to an increase in the disorder present in the sample. The ZE where the rapid drop in energy gap begins in all of these measurements is $\sim 0.3 - 0.5$ K which is small when compared with typical LL widths of several kelvin but the results from the different techniques are remarkably consistent. It would also seem surprising that the disorder could have such a dramatic effect on the exchange energy at $\nu = 1$ without destroying the correlations responsible for the FQHE at $\nu = 2/3$ and $\nu = 2/5$ which are still obvserved clearly as $\tilde{g} \to 0$ [37, 55].

Another possibility that we should consider is the phase separation at $g = 0$ of a spin polarised $\nu = 1$ system into an unpolarised system of two half filled Landau levels, similar to that observed in bilayers where the phases correspond to either $\nu = 1$ in a single layer or $\nu = 1/2$ in two layers or into some of the more complex phases suggested Fal'ko and Iordanskii [22]. Evidence for this sort of domain formation has been observed at around $\nu = 2/3$ by Kronmuller et al [56]. The decrease in gap might then be explained as a transition from a one component to two component phase. Against this is the fact that in the bilayer system the two component phase is destroyed when the layer separation is small and for the single layer 2-D systems described above there is no physical separation of the spin up and spin down electrons. The strong ferromagnetic interactions are acting against any phase separation. However the possiblity remains of forming spatially separated spin up and spin down domains with the cost of forming domain boudaries being paid by the disorder potential. This would certainly limit the size of any Skyrmions, but causes problems in producing well separated Skyrmion–anti-Skyrmion pairs which then contribute to conduction.

Overall therefore it seems reasonable to conclude that there is good evidence for the existence of Skyrmion-like excitations in the quantum Hall ferromagnet although the precise explanations for the magnitude of the energy gaps and the role of disorder have not been fully established yet.

## Acknowledgements

I would like to acknowledge many stimulating discussions on the topic of Skyrmions with David Leadley, Duncan Maude and Victor Zhitomirsky.

## References

[1] K. von Klitzing, G. Dorda and M. Pepper, Phys. Rev. Lett. **45**, 494 (1980); T. Chakraborty & P. Pietilainen, *The Quantum Hall Effects* (Springer-Verlag, New York, 1988); S. Das Sarma & A. Pinczuk, *Perspectives in Quantum Hall Effects* (Wiley, New York, 1997)

[2] S.L. Sondhi, A. Karlhede, S.A.. Kivelson and E.H. Rezayi, Phys. Rev. B **47**, 16419 (1993)

[3] H.A. Fertig *et al.* Phys. Rev. B **50**, 11018 (1994)

[4] A. Usher, R.J. Nicholas, J.J. Harris and C.T. Foxon, Phys. Rev. B **41**, 1129 (1990)

[5] C. Kallin and B.I. Halperin, Phys. Rev. B **30**, 5655 (1984)

[6] M. Dobers, K. von Klitzing and G. Weimann, Phys. Rev. B **38**, 5453 (1988)

[7] In this work we use the convention that $\tilde{S}=\tilde{\mathcal{A}}=0$ in the non-interacting limit, instead of the original definition [8], $S=\mathcal{A}=1$, in the same limit. This prevents any amiguities when comparing the results away from $\nu = 1$ with those where Skyrmion-AntiSkyrmion pairs are created.

[8] S.E. Barret et al. Phys. Rev. Lett. **74**, 5112 (1995)

[9] E.H. Aifer et al. Phys. Rev. Lett. **76**, 680 (1996)

[10] H.A. Fertig et al. Phys. Rev. B **55**, 10671 (1997)

[11] K. Lejnell, A. Karlhede and S.L. Sondhi, Phys. Rev. B **59**, 10183 (1999)

[12] R.K. Kamilla, X.G. Wu and J.K. Jain, Solid State Commun. **99**, 289 (1996)

[13] P. Khandelwal et al. Phys. Rev. Lett. **81**, 673 (1998)

[14] L. Brey, H.A. Fertig, R. Cote and A.H. MacDonald, Phys. Rev. Lett. **75**, 2526 (1995)

[15] A.G. Green, I.I. Kogan, and A.M. Tsvelik, Phys. Rev. B **54**, 16838 (1996)

[16] M. Rao, S. Sengupta, and R. Shanar, Phys. Rev. Lett. **79**, 3998 (1997)

[17] C. Timm, S.M. Girvin, and H.A. Fertig, Phys. Rev. B **58**, 10634 (1998)

[18] K. Moon, and K. Mullen, Phys. Rev. Lett. **84**, 975 (2000)

[19] A.H. MacDonald, H.A. Fertig, and L. Brey, Phys. Rev. Lett. **76**, 2153 (1996)

[20] V. Melik-Alaverdian, N.E. Bonesteel and G. Ortiz, Phys. Rev. B **60**, R8051 (1999)

[21] I. Mihalek and H.A. Fertig, Phys. Rev. B **62**, 13573(2000)

[22] V.I. Fal'ko and S.V. Iordanskii, Phys. Rev. Lett. **84**, 127 (2000)

[23] X.-G. Wu and S.L. Sondhi, Phys. Rev. B **51**, 14725 (1995)

[24] N.R. Cooper, Phys. Rev. B **55**, R1934 (1997)

[25] A.J. Nederveen, Y.V. Nazarov, Phys. Rev. Lett. **82**, 406 (1999)

[26] J. Sinova, A.H. MacDonald, and S.M. Girvin, Phys. Rev. B **62**, 13579 (2000)

[27] M.M. Fogler and B.I. Shklovskii, Phys. Rev. B **52**, 17366 (1995)

[28] M.J. Manfra et al. Phys. Rev. B **54**, R17327(1996)

[29] N. Read and S. Sachdev, Phys. Rev. Lett. **75**, 3509(1995)

[30] N.R. Cooper, and D.B. Chklovskii, Phys. Rev. B **55**, 2436 (1997)

[31] T. Portengen, J.R. Chapman, V. Nikos Nicopolous, and N.F. Johnson, Phys. Rev. B **55**, R7367 (1997)

[32] I.V. Kukushkin, K.v. Klitzing and K. Eberl, Phys. Rev. B **55**, 10607 (1997)
[33] I.V. Kukushkin, K.v. Klitzing and K. Eberl, Phys. Rev. B **60**, 2554 (1999)
[34] S.W. Brown, T.A. Kennedy, D. Gammon, and E.S. Snow, Phys. Rev. B **54**, R17339 (1996)
[35] V. Bayot *et al.*, Phys. Rev. Lett. **76**, 4584 (1996); V. Bayot *et al.*, Phys. Rev. Lett. **79**, 1718 (1997).
[36] S. Melinte, E. Grivei, V. Bayot, and M. Shayegan, Phys. Rev. Lett. **82**, 2764 (1999)
[37] D.R. Leadley *et al.* Semicond.Sci.Technol. **13** 671 (1998)
[38] D. Shoenberg, *Magnetic oscillations in metals* Cambridge University Press, (1984)
[39] R.J. Nicholas, R.J. Haug, K. von Klitzing and G. Weimann, Phys. Rev. B **37**, 1294 (1988) and refs. therein.
[40] F.C. Zhang and S. Das Sarma, Phys. Rev. B **33**, 2903 (1986); T. Chakraborty *et al.* Phys. Rev. Lett. **57**, 130 (1986)
[41] D.R. Leadley, R.J. Nicholas, J.J. Harris and C.T. Foxon, Phys. Rev. B, **58**, 13036 (1998)
[42] R.J.Nicholas *et al.* Semicond.Sci.Technol. **11**, 1477 (1996)
[43] A. Schmeller *et al.* Phys. Rev. Lett. **75**, 4290 (1995)
[44] S.P. Shukla *et al.*, Phys. Rev. B **61**, 4469 (2000).
[45] B. Králik, A. M. Rappe, and S. G. Louie, Phys. Rev. B **52**, R11626 (1995).
[46] D.K. Maude, M. Potemski, J.C. Portal, M. Henini, L. Eaves, G. Hill and M.A. Pate, Phys. Rev. Lett. **77**, 4604 (1996)
[47] D.K. Maude *et al.*, Physica B **251**, 251 (1998)
[48] Y.Q. Song, B.M. Goodson, K. Maranowski, and A.C. Gossard, Phys. Rev. Lett. **82**, 2768 (1999)
[49] G.C. Kerridge, A.S. Plaut, M.C. Holland, C.R. Stanley and K. Ploog, Phys. Rev. B **60**, R5141 (1999)
[50] D.C. Tsui *et al.* Phys. Rev. Lett. **48**, 1559 (1982); R.L. Willett *et al. ibid.* **59**, 1776 (1987); T. Chakraborty & P. Pietilainen, *The Quantum Hall Effects* (Springer-Verlag, New York, 1988); S. Das Sarma & A. Pinczuk, *Perspectives in Quantum Hall Effects* (Wiley, New York, 1997)
[51] J.K. Jain, Adv. Phys. **41**, 105 (1992); B.I. Halperin, P.A. Lee and N. Reed, Phys. Rev. B **47**, 7312 (1993)
[52] D.R. Leadley *et al.* Phys. Rev. Lett. **72**, 1906 (1994); R.R. Du *et al.* Phys. Rev. Lett. **73**, 3274 (1994)
[53] R.R. Du *et al.* Phys. Rev. Lett. **75**, 3926 (1995)

[54] R.G. Clark *et al.* Phys. Rev. Lett. **62** 1536 (1989); J.P. Eisenstein *et al. ibid.* 1540; L.W. Engel *et al.* Phys. Rev. B **45**, 3418 (1992)

[55] D.R. Leadley *et al.* Phys. Rev. Lett. **79** 4247 (1997)

[56] S. Kronmuller *et al.*, Phys. Rev. Lett. **81** 2526 (1998)

# Chapter 5

# SUPERCONDUCTING DEVICES FOR QUANTUM COMPUTATION

James F. Annett, Balazs L. Gyorffy
*University of Bristol, H.H. Wills Physics Laboratory, Royal Fort, Tyndall Ave., Bristol BS8 1TL, UK.*

Timothy P. Spiller
*Hewlett-Packard Laboratories, Filton Road, Stoke Gifford, Bristol BS34 8QZ, UK.*

**Abstract**

The ideas of *quantum information* have led to exciting new developments combining concepts from quantum physics with those of computer science and information technology. The key idea is to use 'bits' of information which can be manipulated using the rules of quantum physics, *qubits*, rather than standard rules of classical Boolean logic. A *quantum computer* which could process information stored in these qubits, could make possible calculations which would be impractical on any conventional classical computer. However any actual implementation of these ideas requires a physical system which can represent the qubits. In this article we shall argue that superconducting devices have some unique properties, which may make them ideally suitable. Superconductors and superfluids have the unique property of *macroscopic quantum coherence*. Mesoscopic superconducting devices can be fabricated in any desired geometry using modern lithographic techniques. Recently quantum coherent operation has been clearly demonstrated in two different types of mesoscopic superconducting devices, relying on quantization of charge and magnetic flux, respectively. Precise *quantum state control* has been achieved, showing that a quantum state can be manipulated at will, and the decoherence effects of the environment are not too severe. We review these experiments, and discuss prospects for exploiting this quantum coherence as a possible practical realization of qubit devices.

## 1. Introduction

Every physics undergraduate knows about the paradox of *Schrödinger's cat* [1]. It is important because it brings home just how strange the rules of quantum mechanics are compared to those of ordinary classical physics. The key principle which it illustrates is the idea of a *quantum superposition*, i.e. the fact that if $|\psi_1\rangle$ and $|\psi_2\rangle$ are two different quantum states, then superpositions such as

$$|\psi\rangle = c_1|\psi_1\rangle + c_2|\psi_2\rangle \qquad (1.1)$$

are also valid quantum states, where $c_1$ and $c_2$ are any complex numbers. Mathematically this is a trivial consequence of the fact that Schrödinger's equation is linear, or one could say that this follows from the mathematical axiom that quantum states are rays in a Hilbert space. The fact that some abstract quantum states can be added together in this way is not particularly worrying at first sight, but the brilliance of Schrödinger's argument is to show that this is far from trivial, indeed it calls into question the very nature of what physical reality actually means. If we are allowed to associate the state $|\psi_1\rangle$ with the everyday statement "the cat is dead", and $|\psi_2\rangle$ with "the cat is alive", then what can $c_1|\psi_1\rangle + c_2|\psi_2\rangle$ possibly represent?

Physicists have been aware of this dilemma since Schrödinger's paper was published in the 1930's, and nowadays students are usually told that the paradox is resolved by the effects of *decoherence*. The pure quantum states $|\psi_1\rangle$ and $|\psi_2\rangle$ interact with the environment, in such a way that any effects of quantum interference are washed out. The expectation value of any physical observable $\hat{O}$ becomes

$$\begin{aligned}\langle \hat{O} \rangle &= \langle \psi|\hat{O}|\psi\rangle \\ &\sim |c_1|^2 \langle \psi_1|\hat{O}|\psi_1\rangle + |c_2|^2 \langle \psi_2|\hat{O}|\psi_2\rangle\end{aligned} \qquad (1.2)$$

because the cross terms like $\langle \psi_1|\hat{O}|\psi_2\rangle$ are destroyed by the decoherence. Therefore the expectation values of any observables become determined purely by the classical probabilities $|c_1|^2$ and $|c_2|^2$ of finding the system in each of the two states. Thus there is no physically meaningful possibility of having a cat which is both dead and alive simultaneously. We must distinguish between microscopic objects (atoms, molecules, elementary particles etc.) for which we must apply quantum mechanics, and macroscopic objects (cats, people, peoples' brains etc) for which only classical physics is meaningful. Working physicists and undergraduates usually accept something like this division into micro- and macro- properties, to which they happily apply different laws of physics, and leave further debate to the philosophers.

Unfortunately, this division into the microscopic and macroscopic is not as clear cut one might expect. This particularly comes to light in experiments which demonstrate *macroscopic quantum coherence* [2]. This directly implies that we can indeed apply quantum mechanics to macroscopic objects, and it is not simply the physical size of a system which determines whether or not it exhibits quantum coherence. Quantum superpositions such as Eq. 1.1, have been directly observed in macroscopic superconducting devices [3, 4, 5], of sizes at least of order $0.1\mu$m, i.e. containing of order $10^8 - 10^{10}$ atoms. As well as having implications for the philosophy of quantum physics, these experiments open up many possibilities for directly exploring the actual sources and effects of decoherence in quantum systems. Also there is the very exciting possibility of using the existence of macroscopic quantum states to produce new electronic devices which are inherently quantum mechanical in their operation. Most prominently there is the idea that one could build a *quantum computer* using these devices.

In standard mathematical logic we have the binary possibilities that any statement is either **TRUE** or **FALSE**. Similarly modern computers all operate on binary digits in which any given bit is either a 0 or a 1. The computer carries out computation by manipulating these bits using the standard rules of Boolean logic and binary arithmetic. *Quantum computation* also uses binary digits, but with the key difference that these are viewed as two quantum states,

$$|0\rangle, \quad |1\rangle.$$

Because these are quantum states, the system need not be in either one or the other state at any given time, but it can also be in superposed states,

$$c_0|0\rangle + c_1|1\rangle.$$

It turns out, perhaps surprisingly, that if one could build a machine for manipulating these quantum bits, or *qubits*, then for many problems it would be immensely faster than any conventional classical computer. This remarkable speed up occurs in many important algorithms, including factorization of integers [6], searching [7] and others. Interest in this field grew enormously after 1994 when Shor [6] showed that quantum computation would provide an efficient way to factorize large composite integers. As well as having uses in number theory, if one could physically implement Shor's algorithm then one could also break most modern computer encryption algorithms such as RSA. Suddenly national governments and large companies became interested in the laws of quantum physics!

In this paper we shall review the recent experiments which demonstrate macroscopic quantum coherence in superconducting devices. We shall particularly focus on the question of whether or not it will eventually be possible to use such devices to implement the qubit in a possible quantum computer. Given that coherent macroscopic superposition states have now been observed, it is certainly possible in principle to build a quantum computer using superconducting devices. Of course the devil is in the details, and there are still many technical challenges to overcome before this ever becomes a practical possibility.

This chapter is organized as follows. First, in section 2, we discuss the fundamental concepts of classical and quantum computation, especially viewed as physical processes. In section 3 we discuss how the qubits of quantum computers may be implemented in various physical systems, especially focusing on the specific case of superconducting devices and the fundamental concepts of macroscopic quantum coherence. We then discuss two specific devices which have explicitly been used to demonstrate quantum superpositions in superconducting devices, the *charge* and *flux* devices. We analyse their advantages and disadvantages as possible qubit devices. Finally, in section 4 we discuss the effects of decoherence, and describe some possible experiments which could perhaps demonstrate quantum entanglement in a pair of coupled superconducting qubits. If this could be achieved then there would be a clear route towards developing superconducting quantum logic gates, and perhaps eventually a superconducting quantum computer.

The article is intended to introduce much of the same material as the lectures given at the Windsor summer school by of Professor G. Schön, however we have taken the liberty of describing the material from our own perspective. We have also included a brief review some of the more recent experimental work which has taken place since the summer school. Much of the material originally presented in Prof Schön's lectures has since appeared in an excellent review article [8].

## 2. Classical and Quantum Computation as Physical Processes

### 2.1. The Universal Turing Machine

All modern computers are in one way or another realizations of a *universal Turing machine*. This is an idealized computer, first described by the mathematician Alan Turing in the 1930s [9]. It is simply a 'black box' which is characterized by having a finite set of possible internal states $q \in \{A, B, C, \ldots Z\}$, as sketched in Fig. 1.1. The box has a simple 'input-output' mechanism: it can read and write characters to a memory

*Superconducting Devices for Quantum Computation* 169

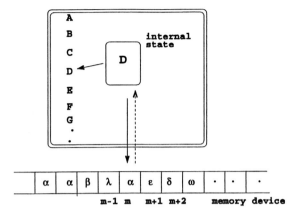

*Figure 1.1.* A schematic representation of a universal Turing machine. The machine has a finite set of possible internal states, $q \in \{A, B, C, \ldots Z\}$, and it reads and writes a finite set of characters $s_m \in \{\alpha, \beta, \ldots \omega\}$ to a memory device (or paper tape), which is assumed to be infinitely big. At each step the current character $s_m$ at location $m$ is read in, the machine changes internal state and writes a new character to location $m$, before moving to a new memory location one step to the right or left ready to start the next step. The machine stops when the computation has completed.

device, (originally described as a 'paper tape' when the Turing machine was first introduced in the 1930s!). The the memory has a finite set of possible characters, $s_m \in \{\alpha, \beta, \ldots \omega\}$, at memory cell $m$. But the memory itself is assumed to be effectively infinitely large, so that the machine can never 'run out of memory' (unlike our own computers!).

A *computation* with the Turing machine proceeds by a sequence of discrete steps. In each step the machine first reads the current memory character, $s_m$. Then, depending on its current internal state, $q$ it will change to a new internal state and write a new character to the current memory location, $m$. The machine them may moves on to a new memory location either one step to the left or right of the current position. The machine then repeats this sequence. Eventually the calculation should come to an end, at which point the which machine stops with the final result of the calculation written to the memory.

The idea of a Turing machine was originally introduced in order to prove certain fundamental theorems in the foundations of mathematics, not as a practical computer! A *computable function* in mathematics is one which could be calculated by a Turing machine in a finite number of steps [9]. However, the Turing machine is in fact rather close to a modern digital computer. The internal states of the machine can be thought of as the registers in the CPU of a modern computer, and the memory device is the RAM or hard disk. Turing specified that the number of internal states of the machine and the character set of the memory must both be finite. Modern digital computers use 'words' of binary digits which obviously satisfy this requirement. Therefore in a mathematical sense any given modern computer is equivalent to a Turing machine, at least if the memory is effectively infinite. If a given function is Turing computable then it can also be calculated on any real computer, provided only that the memory is sufficiently large. (Of course a massively parallel CRAY-T3E will find the result faster than a single old Intel 386, but they will both get the same result eventually!).

## 2.2. Computation as a Physical Process

While the Turing machine is a good mathematical model for the concept of 'computation', let us briefly consider the physics of how computation is actually realized. It is clear that the Turing machine, or its equivalent, is completely deterministic. The final state of the machine is precisely specified given its initial input in the memory and the rules of operation of the machine. The machine and memory elements are also always in one of a finite number of discrete states. (A classical analogue computer would have a continuum of states, and is therefore not equivalent to a Turing machine [10].)

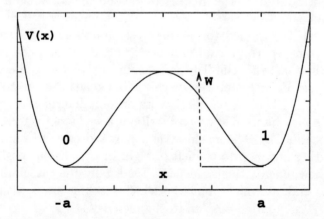

*Figure 1.2.* A typical physical two state system, corresponding to a physical memory 'bit'. Here $x$ represents a generalized coordinate of the system, and $V(x)$ a cross section of the potential energy surface.

Obviously it is convenient to represent the discrete state of the Turing machine and memory elements in terms of binary bits. Physically each bit must be some sort of a simple two state physical system. For example, the double well potential shown in Fig. 1.2 has two well defined minima, which could correspond to the two values 0 and 1 of the memory element. The variable $x$ represents some generalized coordinate, which describes the two distinct states and the potential barrier between them. The physical requirements of classical computation are (i) the bit must be able to remain in either minimum indefinitely (i.e. spontaneous hopping over the barrier must be forbidden) and (ii) one must be able to switch the bit from 0 to 1 or vice versa reliably. In practice it is possible to tolerate some small error rate, such as events where the device spontaneously switches from one state to the other by thermal excitation over the barrier or from random external events, such as hits by cosmic rays. The methods of error correction can detect and recover from such events, provided they are sufficiently rare.

To be specific, let us briefly consider a classical dynamical equation which could model the two state device shown in Fig. 1.2. For example, we might model the bit with the classical Langevin type equation of motion [11]

$$m\ddot{x} + b\dot{x} + \frac{dV(x)}{dx} = F(t) + \eta(t), \tag{1.3}$$

where the generalized coordinate $x$ is assumed to have some inertia, $m$, and also some frictional damping, $b$, in addition to the potential energy $V(x)$. There are two time dependent forces on the right hand side. The first, $F(t)$, is the switching field, which is an externally applied force that is turned on or off in order to switch the device from one state to the other. The second force, $\eta(t)$, is a stochastic force which leads to Brownian motion of $x$. It represents the interaction of the device with the thermal environment.

The requirement for reliable and fast operation of the device will place a number of limits on the parameters of this simple model. The device is in contact with its environment. For example, if the noise term is too strong compared to the barrier height ($W$ in Fig. 1.2), if $W \sim k_B T$, then the bit will have a large error rate due to spontaneous transitions over the barrier. Similarly if the noise term is too large compared to the friction term, $b$, then the device may fail to switch correctly from one state to the other when the switching field $F(t)$ is turned on. Both of these conditions would constitute errors in the computational process. Simple correction techniques can correct for these errors, by using additional bits to store the actual information redundantly. For example most modern RAM devices check 'parity' for each byte stored, so simple parity errors corresponding to a single incorrect bit can be detected and then automatically corrected.

Notice that both the coupling to the environment (the $\eta(t)$ term), and the presence of dissipation ($b$) are necessary in any realistic physical system. The device is necessarily *dissipative* and *irreversible*. The switching force $F(t)$ does work on the system, and so heat and entropy are generated by switching. The force $F(t)$ moves the coordinate $x$ by a distance $2a$, doing work $2aF \sim 2a(dV/dx)_{max} \sim W$ where $W$ is the barrier height. The work done is eventually dissipated as heat, $Q = W$, and entropy $S \sim Q/T$ is generated. But we have seen that the barrier height must be much greater than $k_B T$, and so the entropy generated is $S \gg k_B$ every time the device is switched.

The above simple physical model is probably reasonably accurate to describe the elements in modern computers. The memory bits are basically magnetic domains of size $\sim 1\mu$m, while the RAM consists of charges on small capacitors, fabricated nowadays on silicon wafers with a feature size approaching $0.1\mu$m. These systems are indeed dissipative and irreversible. They are obviously microscopic in size, but they are also sufficiently macroscopic that classical dynamics is appropriate.

During by late 1970's it was already clear that the device sizes on the semiconductor chips were steadily becoming smaller and smaller, and computers were becoming correspondingly more and more powerful. This trend, first noted by Gordon Moore (a founding director of Intel) became known as *Moore's Law*. It states that computer speed increases exponentially with time, doubling approximately every 18 months. Remarkably this trend has continued more or less unbroken from the 1970's to the present.

However, as Malthus [12] realized in the eighteenth century, exponential growth cannot continue forever! There has been a great deal of speculation about what will eventually end this exponential trend. Extrapolating Moore's law into the future, one sees that by 2020 devices would operate with a single electron. In a famous paper Richard Feynman [13] speculated about the ultimate limits that might prevent us from building smaller and smaller computing devices down to the atomic scale. Others [15] discussed the possibility of constructing a *reversible Turing machine*. This would operate in a completely reversible way (in the thermodynamic heat engine sense), so that there is no entropy generated by the machine's operation. This would overcome the $S \gg k_B$ entropy cost per switching operation that we have seen is inevitable with our simple classical model of two state memory bit devices.

Feynman also raised the point that as devices become smaller and smaller, then inevitably it will be necessary to consider the effects of quantum mechanics. In the current generation of semiconductor devices the basic elements of the computer are already fabricated on the $0.1-0.2\mu$m scale, so they have perhaps as few as $10^{10}$ atoms. Modern memory elements also include the magnetic hard disc, where again the physical size of a single stored bit is approaching the micron scale of a single magnetic domain. These are both large enough that the device can be considered classical, as we have assumed in our simple model described earlier. However as the devices become smaller and more isolated from their environments, then ultimately quantum mechanics will come to dominate. For example, in our simple two state model of Fig. 1.2 we see that quantum effects may become important when the thermal de Broglie wavelength,

$$\lambda = \frac{h}{\sqrt{2mk_BT}}, \tag{1.4}$$

becomes comparable to the potential well width, $a$. Also tunnelling through the potential barrier will become possible unless

$$a\sqrt{2mW/\hbar^2} \gg 1.$$

From the point of view of classical computation, this would obviously be viewed as an error. However, as we shall see in the next section, it is possible to devise an entirely new concept of computing, which is entirely consistent with the rules of quantum mechanics.

## 2.3. Quantum Information: Qubits

The implicit assumption of the simple model above is that the dynamics is purely classical. This is consistent with the Turing machine ideal in which of the machine is one of the discrete states such as $A$ or $B$ at any given time, and never in a superposition state such as

$$\frac{1}{\sqrt{2}}(|A\rangle + |B\rangle).$$

However, as Feynman first noted [13], if the physical system representing the bit is sufficiently small, at some microscopic size eventually the classical physics model will no longer be valid, and it will be necessary to consider the system fully quantum mechanically.

Let us retain the simple model of a bit, shown in Fig. 1.2, but now interpret this as a quantum system. The system will obviously have a wave function $\psi(x,t)$ determined by the Schrödinger equation

$$\hat{H}\,\psi(x,t) = i\hbar\,\frac{\partial \psi(x,t)}{\partial t} \tag{1.5}$$

where the Hamiltonian is

$$\hat{H} = -\frac{\hbar^2}{2m}\frac{\partial^2}{\partial x^2} + V(x) + F(t)x \tag{1.6}$$

(for now neglecting any effects of decoherence or interaction with the environment). If we prepare an initial state, $\psi(x,0)$ and only change the external force $F(t)$ very slowly then the state evolves adiabatically. For example, if the initial state is the ground state of the system for the the initial value of the force $F(0)$, so $\psi(x,0) = u_0(x, F(0))$, the final state will be the ground state for the final value of the force,

$$\psi(x,t) = \exp(-i\gamma) u_0(x, F(t)). \tag{1.7}$$

The phase $\gamma$ is the usual dynamical factor $\frac{i}{\hbar} \int_0^t dt' E_0(F(t'))$ from the ground state energy $E_0$ as it varies with $F$, although in principle there may also be a geometric contribution [14] when the parameter $F$ is a vector.

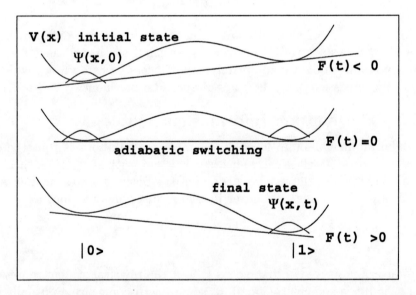

*Figure 1.3.* Adiabatic switching in a quantum mechanical two state system. The system is prepared in initial state $\psi(x,0)$, in the left hand well, and then adiabatically moved to the final state $\psi(x,t)$ in the right hand well.

By suitably changing the external switching field $F(t)$ we can manipulate the state in any manner we choose. For example, we can adiabatically move the particle from one minimum to the other. Supposing that we initially have the field $F(t)$ small and negative, so that the left minimum is slightly lower than the right hand side one, as shown in Fig. 1.3. The system will therefore adopt an initial state, $\psi(x,0)$, in which the system is localized in the left hand minimum. Now increase $F(t)$ slowly until it changes sign. The wave function will evolve adiabatically as shown in Fig. 1.3. The final state of the system is now localized in the right hand potential well.

If we define $|0\rangle$ to mean that the system is localized in the left hand minimum of Fig. 1.3, and $|1\rangle$ to mean that is localized on the right hand side, then the simple adiabatic process shown has *reversibly and coherently* switched the system from state $|0\rangle$ to $|1\rangle$. This obviously contrasts strongly with the irreversible and dissipative switching in the classical case described in the previous section.

In addition to switching, we can obviously also use adiabatic evolution to prepare the system in any quantum state we choose, such as the superposition

$$\frac{1}{\sqrt{2}}(|0\rangle + |1\rangle).$$

For example, this occurs at the half way point of the switching process shown in Fig. 1.3.

The key idea of *quantum information* is to replace the classical information 'bit', with two classical values 0 and 1, with a quantum mechanical two state system, or *qubit*. The qubit has two distinct (orthogonal) quantum states $|0\rangle$ and $|1\rangle$ but, unlike the classical bit, it can exist in any quantum superposition

$$|\psi\rangle = c_0|0\rangle + c_1|1\rangle. \qquad (1.8)$$

The physical model of the two state system, Fig. 1.2, would be one possible approximation to a qubit, (approximate because in the quantum case there are also many other excited states, so $|0\rangle$ and $|1\rangle$ in the wells do not span the full Hilbert space). Another obvious realization of a qubit would be a spin-1/2 particle; here there is no approximation as the Hilbert space really is just two dimensional.

## 2.4. Quantum Computation

To perform a *quantum computation* [16] we assume that we have a set of qubits, called a register. We first *prepare* the register of qubits in a given quantum state. Secondly, we *evolve* the state (adiabatically or otherwise) coherently. Finally we *measure* the final state of the system, by performing measurements on one or more of the qubits. The measurement collapses the wave function according to the standard rules of quantum measurement theory.

The initial preparation of the state defines a wave function at time $t_0 = 0$, $|\psi_0\rangle$. The state will evolve in time according to a unitary evolution

$$|\psi_{t'}\rangle = \hat{U}(t',t)|\psi_t\rangle. \qquad (1.9)$$

The evolution must be unitary in order to satisfy the fundamental quantum mechanical axioms of linearity, and conservation of probability (i.e. each state is normalized to unity).

If the register contains $M$ qubits, then the system has $2^M$ basis states

$$|00\ldots00\rangle\ldots|11\ldots11\rangle.$$

The general state of the state at any given time is a superposition

$$|\psi_t\rangle = \sum_{\{\sigma_i\}} c_{\sigma_1\ldots\sigma_M}(t)|\sigma_1\sigma_2\ldots\sigma_n\rangle \qquad (1.10)$$

where $\sigma_i \in \{0,1\}$. Expressed in this basis the unitary evolution Eq. 1.9 becomes

$$c_{\sigma_1\ldots\sigma_M}(t') = \sum_{\{s_i\}} U_{\sigma_1\ldots\sigma_M;s_1\ldots s_M}(t',t) c_{s_1\ldots s_M}(t) \qquad (1.11)$$

where $s_i, \sigma_i \in \{0,1\}$. The unitary operators acting on these states are therefore represented by $2^M \times 2^M$ unitary matrices, $U_{\sigma_1\ldots\sigma_m;s_1\ldots s_M}$.

It is convenient, as we shall see below, to consider the evolution as a sequence of discrete steps, defined by some discrete set of time intervals $t_n$, $n = 0,\ldots N$. Hence the final wave function is given by

$$|\psi_{t_N}\rangle = \left(\prod_{n=0,N-1} \hat{U}(t_{n+1},t_n)\right)|\psi_0\rangle, \qquad (1.12)$$

or by the matrix product

$$c_{\sigma_1\ldots\sigma_M}(t_N) = \left(\prod_{n=0,N-1} U(t_{n+1},t_n)\right) c_{\sigma_1\ldots\sigma_M}(0). \qquad (1.13)$$

The "output" of the machine is simply its state at the last step, $|\psi_{t_N}\rangle$. Because this is a quantum state, we must apply the usual rules of quantum measurement in order to "collapse" the wave function. This will require us to make measurements of the system. For example we could measure the state of one of the qubits, $\sigma_i$, or make joint measurements on more than one qubit (e.g. EPR Bell state type measurements, projecting onto entangled states). By the usual rules, quantum measurement is projective, and we will find the system in state $|\Phi\rangle$ after the measurement with probability

$$P(\Phi) = |\langle\Phi|\psi_{t_N}\rangle|^2 \qquad (1.14)$$

The resulting answers from the computation are inherently probabilistic, and so in principle one should repeat the calculation many times.

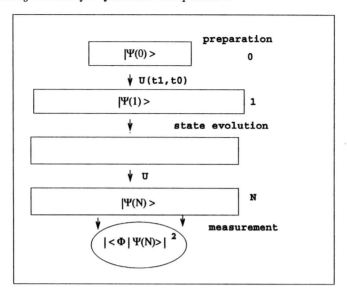

*Figure 1.4.* A quantum computation. A system is prepared in an initial quantum state, $|\psi(t=0)\rangle$. This state is evolved by a sequence of unitary operations $\hat{U}(t_{n+1}, t_n)$. At the end of the computation the final state $|\psi_{t_N}\rangle$ must be 'collapsed' by measurement of the state. The observable results of the calculation are the respective probabilities of projective measurements into states $|\Phi\rangle$, $|\langle\Phi|\psi_{t_N}\rangle|^2$.

However, in some cases it may be possible to reduce the set of possible outcomes to probabilities of 0% or 100% and so only one measurement is sufficient.

## 2.5. Quantum Logic Gates

As we have seen the evolution of the quantum state in a quantum computation is simply described in terms of a set of $2^M$ dimensional square matrices, $U(t_{n+1}, t_n)$, representing the evolution in each time step. The natural question then arises: how can one implement a general $2^M$ dimensional unitary operator? The answer to this problem is to introduce a *complete set of logic gates*.

A 'logic gate', is simply a device for manipulating qubits singly or in pairs, and is the natural quantum extension of the idea of the gates in classical binary logic. The simplest logical operation in classical logic is the **NOT** gate, which simply inverts a given bit $0 \to 1$, $1 \to 0$. We can apply the same rules to a qubit

$$\begin{aligned} |0\rangle &\to |1\rangle \\ |1\rangle &\to |0\rangle. \end{aligned} \quad (1.15)$$

Because of the superposition principle, when we apply this operator on a general state we obtain

$$c_0|0\rangle + c_1|1\rangle \to c_1|0\rangle + c_0|1\rangle. \quad (1.16)$$

This can obviously be represented as a simple $2 \times 2$ unitary matrix

$$\begin{pmatrix} c_0 \\ c_1 \end{pmatrix} \to \begin{pmatrix} 0 & 1 \\ 1 & 0 \end{pmatrix} \begin{pmatrix} c_0 \\ c_1 \end{pmatrix}, \quad (1.17)$$

which is just the Pauli matrix, $\sigma_x$.

Another single qubit operation is the Hadamard gate, defined by

$$\begin{pmatrix} \alpha \\ \beta \end{pmatrix} \to \frac{1}{\sqrt{2}} \begin{pmatrix} 1 & 1 \\ 1 & -1 \end{pmatrix} \begin{pmatrix} c_0 \\ c_1 \end{pmatrix} = \frac{1}{\sqrt{2}}(\sigma_x + \sigma_z) \begin{pmatrix} c_0 \\ c_1 \end{pmatrix}. \quad (1.18)$$

Obviously a quantum gate such as this has no classical analogue.

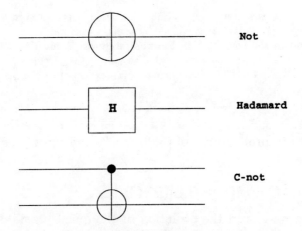

*Figure 1.5.* Some of the elementary quantum logic gates. Each corresponds to a particular unitary matrix, as defined in the text.

There is no direct quantum version of the classical **AND**, **OR** and **XOR** gates, since they each have two inputs and only one output and cannot be represented by unitary matrices. However a number of elementary quantum operations can be defined on two qubits, and which can be represented by $4 \times 4$ unitary matrices. The *controlled-not* gate, **C-NOT**, takes two input qubits, $\sigma_1$ and $\sigma_2$. The target qubit $\sigma_2$ is inverted, depending on the state of a control bit, $\sigma_1$,

$$\sigma_2 \to \begin{cases} \sigma_2 & \text{if } \sigma_1 = 0 \\ 1 - \sigma_2 & \text{if } \sigma_1 = 1 \end{cases}$$

This operation acting on the the four basis states give $|00\rangle \to |00\rangle$, $|01\rangle \to |01\rangle$, $|10\rangle \to |11\rangle$, and $|11\rangle \to |10\rangle$, which can be represented as a unitary $4 \times 4$ matrix

$$\begin{pmatrix} 1 & 0 & 0 & 0 \\ 0 & 1 & 0 & 0 \\ 0 & 0 & 0 & 1 \\ 0 & 0 & 1 & 0 \end{pmatrix} \qquad (1.19)$$

A standard notation has been developed in order to represent these quantum logic gates, as shown in Fig. 1.5.

A number of other two qubit logic gates have also been defined. For example, the Ideal-Model gate, with operator

$$U_{2b}(\phi) \begin{pmatrix} 1 & 0 & 0 & 0 \\ 0 & \cos\phi & i\sin\phi & 0 \\ 0 & -i\sin\phi & \cos\phi & 0 \\ 0 & 0 & 0 & 1 \end{pmatrix}, \qquad (1.20)$$

can be used to swap two qubits ($\sigma_1 \sigma_2 \to \sigma_2 \sigma_1$) when $\phi = \pi/2$, or to perform a "square root of swap" for $\phi = \pi/4$.

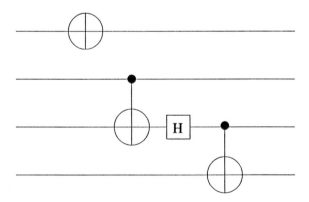

*Figure 1.6.* A quantum computation as a sequence of quantum logic gates acting on one or two qubits at any given time.

It has been shown that any possible operation on an $M$ qubit register can be represented in terms of a suitable sequence of actions of such elementary logic gates, provided that a sufficiently rich set of elementary operations are available [17]. In fact it is sufficient to have arbitrary single-qubit rotations (which, for example, can be built from Hadamard and simple phase shifts $|0\rangle \to |0\rangle$ and $|1\rangle \to e^{i\phi}|1\rangle$, for arbitrary phase angle $\phi$) and at least one suitable two qubit operation, such as C-NOT or "square root of swap", capable of generating entanglement. Any general unitary operator can be generated by successive applications of these elementary gates. Therefore any quantum computation on an $M$ qubit register can be achieved by some repeated application of these elementary logic gates. Any quantum computation can then be represented diagrammatically as a sequence of these elementary gates acting on single qubits and pairs of qubits, as illustrated in Fig. 1.6.

## 2.6. Computational Complexity and Speedup

The great interest in quantum computation is because of the immense speed up it offers relative to standard classical computation [6, 7]. In classical computation the *complexity* of a problem is related to the number of steps that a Turing machine would require before finding the solution. Usually the number of steps scales in a simple way with the size of some input parameter. For example to multiply an $N$ dimensional vector by a $N \times N$ matrix clearly takes $N^2$ multiplication and addition steps. To multiply two $N \times N$ matrices takes (using the naive algorithm) $N^3$ steps. These would be classified as polynomial time algorithms, since the number of steps is a polynomial of $N$.

On the other hand, many important computational problems are N-P or non-polynomial. The required computer time may grow exponentially or faster with $N$. Common examples are optimization problems (e.g. the 'travelling salesman' problem, where the salesman must find the shortest path which visits $N$ cities), and searching problems. Another key example is factorization of large integers. Given an integer $N$ there is no known simple test to discover whether or not it is prime, and if it is not prime to find the prime factors,

$$N = p_1 p_2 \ldots p_n.$$

Naively, in the worst case, one would have to divide $N$ by all the odd numbers up to $\sqrt{N}$ in order to be sure that it is not prime. If $N$ is a binary digit of $M$ bits, $2^{M-1} < N < 2^M$, then this clearly involves a time $\sim 2^{M/2}$, and the complexity is therefore exponential in $M$.

Many modern 'secure' communication systems use *public-key cryptography*, in which, essentially, you publish a large integer $N$. You know that it is a product of two prime factors, $p_1$ and $p_2$, but you keep these secret. If someone wants to send you a secure message they encode it using the number $N$ which you have published. When you receive the message you can easily decode it using the prime factors $p_1$ and $p_2$. But only you can do this, because only you have access to the secret prime numbers. The security of the system relies entirely on the exponential difficulty of finding the factors of the public number $N$. If $N$ has 100 or more decimal digits (say 1024 binary bits), then finding the factors should be essentially impossible on any currently available (or even foreseeable) computer, except a quantum one [6, 18]!

In order to properly define the computational complexity of a quantum computation, one must have some analogue of the universal Turing machine as a reference. Deutsch [19, 16] has formulated the idea of a *Quantum Turing Machine*, QTM. This simply replaces the classical, deterministic, motion of the Turing machine with a quantum evolution. The classical Turing machine has a discrete set of possible states at any given time, determined by the internal state, $q$, the characters in each memory position, $s_i$, and the current reading position in memory of the machine, $m$. The deterministic evolution of the machine is therefore simply a map: $\{q, \{s_i\}, m\} \to \{q', \{s'_i\}, m'\}$ which evolves the system from its state at time step $n$ to that at step $n+1$.

In the quantum Turing machine we change from a classical dynamics to a quantum one. We allow the system to be in an arbitrary superposition state $|\psi_n\rangle$ at time step $n$,

$$|\psi_t\rangle = \sum_{q\{s_i\}m} c_{q\{s_i\}m}(t)|q, \{s_i\}, m\rangle, \quad (1.21)$$

where $c_{qs_im}(t)$ is the probability amplitude that at time $t$ the machine is in internal state $q$ at memory position $m$ and the memory characters are $s_i$ for $i = -\infty \cdots +\infty$. The set of basis states, $|q, \{s_i\}, m\rangle$, includes all possible internal states of the machine ($q = A, B, \ldots Z$), all possible characters in each memory location ($s_i = \alpha, \beta \ldots \omega$ for memory cell $i = -\infty \ldots \infty$) and the machine being at any possible memory position $m$. These states constitute $|q, s_i, m\rangle$ constitute a countably infinite set of orthogonal basis states in the Hilbert space of the system (or finite if we restrict attention to machines with a finite memory).

The quantum Turing machine proceeds by a series of discrete steps, just like the classical version. The "computational complexity" of the Quantum Turing Machine is therefore essentially the number of unitary operators $\hat{U}(t_{n+1}, t_n)$ which must be applied. It is assumed that these operators must be 'local', i.e. they only act on the internal state of the machine $q$ the current memory cell, $s_m$, and move the machine one step left or right in memory $m' = m \pm 1$.

With this definition we can directly compare the number of such unitary evolutions required to solve a given problem with the number of steps taken by a classical Turing machine. It turns out that the computational complexity of a given problem can be quite different in the quantum and classical cases. The example of factoring large integers [6] is one example of a quantum algorithm which is known to offer *exponential* speed up over the (best known) classical algorithm. Shor's algorithm reduces the factorization problem to a number of unitary operators which is a polynomial in the number of bits, $M$. A similar result holds for the related discrete log problem [20].

It is interesting to speculate about the deep physical reasons for this speedup. One can say that a quantum state of $M$ bits is a superposition of $2^M$ states. In some sense the quantum computation is a parallel computation in which all $2^M$ basis vectors are acted upon at the same time. If one wanted to simulate a quantum computer using a classical computer one would need to multiply together $2^M$ dimensional unitary matrices, to simulate each step. So the classical simulation would be about $2^{2M}$ times longer than the corresponding quantum calculation. Interestingly, the fact that one could simulate a quantum computer using a classical one shows that both quantum and classical computers are limited to the same set of possible 'computable' functions as an ordinary Turing machine.

## 3. Superconducting Qubit Devices

### 3.1. Possible Qubit Realizations

The ideas of quantum computation of course immediately raise the issue of how to physically implement the qubits and realize gates between them. In principle any quantum mechanically coherent system could be used to implement the ideas of quantum computation. However in practice this has not proved so easy as one might have thought. The prospect of actually achieving quantum computation for useful size problems will remain unattainable unless one has suitable devices which can implement the fundamental logic operations necessary while maintaining quantum mechanical coherence. A wide variety of different possible implementations have been proposed [21]. However, at present no single option has a clear cut lead over the alternatives. Manipulations of single photons, nuclear spins, and trapped ions have been shown to have the necessary level of quantum coherence, at least for small number of qubits. However these realizations will be difficult to scale up to the large numbers of qubits (at least hundreds, probably thousands) required for serious computation. At present liquid state NMR has achieved coherent operation with the largest number of qubits (7 at present) but there are fundamental obstacles which make it unlikely that this can ever be extended to sufficiently large systems.

Five detailed criteria have been proposed by David DiVincenzo [22, 21], which can be used to judge the suitability of the various proposals which have been made. Obviously one must be able to initialize the qubits to a given state at the beginning, manipulate the qubits in order to implement some set of quantum logic gates during the computation and read out (i.e. measure) its value at the end of the computation. Clearly it must be possible for qubits to interact with each other, and quantum mechanical coherence must be maintained during the whole computation process. Another important criterion is *scalability*, i.e. the ability to work with large registers of qubits, say of order 100 − 1000 qubits, rather than with just one or two.

It is the possibility of a *scalable* implementation of the qubits which is the main advantage of solid state implementations, compared to liquid state NMR, photon optics or ion trap technology. While one can manipulate single photons, or small numbers of ions in a trap, it is not going to be easy to manipulate a hundred of them arbitrarily. On the other hand, if it is possible to fabricate devices 'on a chip', just like modern transistors on a silicon wafer, then in principle one could make any number of such devices and address each of them at will.

On the other hand, any solid state device is necessarily coupled to an environment (e.g. the wafer on which it is fabricated, the electrical contacts and leads into and out of the device etc.). This environmental coupling necessarily leads to *decoherence*. If this decoherence is too strong then the qubit device will not be able to operate coherently for a sufficient length of time, and the quantum computation will fail. A certain amount of decoherence can be tolerated, using the techniques of quantum error correction [23] and fault-tolerance [24]. DiVincenzo and others have estimated that, roughly, one would need to be able to apply at least of order $10^4$ quantum logic gate operations before the decoherence time in order to have a viable chance of implementing error correction. In practice the relevant number may even be significantly higher than this.

Therefore, with the range of possible solid state devices which could implement quantum computation, one should look for those which have the minimum levels of decoherence. Below, we shall argue that superconducting devices could well come closest to satisfying the requirements. Possible devices have been proposed making use of superconductors [25, 26, 27, 28, 29], semiconductor quantum dots [30], nanoscopic magnetic systems [31], and implanted ions in semiconductors [32]. The relative decoherence properties of these different systems are at the moment quite poorly understood experimentally [33]. However, in one of the first experiments carried out with superconducting qubits [3] (which we describe below) the device was found to operate coherently for more than 36 cycles. While this is still a long way from the DiVincenzo target, this was only the very first experiment of its kind and it should be possible to improve this number considerably. Whether or not one can ever raise this to the required figure of merit will only be determined after many more years of effort. However, the fact that 36 cycles have indeed been observed immediately places the superconducting devices in the front running position among solid state qubit implementations (although still far behind liquid state NMR in the same race [34]). There does not appear to be any truly fundamental limit to the level of coherence that can be obtained in the solid state, and at present the difficulties appear to be ones of implementation, rather than fundamental physics. It remains to be seen how far the devices can be improved before we reach any fundamental limits.

## 3.2. The Condensate of Cooper Pairs

The reason why superconductors are one of the leading candidates for physical realizations of a qubit is because they possess *macroscopic quantum coherence*. I.e. they exhibit quantum phenomena even in systems on a macroscopic or essentially infinite scale [2]. This quantum nature is apparent in phenomena such as the a.c. Josephson effect in which a d.c. voltage $V$ leads to an a.c. current with frequency exactly $\nu = 2eV/h$ [35, 36, 37]. The presence of the Planck constant, $h$, in the Josephson frequency clearly indicates that this is a quantum effect. Evidence for the existence of macroscopic quantum phenomena has emerged over the past decade or more [38, 39, 40, 41, 42]. However the clearest evidence for macroscopic quantum superposition states arose only very recently [3, 4, 5]. In this section we briefly introduce the nature of the quantum state in superconductors, which is necessary to set the scene before going on to describe the role of macroscopic quantum coherence in possible superconducting qubit implementations.

The origins of macroscopic quantum coherence derive fundamentally from the BCS theory of superconductors. There are two key aspects of the BCS theory which are essential in this. First the electrons in the superconductor are *paired* into Cooper pairs. Secondly the pairs form a *condensate*.

The pairing is vitally important because it means that there is an energy scale $\Delta$, the BCS energy gap, below which there are no single electron excitations (quasiparticles). If we work at temperatures where $k_B T \ll \Delta$ then all the unpaired electron states are effectively frozen out. Therefore all of the accessible low energy quantum states involve only states which can be described in terms of the Cooper pairs.

The second idea, the condensate, is equally important. The Cooper pairs do not just form a gas of bound electron pairs (like helium gas), but form a collective many-body condensed state (like superfluid helium, or a bose condensate). The BCS theory is the mean-field theory for this condensed state. The key point is that condensate possesses an order parameter which can be represented by a complex scalar field

$$\psi(\mathbf{r}) = |\rho_s|^{1/2} e^{i\varphi}. \quad (1.22)$$

The presence of the condensate is signalled by the presence of the definite phase, $\varphi$. As we shall show below, it is the presence of this phase which directly leads to the macroscopic quantum phenomena.

The phase in the order parameter, Eq. 1.22 arises from the BCS many-body wave function (although its presence was originally deduced by Ginzburg and Landau long before the BCS theory). Omitting many details, the important fact which we need about the BCS state is that it is a *coherent state* with a definite phase and an indefinite number of particles. The BCS wave function is of the general form

$$|\varphi\rangle = \sum_n e^{in\varphi} a_n |n\rangle, \quad (1.23)$$

where $|n\rangle$ is a many-body wave function with $n$ Cooper pairs ($2n$ electrons). The coefficients $a_n$ are probability amplitudes for being in each of the different states, $n$, of the Cooper pairs (remembering that all single electron states are frozen out at low temperatures). The BCS theory gives specific numerical values for the $a_n$, but these values will not be important for our argument.

*Figure 1.7.* The Josephson tunnelling current between two superconductors.

The d.c. Josephson effect is the first important consequence of this coherent many-body state. This is a result of tunnelling processes between two superconductors which are separated by a thin insulating layer, as shown in Fig. 1.7. Suppose that the Hamiltonian of the two superconductors and junction is of the form

$$\hat{H} = \hat{H}_L + \hat{H}_R + \hat{H}_t \quad (1.24)$$

where the terms represent the left hand side, right hand side and the tunnelling, respectively. $\hat{H}_t$ is the effective tunnelling Hamiltonian for

Cooper pairs to cross the junction (involving two actual single electron tunnelling events [35, 37]).

In the absence of tunnelling we have uncoupled condensates on the left and right hand sides, with condensate phases $\varphi_L$ and $\varphi_R$. Writing this combined state $|\varphi_L, \varphi_R\rangle$, then first order perturbation theory leads us to expect an energy change associated with the junction of

$$E = \langle \hat{H}_t \rangle = \langle \varphi_L, \varphi_R | \hat{H}_t | \varphi_L, \varphi_R \rangle. \quad (1.25)$$

Substituting in the coherent state representation, Eq. 1.23 for each condensate gives

$$E = \sum_{n_L n_R} a^*_{n_R} a_{n_R+1} a^*_{n_L} a_{n_L-1} e^{i(\varphi_L - \varphi_R)} \langle n_L, n_R | \hat{H}_t | n_L - 1, n_R + 1 \rangle + c.c. \quad (1.26)$$

where c.c. denoted the complex conjugate. The phase factors are constant and can be taken outside of the sums, giving a junction energy of the form

$$E = -E_J \cos(\varphi_L - \varphi_R). \quad (1.27)$$

The corresponding d.c. Josephson current is ( $I = (2e/\hbar) dE/d\varphi$)

$$I = I_c \sin(\varphi_L - \varphi_R) \quad (1.28)$$

where the junction critical current is $I_c = (2e/\hbar) E_J$.

## 3.3. Quantization of Superconducting Circuits

The BCS theory is a mean-field theory in which the order parameter phase, $\varphi$ in Eq. 1.22, is a constant. It has an arbitrary value, corresponding to the spontaneous symmetry breaking that occurs at the superconducting transition temperature $T_c$. Once the transition has occurred the value of $\varphi$ is frozen, as required by the thermal equilibrium of the condensate.

Therefore we must be quite cautious in interpreting $\varphi$ as the phase of the 'wave function' of the condensate. Although we have described the condensate in terms of a coherent state wave function $|\varphi\rangle$, it *does not*, follow that a superposition state of the form

$$c_1 |\varphi_1\rangle + c_2 |\varphi_2\rangle \quad (1.29)$$

would be a valid many-body wave function. In fact it would not be. The condensate is a strongly interacting many-body system. The Ginzburg Landau equation describing the condensate order parameter, Eq. 1.22 is a *non-linear* Schrödinger equation for which the superposition principle is *not valid*.

The problem is that the order parameter Eq. 1.22 is describing the density matrix of a system in thermal equilibrium, and not a simple wave function. It implicitly includes interactions with heat baths, particle baths, normal electrons outside the condensate and so on. Any pure superposition state, such as Eq. 1.29 would interact with these external environmental degrees of freedom in such a way as to collapse the wave function down to a single coherent state with a single definite value of the phase. In this sense, the phase $\varphi$ is essentially a classical variable in the BCS state.

In order to derive an effective quantum mechanical Hamiltonian of a superconductor it is necessary to start with the mean-field BCS or Ginzburg Landau equations for the condensate. We can then assume that there is a well defined order parameter $\psi(\mathbf{r})$ at each point in the superconductor, so that there is a well defined condensate phase $\varphi(\mathbf{r})$ at each point.

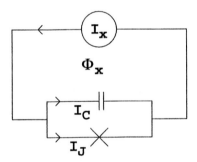

Figure 1.8. Schematic circuit diagram of a Josephson junction connected to an external current source $I_x$. The junction tunnelling current is $I_J$ and the junction capacitance charging current is $I_C$. The circuit encloses a total magnetic flux $\Phi$ which contains an externally applied part $\Phi_x$.

Starting from this BCS state we can derive an effective quantum mechanical Hamiltonian to describe a superconducting circuit. Consider the schematic circuit shown in Fig. 1.8 As a circuit the system has three elements, the ring itself, the Josephson tunnel junction, and the capacitance of the Josephson junction. From Kirchoff's law the currents in the three elements add as follows

$$I_x = I_C + I_J. \tag{1.30}$$

The Josephson current depends on the order parameter phase difference across the junction,

$$I_J = I_c \sin(\varphi). \tag{1.31}$$

The order parameter phase gradient in the closed superconducting loop leads to a current,

$$\begin{aligned} I_x &= -\frac{1}{L}\left(\frac{\hbar\varphi}{2e} - \Phi_x\right) \\ &= -\frac{1}{L}(\Phi - \Phi_x) \end{aligned} \quad (1.32)$$

in terms of an effective loop inductance, $L$, and an effective magnetic flux through the loop defined by, $\Phi = \hbar\varphi/2e$. The external magnetic flux is $\Phi_x$. The capacitor current is

$$I_C = \frac{dQ}{dt} = C\frac{dV}{dt} \quad (1.33)$$

where $C$ is the junction capacitance and $V$ is the potential drop across the junction. In order to close the set of equations we use the the a.c. Josephson relation to relate the junction voltage to the phase difference $\varphi$.

$$V = \frac{\hbar}{2e}\frac{d\varphi}{dt}. \quad (1.34)$$

Combining the above, we obtain an equation of motion for the flux $\Phi$,

$$C\frac{d^2\Phi}{dt^2} + I_J \sin(2\pi\Phi/\Phi_0) + \frac{1}{L}(\Phi - \Phi_x) = 0, \quad (1.35)$$

where $\Phi_0 = h/2e$ is the superconducting magnetic flux quantum. Notice that this is a purely classical equation of motion, which contains no explicit $\hbar$ factors. It would apply to any size superconducting ring, whether $1\mu$m in diameter or 1km.

In order to find the corresponding quantum dynamical equation we must first find an appropriate classical Hamiltonian, and apply the rules of canonical quantization. We first write down the classical Lagrangian,

$$\mathcal{L} = \frac{1}{2}C\dot{\Phi}^2 + E_J \cos(2\pi\Phi/\Phi_0) - \frac{(\Phi - \Phi_x)^2}{2L}, \quad (1.36)$$

for which the equation of motion is the Euler-Lagrange equation

$$\frac{d}{dt}\left(\frac{\partial \mathcal{L}}{\partial \dot{\Phi}}\right) - \frac{\partial \mathcal{L}}{\partial \Phi} = 0. \quad (1.37)$$

The classical momentum conjugate to $\Phi$ is

$$\frac{\partial \mathcal{L}}{\partial \dot{\Phi}} = C\dot{\Phi} = Q. \quad (1.38)$$

Therefore the junction capacitance charge $Q$ is the classical momentum conjugate to the flux $\Phi$. In terms this conjugate pair of variables the classical Hamiltonian is

$$\mathcal{H} = Q\dot{\Phi} - \mathcal{L} \tag{1.39}$$

giving

$$\mathcal{H} = \frac{Q^2}{2C} - E_J \cos(2\pi\Phi/\Phi_0) + \frac{(\Phi - \Phi_x)^2}{2L}. \tag{1.40}$$

We can accomplish canonical quantization by introducing commutator

$$[\hat{Q}, \Phi] = \frac{\hbar}{i} \tag{1.41}$$

and the standard substitution

$$Q \to \hat{Q} = \frac{\hbar}{i}\frac{\partial}{\partial \Phi}. \tag{1.42}$$

The quantized Hamiltonian to describe the circuit is [8],

$$\hat{H} = -\frac{\hbar^2}{2C}\frac{\partial^2}{\partial \Phi^2} - E_J \cos(2\pi\Phi/\Phi_0) + \frac{(\Phi - \Phi_x)^2}{2L}. \tag{1.43}$$

Equation Eq. 1.43 defines a quantum mechanical Hamiltonian for a superconducting circuit in terms of the flux $\Phi$. In principle this will allow a quantum wave function $\psi(\Phi)$ which is a superposition state of different flux states. However, such superposition states will eventually become decoherent and collapse because of coupling to the external thermal environment, consistent with the absence of superposition states in the non-linear BCS or Ginzburg-Landau mean-field equations. But it is still possible to observe superposed states, provided the eventual wave function collapse is sufficiently slow, i.e. the coupling to the environment is sufficiently weak, then it will be possible to observe macroscopic quantum superpositions in superconducting circuits.

We shall discuss the main effects leading decoherence in more detail below. However two requirements are obviously necessary to prevent this wave function collapse. Firstly it is vital to isolate the condensate from external interactions. Clearly this can only be achieved at low temperatures ($k_B T \ll 2\Delta$ or $T \ll T_c$), so that there are essentially no interactions with non-condensed electrons. It is also necessary to isolate the system as much as possible from external sources of decoherence. This in turn implies that one should consider small, or mesoscopic, superconductors rather than bulk superconductors. These mesoscopic superconductors may still be considered macroscopic, since they may contain of order $10^{10}$ electrons in the condensate.

There are two basic designs which have been shown to exhibit macroscopic quantum coherence. They differ primarily in whether the charge, $Q$, or flux, $\Phi$, is directly manipulated. We also call them the 'charge' and 'flux' devices, respectively.

### 3.4. The Cooper Pair Box Qubit

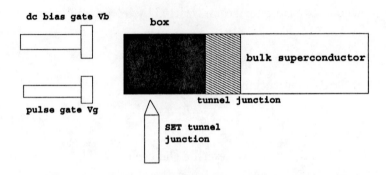

*Figure 1.9.* A schematic representation of the Cooper-pair box.

The 'charge' superconducting qubit, discussed theoretically in [25, 26, 27], is the Cooper pair box developed experimentally by Nakamura, Pashkin and Tsai [3, 45]. The key element in the device is a small metallic island, or box, of dimensions 700 × 50 × 15nm. The island is aluminium, which becomes a superconductor below about 1K. This box contains of order $10^8$ conduction electrons. Because the operating temperature (nominally 30mK) is substantially less than the BCS gap in Al ($2\Delta \sim 3K$), essentially all of the conduction electrons will be paired into Cooper pairs an inside the condensate. We can therefore speak of this system as a Cooper pair box.

In the Cooper pair box, Fig. 1.9, the superconducting island is isolated from its surroundings, except for the Josephson junction. If the Josephson tunnelling is weak, then the number of Cooper pairs on the island will remain constant. But, on the general principles outlined above we know that circuit charge and flux do not commute. Converting to variables of Cooper pair number, $n = Q/2e$, and Josephson junction phase, $\varphi = 2e\Phi/\hbar$, the commutator becomes

$$[\hat{n}, \varphi] = \frac{1}{i}, \qquad (1.44)$$

where the number operator is

$$\hat{n} = \frac{1}{i}\frac{\partial}{\partial \varphi}. \qquad (1.45)$$

In a completely isolated superconducting island the number of Cooper pairs must be conserved, and so the quantum state will be an eigenstate of number, $\hat{n}$. Uncertainty implies that the phase must be indefinite in this case. The presence of a small Josephson coupling to a second large superconductor allows quantum transitions to occur between states of different $n$.

Because the Cooper pair box states are close to being eigenstates of $\hat{n}$, it is most convenient to write the Hamiltonian in terms of eigenstates of particle number, $|n\rangle$. As shown in the figure, the Cooper pair box is coupled to several external leads. Two of these are external gates, which provide external electrostatic potentials $V_b$ and $V_g$, which we shall call the bias and the gate voltages, respectively. The former is used to provide a d.c. bias voltage to set the device at the correct operating point, and the second is used to input fast voltage pulses into the device. The Cooper pair box has a finite total capacitance, $C = C_J + C_b + C_g$, of order 600aF. We can express the energy of the $n$ Cooper pair state as [25, 27, 43, 44]

$$E(n) = \frac{(2en - C_b V_b)^2}{2C} \qquad (1.46)$$

when the gate voltage is turned off, $V_g = 0$. This Hamiltonian is clearly diagonal in the particle number operator.

The third gate to the Cooper pair box is a Josephson junction which couples it to a macroscopic superconductor. As discussed above, Josephson tunnelling is a coherent tunnelling of Cooper pairs, and so this junction introduces an additional tunnelling term into the Hamiltonian which becomes

$$\hat{H} = \sum_n \frac{E_J}{2} \left( |n\rangle\langle n+1| + |n+1\rangle\langle n| \right) + \frac{(2en - C_b V_b)^2}{2C} \qquad (1.47)$$

With this Hamiltonian it is clear that the system is generally in a quantum superposition of one or more different $n$ eigenstates. In the Nakamura device the Josephson energy was $51.8\mu eV$, which is significantly smaller than the Cooper pair charging energy, $E_C = (2e)^2/2C = 468\mu eV$.

The quantum state of the Cooper pair box can be controlled by suitable manipulations of the bias and gate voltages. It is possible to move the number eigenstates $|n\rangle$ to have two energies, say $E(n)$ and $E(n+1)$, which cross on changing the gate voltage. This is shown in Fig. 1.10, where the dotted lines represent $E(n)$ and $E(n + 1)$. However, because of the Josephson tunnelling between states $|n\rangle$ and $|n + 1\rangle$ these are not eigenstates. The true eigenstate energies are shown in the continuous curves, and these have a typical quantum mechanical avoided crossing

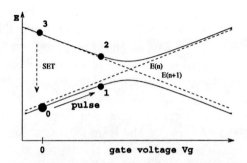

*Figure 1.10.* The energy levels of the Cooper pair box as a function of gate voltage $V_g$. The initial state is prepared in state 0. The gate pulse leads to a superposition of 1 and 2. After the pulse the state is a superposition of 0 and 3. Incoherent transitions from 3 to 0 lead to detected current in the SET.

behaviour. Clearly, precisely at the avoided crossing point the energy eigenstates are not $|n\rangle$ and $|n+1\rangle$, but the even and odd superpositions

$$|\psi_+\rangle = \frac{1}{\sqrt{2}}(|n\rangle + |n+1\rangle)$$
$$|\psi_-\rangle = \frac{1}{\sqrt{2}}(|n\rangle - |n+1\rangle).$$

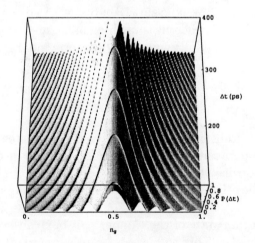

*Figure 1.11.* Calculated Cooper pair box SET current $I(\Delta t)$. $I$ is proportional to the probability of excitation after the pulse $P(\Delta t)$ given in from Eq. 1.51. The result is remarkably similar to the actual data obtained by Nakamura, Pashkin and Tsai [3]. The horizontal axis is proportional to the pulse gate voltage, $V_g$, represented in units of as an effective number of Cooper pairs, $Q$. The observed coherent oscillation persists for longer than 36 cycles.

Nakamura and co-workers [3, 45] explicitly demonstrated in their experiment that these superposition states do indeed occur. Firstly the system is prepared with the gate voltage, $V_g$, turned off, and a steady d.c. bias voltage $V_b$ applied to prepare the state in the eigenstate 0 indicated on Fig. 1.10. Call this initial eigenstate $|\psi_0\rangle$, which is an eigenstate of $\hat{H}_0$. Then a sudden square pulse is applied to the gate $V_g$. Assuming an infinite pulse rise time we can apply the sudden approximation and write the initial state $|\psi_0\rangle$ as a superposition of the two eigenstates of $\hat{H}$, say $|\psi_1\rangle$ and $|\psi_2\rangle$

$$|\psi\rangle = |\psi_1\rangle\langle\psi_1|\psi_0\rangle + |\psi_2\rangle\langle\psi_2|\psi_0\rangle. \qquad (1.48)$$

These two eigenstates are indicated as points 1 and 2 in Fig. 1.10. The square gate pulse lasts for some time $\Delta t$, and so at the end of this time the wave function is

$$|\psi\rangle = e^{-iE_1\Delta t/\hbar}|\psi_1\rangle\langle\psi_1|\psi_0\rangle + e^{-iE_2\Delta t/\hbar}|\psi_2\rangle\langle\psi_2|\psi_0\rangle, \qquad (1.49)$$

where $E_1$ and $E_2$ are the energy eigenvalues of states 1 and 2. At the end of the pulse the gate voltage $V_g$ returns to zero. We must therefore project the state back onto the ground state $|\psi_0\rangle$,

$$\begin{aligned}\langle\psi_0|\psi\rangle &= e^{-iE_1\Delta t/\hbar}|\langle\psi_0|\psi_1\rangle|^2 + e^{-iE_2\Delta t/\hbar}|\langle\psi_0|\psi_2\rangle|^2 \\ &= e^{-iE_1\Delta t/\hbar}\cos^2\frac{\eta}{2} + e^{-iE_2\Delta t/\hbar}\sin^2\frac{\eta}{2}\end{aligned} \qquad (1.50)$$

The probability that the state does not return to the the ground state, 0 in Fig. 1.10, after the pulse has finished is

$$\begin{aligned}P(\Delta t) &= 1 - |\langle\psi_0|\psi\rangle|^2 \\ &= 4\cos^2\frac{\eta}{2}\sin^2\frac{\eta}{2}\sin^2\Delta E\Delta t\end{aligned} \qquad (1.51)$$

where $\Delta E = (E_2 - E_1)/2$. This is the probability that the system is left in the excited state 3 in Fig. 1.10 after the pulse. A plot of this probability function is shown in Fig. 1.11.

Nakamura and co-workers measured the probability $P(\Delta t)$ for their device, Fig. 1.12, by measuring the transition rate back down to the ground state (the dashed arrow in Fig. 1.10). This is accomplished with the fourth electrode on the device, Fig. 1.9, which is a single electron transistor (SET). The current in this transistor represents *incoherent* processes where the device makes the transition down to the ground state by emitting two unpaired electrons into the SET.

Their results for the SET current, reproduced in Fig. 1.12, as a function of pulse length, $I(\Delta t)$, are very similar to the those calculated from

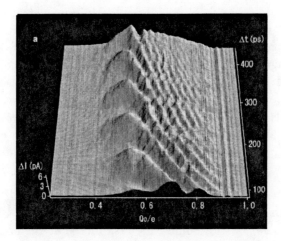

*Figure 1.12.* Measured Cooper pair box SET current $I(\Delta t)$. obtained by Nakamura, Pashkin and Tsai [3]. The horizontal axis is proportional to the pulse gate voltage, $V_g$, represented in units of as an effective number of Cooper pairs, $Q$. The observed coherent oscillation persists for longer than 36 cycles. Figure reproduced from: Y. Nakamura, Yu. A. Pashkin, J. S. Tsai, Nature **398** 786 (1999).

Eq. 1.51, shown in Fig. 1.11. The horizontal axis represents the pulse height, converted to units of fractions of a Cooper pair on the box (using the box capacitance $C$). The crossing point in Fig. 1.10 corresponds to $Q = 0.5$ in this representation, because this is the point where the basis states $|n\rangle$ and $|n+1\rangle$ cross in Fig. 1.10. One can clearly see that the coherent oscillations are particularly strong at this point. Indeed Nakamura and co-workers [3] observed that these coherence oscillations persist in excess of 36 coherent cycles. In fact the precise decoherence time was not reported, and the 36 cycle figure is only a lower bound. Clearly this is not yet close to the qubit figure of merit of $10^4$ device cycles, proposed by DiVincenzo [22], but this is only the first such experiment of its kind and there may be many possible improvements to be made, which we discuss below.

## 3.5. The Flux Qubit

There is a second fundamental design which has been proposed [28, 29] to implement superconducting qubits. With this second design, experiments have also been able to explicitly demonstrate quantum mechanical coherence and superposition states [4, 5]. This is the 'flux qubit'. The

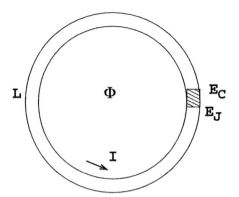

*Figure 1.13.* A schematic flux qubit. The Josephson junction is shorted by a thick superconducting ring of inductance $L$, which encloses a total magnetic flux $\Phi$, due to the current $I$ and an external contribution $\Phi_x$.

basic design is shown in Fig. 1.13. It is essentially a superconducting ring, not an island. The ring contains one (or more) Josephson junctions which act as a weak link. Here it is the flux quantization in superconducting rings which provides the basic set of basis functions.

Ignoring the Josephson junction for the moment, the energy levels of an ideal superconducting ring are quantized in units of flux quanta, $\Phi = n\Phi_0$, with energies analogous to Eq. 1.46,

$$E(n) = \frac{(n\Phi_0 - \Phi_x)^2}{2L} \qquad (1.52)$$

where $L$ is the self-inductance of the ring, $\Phi_0 = h/2e$ is the superconducting flux quantum and $\Phi_x$ is the externally applied flux. If the externally flux is chosen as half a flux quantum, $\Phi_x = \Phi_0/2$, then two of the states, say $n = 0$ and $n = 1$, cross and again there is a two level system analogous to Fig. 1.10.

The full potential energy surface $V(\Phi)$ for the Hamiltonian in Eq. 1.43 is a system of potential wells with minima near $\Phi = n\Phi_0$, superimposed upon a larger scale parabolic curve. However, for values of the external flux a $\Phi_x = \Phi_0/2$ there are two degenerate ground state minima, very similar to Fig. 1.2. Quantum tunnelling between these two states occurs under the action of the kinetic energy term

$$-\frac{\hbar^2}{2C}\frac{\partial^2}{\partial \Phi^2},$$

with the junction capacitance $C$ acting as the effective particle mass.

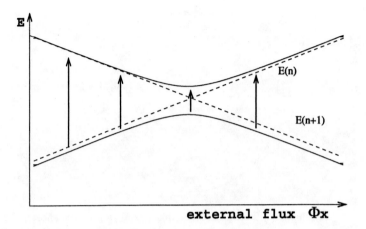

*Figure 1.14.* Energy level crossing as a function of external flux, $\Phi_x$ in a flux qubit. The two flux states $n$ and $n+1$ cross as a function of $\Phi_x$ (dashed lines). The presence of the quantum tunnelling leads to an avoided crossing in the true spectrum (solid lines). The avoided crossing can be measured by detecting the frequency of microwave absorption (vertical arrows) as a function of $\Phi_x$.

In the experiments of Friedman *et al.* [4] and van der Wal *et al.* [5] the eigenvalues of this effective Hamiltonian were measured, directly demonstrating the existence of macroscopic superposition states. As illustrated in Fig. 1.14, the external flux, $\Phi_x$ was tuned to a point where two flux states, say $n$ and $n+1$, cross. Quantum coherent tunnelling between these eigenstates leads to an avoided crossing, as illustrated in the figure. By measuring the microwave transition energy between the two eigenstates as a function of $\Phi_x$, Friedman *et al.* [4] and van der Wal *et al.* [5] were able to show directly that an avoided crossing does indeed occur, and hence there is a quantum coherent tunnelling between the two flux states. The Friedman *et al.* [4] experiment showed this coherent tunnelling between two excited states of the double well potential, Fig. 1.2, while the van der Wal *et al.* experiment showed a similar avoided crossing in the ground state of the double well potential.

The flux qubits used in these experiments are about $20\mu$m in circumference, and about $450 \times 80$nm$^2$ in cross sectional area. Therefore they contain about $10^{10}$ atoms. The tunnelling observed can be considered a truly *macroscopic quantum tunnelling*, because the state of the whole superconducting condensate is changed coherently. These experiments therefore clearly vindicate the laws of quantum mechanics, even when applied to objects of macroscopic size, as suggested originally by Leggett [2]. The existence of quantum tunnelling in macroscopic objects is only limited by the importance of decoherence due to dissipative coupling to the environment [46]. Below, in Sec.4.3, we shall briefly discuss the role of decoherence and dissipation in these superconducting qubit devices.

## 3.6. Tunable Josephson Energy

It is very useful to be able to tune the Josephson coupling $E_J$. This has been noted by some of the superconducting qubit proposers [27, 29] and was implemented in the charge qubit experiments [3, 45]. The variable $E_J$ is achieved by constructing a composite junction, made from two equal junctions (of capacitance $C_J/2$ and tunnelling energy $E_J/2$) mounted symmetrically in a small loop (of inductance $L_J$), as shown in Fig. 1.15. An external magnetic flux $\Phi_{xJ}$ is applied to this ring. This system has two pairs of conjugate variables, the discrete charge $2ne$ with the phase difference $\varphi$ across the whole system and the continuous electric flux $Q$ with the total magnetic flux $\Phi$ in the loop. The full

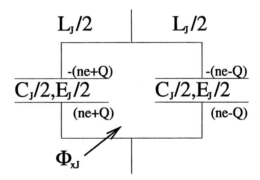

*Figure 1.15.* Schematic circuit for a composite, tunable Josephson junction, controlled by an external magnetic flux $\Phi_{xJ}$.

Hamiltonian for this system is

$$H_{tot} = \frac{(2ne)^2}{2C_J} + \frac{Q^2}{2(C_J/4)} + \frac{(\Phi - \Phi_{xJ})^2}{2L_J} - \frac{E_J}{2}\cos\left(\varphi - \frac{\pi\Phi}{\Phi_0}\right) - \frac{E_J}{2}\cos\left(\varphi + \frac{\pi\Phi}{\Phi_0}\right). \quad (1.53)$$

Various parameter conditions need to be arranged:
(i) $L_J$ is small (so $L_J\dot{Q} \ll \Phi_{xJ}$ and $\Phi \approx \Phi_{xJ}$),
(ii) $\Phi_0^2/L_J \gg e^2/C_J$ , $(2\pi)^2 E_J$ (so the $\Phi$ wave function peaks sharply around $\Phi \approx \Phi_{xJ}$) and
(iii) $\hbar\omega_J \equiv \hbar(L_J C_J/4)^{-1/2} \gg E_J$ , $e^2/C_J$ , $k_B T$.

Then the $\Phi,Q$ system exhibits ground state behaviour with $\Phi \approx \Phi_{xJ}$ and the effective Hamiltonian for the $n,\varphi$ system is

$$H_{varE} = \frac{(2ne)^2}{2C_J} - E_J \cos\left(\frac{\pi\Phi_{xJ}}{\Phi_0}\right) \cos\varphi . \quad (1.54)$$

The capacitances add in parallel for $n$ and there is an effective tunnelling energy $E_J \cos\left(\frac{\pi\Phi_{xJ}}{\Phi_0}\right)$ for the phase $\varphi$, tunable through the external applied flux. The ability to vary the tunnelling, in principle to suppress it to zero, is extremely useful for the implementation of superconducting quantum gates.

## 4. Superconducting Qubits and Decoherence

### 4.1. The Advantages of Macroscopic Quantum States

We have argued above that it is *macroscopic quantum coherence*, which is the special advantage of superconductors compared to other possible solid state realizations of qubit devices. However, one can ask why exactly is it such an advantage to have a macroscopic quantum state. One could argue that it is better to have fully microscopic devices, since they may be more isolated from their environments and hence more coherent.

To see how the macroscopic size of the superconductor becomes relevant, consider an idealized model of the flux qubit, as shown in Fig. 1.13. As shown, there is a circulating current $I_x$ driven by the order parameter phase gradient around the loop. We have also seen earlier that the current is related to the total flux $\Phi$ and the external flux $\Phi_x$ by

$$I_x = -\frac{1}{L}(\Phi - \Phi_x) = -\frac{1}{L}(n\Phi_0 - \Phi_x) \quad (1.55)$$

for a ring containing $n$ flux quanta and in zero external field. To estimate the effective inductance, $L$, we can assume that the superflow corresponds to a perfect solid-body rotation of the condensate. The inductive energy is mainly due to to the inertial energy of the rotating condensate,

$$\frac{1}{2}I_x^2 L = \frac{1}{2}mv^2 N_s \quad (1.56)$$

where $v = \omega R$ is the rotational velocity, and there are $N_s$ electrons of effective mass $m$ in the superconducting condensate. If there there are $n_s$ condensate electrons per unit volume of superconductor, then

$$I_x = n_s ev\mathcal{A} \quad (1.57)$$

where $\mathcal{A}$ the cross-sectional area of the superconductor. Solving for the inductance yields

$$L = \frac{2\pi R m}{n_s e^2 \mathcal{A}}. \qquad (1.58)$$

Therefore the macroscopic size parameters of the ring ($R$ and $\mathcal{A}$) enter explicitly in determining the effective ring inductance.

It is even more instructive to estimate the magnetic moment, $M$, associated with the circulating current in the qubit. For a circular loop of radius $R$, the magnetic moment $M$ is defined by $M = \pi R^2 I_x$. For a ring containing $n$ flux quanta and with $\Phi_x = 0$ we therefore obtain

$$\begin{aligned} M &= \pi R^2 \frac{1}{L} n \Phi_0 \\ &= n \frac{e\hbar}{2m} N_s \end{aligned} \qquad (1.59)$$

where $N_s = 2\pi R \mathcal{A} n_s$ is the total number of electrons in the superconducting condensate. The factor $e\hbar/2m$ is the Bohr magneton, $\mu_B$, and so we see that the magnetic moment of the ring with $n$ flux quanta is $N_s \mu_B$ times $n$. Effectively the superconducting ring is a system with a *giant magnetic moment*, of order $10^{10}$ Bohr magnetons.

This simple argument shows the power of the macroscopic quantum state. If the system makes a transition from $n$ to $n+1$ flux quanta, then there is a macroscopic change of magnetic moment, $10^{10}$ times bigger than a for single electron spin. Clearly it is the large magnetic field generated by this macroscopic moment which enables the device to couple well with external electro-magnetic fields, such as the external microwave fields used to detect the quantum tunnelling.

We can contrast this with the corresponding figures for liquid state NMR experiments [34]. In an NMR sample one has an ensemble of perhaps $10^{18}$ nuclear spins. However the ensemble only has a small fractional magnetic polarization, of perhaps $10^{-4}$ because of the high temperature relative to the nuclear Zeeman splitting $\mu_N B$ in the applied field $B$. The nuclear moments, $\mu_N$ are about $10^{-3} \mu_B$, and so the net magnetic signal which is detected from the NMR experiment is also of order $\sim 10^{11} \mu_B$ in magnitude, which is similar to a single superconducting ring. However, unlike NMR, the superconductor case is not an ensemble measurement, because the effective temperature is low (both because $k_B T \ll \Delta$ and because $k_B T \ll E_J, \Phi_0^2/2L$). Therefore the problems of lack of entanglement in NMR, which have been noted recently by Popescu and Linden [47] do not arise in the superconductor case.

The case of the charge qubit device is slightly different. In this case the relevant two states of the system only differ by one Cooper pair, or two electrons. Therefore the change of state of the system does not produce a macroscopic external field. The change in charge state is microscopic, and the effective signal strength does not increase with the size of the Cooper pair box.

However one may make the following observations about the role of macroscopic quantum coherence in the charge qubit case. The Cooper pair box is Josephson coupled to a macroscopic (essentially infinite) second superconductor. That Josephson coupling is uniquely a property of superconductors, and would not be present in a non-superconducting system, such as a semiconductor quantum dot. Remakably, measuring the Josephson frequency, $\nu = 2eV/h$, is the source of the most accurate determination of the fundamental constant $e/h$. In short, the macroscopic nature of the condensate overrides the influence of mechanical, thermal, and electrical imperfections of the any actual Josephson junction. Thus the quantum state control of a Cooper pair box by Josephson coupling it to larger superconductors can be expected to be equally free of decoherence. Moreover, one may recall that impurities in a superconductor do not cause resistance, and hence dissipation. The reason for this is clarified by Anderson's theorem [48] which implies that the energy gap in the quasiparticle spectrum remains absolute even in the presence of impurities. Clearly, this lack of Ohmic dissipation suggests that the influence of more general electromagnetic interactions with noise in the environment will also be weak. Although these general arguments are not necessarily decisive, they do provide strong grounds for optimism concerning small superconducting qubits.

In the specific case of the Nakamura type experiment [3] one further important advantage is the ability to probe the charge state. This is because of the extreme sensitivity of the SET detector. The SET can reliably detect single electrons, and so is sufficient to measure the change in state of the device. One should note, however that the Nakamura experiment is effectively an ensemble measurement, since the gate pulse sequence is applied continuously, leading to a time averaged current in the SET device.

### 4.2. Quantum Logic Gates and Two Qubit Devices

As discussed earlier, any quantum computation with many qubits can be decomposed into just one and two qubit gates between the qubits. For a universal set one must be able to perform arbitrary single qubit rotations along with a two qubit gate capable of generating entanglement. It is possible in principle to perform the required gates with superconducting systems.

**Single Qubit gates.** For the charge qubit, provided that it is operated in the regime where the two charge states $|0\rangle$ and $|1\rangle$ are the dominant quantum states (which requires the external bias charge—$C_b V_b$ in Eq.1.47—to satisfy $Q_x \sim e$), the device is a good approximation to a two state system. Its Hamiltonian can be approximated by

$$H_{cq} \approx \frac{e}{C}(Q_x - e)\sigma_z - \frac{E_J}{2}\sigma_x, \qquad (1.60)$$

omitting unimportant constant terms and with the notation $|0\rangle \equiv |\uparrow\rangle$ and $|1\rangle \equiv |\downarrow\rangle$.

A similar result holds for the flux qubit. Provided that the parameters are such that the curvature of the cosine minima in its potential dominates that due to the inductance and it is operated in the regime where the two flux states which peak around 0 and $\Phi_0$ (denoted by $|\uparrow\rangle$ and $|\downarrow\rangle$ respectively) dominate, this device is also a good approximation to a two state system. Its Hamiltonian can be approximated by

$$H_{fq} \approx \frac{L_t}{LL_k}\left(\frac{\Phi_0}{2}\right)(\Phi_x - \Phi_0/2)\sigma_z - \frac{\Gamma}{2}\sigma_x, \qquad (1.61)$$

again omitting constant terms. Here $L_k \equiv \Phi_0^2/(4\pi^2 E_J)$, $L_t^{-1} = L^{-1} + L_k^{-1}$ and $\frac{\Gamma}{2}$ is the matrix element for tunnelling between the two flux states through the barrier created by the cosine potential.

In each case it is clear that the effective Hamiltonian is sufficient to effect arbitrary single qubit rotations. The coefficients of both Pauli matrices $\sigma_z$ and $\sigma_x$ can be set to zero and turned on for set times to effect desired rotations [27]. The $\sigma_z$ term is controlled by the external bias charge/flux ($Q_x/\Phi_x$) away from the degenerate point and the $\sigma_x$ tunnelling term is controlled by use of a tunable $E_J$, described earlier. For the flux qubit, turning $\Gamma$ off is effected through a large $E_J$ as $\Gamma$ is exponentially damped with increasing barrier height. It is also worth noting that single qubit phase gates can in principle be achieved for a charge qubit using the geometric or Berry phase [14], rather than dynamically [49]. This involves using an *asymmetric* version of the tunable $E_J$ system described earlier, with two unequal junctions in parallel. This generates a 3-dimensional external parameter space for the qubit, in which it is possible to choose paths that produce a non-trivial Berry phase. Given the geometric rather than dynamic nature of such a quantum gate, it is possible that this may be immune to some decoherence effects, although this may be outweighed by the longer gate time if it has to be performed adiabatically.

**Two Charge Qubit gates.** One possibility for coupling charge qubits [25] involves placing the qubits in parallel and connecting a superconducting inductor across the qubit system, as shown in Fig. 1.16. This generates a $\sigma_{yi}\sigma_{yj}$ coupling between qubits $i$ and $j$. This is current-

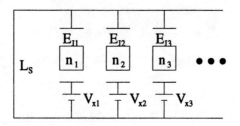

*Figure 1.16.* Schematic circuit for inductively coupled charge qubits. Each of the $E_{Ji}$ shown can be composite, individually tunable through an external flux $\Phi_{xJi}$ as in Fig. 1.15.

current coupling—in the spin picture a qubit tunnelling energy (proportional to $\cos\varphi$) is represented by $\sigma_x$ and so the tunnelling current (proportional to $\sin\varphi$) is represented by $\sigma_y$. Such an interaction term (acting for a set time) can generate a universal two qubit gate of the form Eq. 1.20 [25]. In the original proposal [25] the interaction term is permanently on, as in NMR systems, so two-qubit gates have to be realized by operating qubits at different voltages and bringing them into resonance, either suddenly for a fixed (short) time period, or adiabatically and using an oscillatory external voltage difference pulse.

Inductive qubit coupling can be controlled [27, 51] if the charge qubits in Fig. 1.16 have tunable $E_J$. For a set of qubits (each of total capacitance $C$) in parallel with a shunt inductor $L_s$ (chosen so the oscillation frequency is large compared to any qubit frequencies), the interaction term is $-\frac{1}{2}L_s I^2$ where $I$ is the sum of the qubit currents [27, 51]. The cross terms give qubit interactions of the form $E_{Ji}E_{Jj}\sigma_{yi}\sigma_{yj}$. This is handy because single qubit manipulations only require the turning on of one $E_{Ji}$ (using the external flux $\Phi_{xJi}$), in which case the two qubit interaction remains off. Simultaneous single qubit gates would still be tricky; nevertheless control of single and two qubit gates with the same external sources ($\Phi_{xJi}$ and $Q_{xi}$ for each charge qubit) is certainly useful.

Capacitive coupling between charge qubits is also possible, as illustrated in Fig. 1.17. In this arrangement the couplings are fixed and give

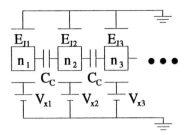

*Figure 1.17.* Schematic circuit for capacitively coupled charge qubits. The boxes are coupled through capacitors $C_C$. Each of the $E_{Ji}$ shown can be composite, individually tunable through an external flux $\Phi_{xJi}$ as in Fig. 1.15.

rise to a $\sigma_{zi}\sigma_{z(i+1)}$ between neighbouring qubits. Such a coupling can be effectively turned on and off by manipulation of the voltages applied to the qubits, using techniques analogous to those used in NMR. This can generate a conditional phase shift gate, another universal two-qubit gate [50].

An alternative adiabatic manipulation of charge qubits has been proposed [26]. Here the interaction is turned on and off through spatial movement of the charges. By using a 1D *array* of superconducting islands to represent each qubit and moving the charges adiabatically, it is possible to turn on a capacitive coupling between qubits. This again gives a $\sigma_{zi}\sigma_{zj}$ coupling which can generate a conditional phase shift.

The geometric phase approach can also be adapted to generate a universal two qubit gate [49]. If two (asymmetric composite) charge qubits are coupled capacitively (as illustrated in Fig. 1.17), the charge state of one (the control) can determine the geometric phase acquired by the other (the target) as its state undergoes adiabatic transport. This generates a conditional phase gate geometrically, as opposed to dynamically.

**Two Flux Qubit Gates.** Flux qubits could be coupled through superconducting inductive loops, or flux transformers [29, 52]. With multi-junction qubits, dependent upon the positioning of the transformer loops, it is possible to generate $\sigma_{zi}\sigma_{zj}$, $\sigma_{zi}\sigma_{xj}$ or $\sigma_{xi}\sigma_{zj}$ coupling terms, as illustrated in Fig. 1.18. The coupling coefficients should be small compared to the qubit energies, as it is proposed that the inductances (self and mutual) are kept small in to avoid pick-up. An estimate for a typical coupling energy is [52] $\sim 0.01 E_J$. Once again, these forms of coupling are able to realize a universal two qubit gate, such as conditional phase. Rather than leaving them on permanently, the flux transformers

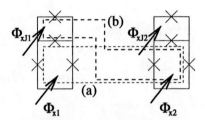

*Figure 1.18.* Schematic circuit for multi-junction (each denoted by ×) flux qubits coupled through flux transformers. Each qubit is subjected to two external flux sources, $\Phi_x$ and $\Phi_{xJ}$; the upper pair of junctions is in effect a composite tunable junction. Scheme (a) produces $\sim \sigma_{z1}\sigma_z$ coupling and scheme (b) $\sim \sigma_{x1}\sigma_{z2}$ coupling.

could be turned on with Josephson switches [52]. However, this switching may introduce additional unwanted noise.

Another proposed approach for coupling flux qubits [51] enables control using the same external sources applied to achieve single qubit gates ($\Phi_{xJi}$ and $\Phi_{xi}$). Each of a 1D array of qubits is mutually coupled to the superconducting inductor $L_s$ of a separate high frequency ($L_s$–$C_s$) oscillator, as illustrated in Fig. 1.19. This generates an interaction term

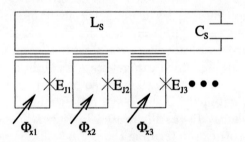

*Figure 1.19.* Schematic circuit for coupling flux qubits through an $L_s$–$C_s$ oscillator mode. Each flux qubit is controlled through the flux $\Phi_{xi}$ and has a tunable junction (×) $E_{Ji}$ implicitly controlled by another flux $\Phi_{xJi}$ as in Fig. 1.15.

$-\frac{1}{2}C_s V^2$ where $V$ is the sum of the qubit voltages induced in the oscillator circuit due to the flux tunnelling in the qubit circuits [51]. Expanding this gives qubit interactions of the form $\Gamma_i \Gamma_j \sigma_{yi} \sigma_{yj}$ As in the charge case, such terms remain off unless both $\Gamma$s are turned on.

**Linear Chain Addressing.** An alternative approach to realizing quantum computing gates is the addressing of a linear chain of qubits $ABABAB...$ through global, rather than qubit-specific, pulses [53, 54]. Such an approach may be applicable to superconducting qubits if all the $A$ qubits can be made very similar (ideally they should be identical) and likewise for all the $B$s. As there are always tolerances in fabricated systems, the approach would probably still need individual qubit addressing. However, as this is just for tuning it could effectively be at dc, so the only pulses needed would be global. For a superconducting qubit array such pulses could be applied from a microwave cavity. This has been suggested for flux qubits [52] but it could also apply to a charge qubit array.

## 4.3. Decoherence Effects

Two of the most important contributions to decoherence of superconducting qubits apply to both charge and flux cases. The first is the presence of fluctuations in the applied external sources, such as the gate voltages, the ring fluxes, or the small loop fluxes $\Phi_{xJ}$ employed to tune the qubit $E_J$s. Clearly such sources are needed to realize gates; however, they are also a source of noise. Even if the stages adjacent to the qubits are cooled in the cryostat along with the qubits, external sources must have some finite coupling back to room temperature apparatus and so will contain some level of thermal noise. Other experimental errors such as pulse shape imperfections will also play a role. It is very interesting to note that the dominant sources of decoherence in the charge qubit experiments [3, 45] are probably of this nature [55]. Rounding the model pulse applied to the charge qubit generates an asymmetry about the central peaks of Fig.1.11, weakening the left side oscillations as was seen experimentally [3, 45] in Fig. 1.12. Furthermore, modelling of Gaussian noise in the pulse height and the Josephson energy $E_J$ respectively damps the oscillations on both sides of the peaks, or weakens the peaks themselves [55]. The former is very much in keeping with the experimental observations [3, 45] in Fig. 1.12. Thus, not only can such simple modelling show that external noise is probably the dominant decoherence effect in the system of Nakamura and co-workers, it can also point to the major source, the voltage pulses. In one sense this is very encouraging, because it is probably easier to reduce such external noise in future experiments, in comparison to reducing intrinsic microscopic sources of decoherence.

The other important general decoherence effect, already mentioned earlier, is the fact that neither charge nor flux qubits are actually two state systems. Even if the two lowest lying energy levels are well below all others, sudden projections or other forms of time-dependent source can excite non-zero amplitudes for the higher states in the full description of the state. As far as the qubit is concerned, these amplitudes are errors and so a source of decoherence. The same is true of unitary quantum leakage to states outside the truncated two state Hilbert space. This effect has been studied in detail for charge qubits [56]. It is possible to optimize the fidelity loss due to leakage in two qubit gates by adjusting the coupling; it is also possible that leakage errors may be corrected through measurement which acts if the system steps outside the qubit space [56]. For single qubit gates it is estimated that, without countermeasures, about 4000 gates might be possible before decoherence due to leakage bites.

The dominant sources of decoherence specific to charge qubits are the effects of external charge or voltage noise [25, 27, 51]. Such fluctuations couple to $\sigma_z$ (in the spin notation) and so cause dephasing in the charge representation. In addition to the noise in the coupled circuits already mentioned, this noise could be due to charged impurities in the fabricated circuit (e.g. in the substrate material). The effects of quasiparticles in the superconducting circuits themselves are estimated to contribute less to the overall charge qubit decoherence [25]. Modelling based on dissipative elements in external circuits connected to charge qubits again suggests around 4000 single qubit operations might be achievable in a decoherence time [27]. However, this still leaves open the question of charged impurities in or around the qubits themselves, which could be a bigger problem.

Some detailed analysis has also been made of decoherence specific to flux qubits [52, 51, 57]. External flux or current noise couples to $\sigma_z$ and so cause dephasing in the flux representation. In addition to the current noise in coupled circuits already mentioned, magnetic impurities in the fabricated circuit (e.g. in the substrate material) could be a factor. It seems unlikely that such impurities will be as damaging as in the charge case. Other possible sources of decoherence for flux qubits are [57]:
(a) quasiparticle effects in the actual superconductor;
(b) magnetic fluctuations due to nuclear spins in the solid;
(c) electro-magnetic radiation by the qubit (generated by time-varying currents during quantum gates);
(d) unavoidable coupling to other flux qubits, i.e. not that deliberately turned on to effect gates.

The discussions of Tian et al. [57] suggest that (d) is potentially the biggest problem. Assuming that desired two qubit couplings are generated through transformers or oscillators, the direct magnetic dipole coupling between qubits is a source of decoherence (unless it could be screened). This leads to an estimate of $10^5$ single qubit operations before this form of decoherence bites.

## 4.4. Measurement

As has already been discussed with reference to the Cooper pair box charge qubit experiments, the charge state of a qubit can be measured using a SET. If the parameter values are suitable, this can form a decent approximation to a projective quantum measurement of charge [58]. Of course it is possible that a quantum computation might require measurement in a different basis. However, this is not a problem because, given that single qubit rotations will have to be pretty routine to effect the computation in the first place, a single qubit rotation can always effectively transform the measurement basis to charge. The same holds for flux qubits. For these, the relevant measurement basis is again $\sigma_z$. In principle, the flux in a qubit can be measured by a SQUID magnetometer [29, 52]. Such measurements will have to be performed if the flux analogues of the charge qubit oscillations [3, 45] are to be seen experimentally.

For both forms of qubit, the actual time scale for the measurement result to appear does not have to be short compared to the typical time scales for gate operations. All that is required is for the qubit evolution to be quickly "frozen", for example by a rapid change in $E_J$. The full measurement process can then be a relatively leisurely affair. This is illustrated in the charge qubit modelling presented in [58], which shows a fast dephasing of the qubit state in the charge basis, but a rather slower development of correlation between the SET current and the qubit charge.

## 5. Conclusions

In summary, it is clear that superconducting devices are a very promising route towards implementation of a scalable quantum computing technology. The field is perhaps less well developed than some alternative approaches, such as NMR or ion traps, but the long term prospects seem good. Because they are solid state devices, superconducting qubits are scalable, and individually addressable. Superconductors are particularly attractive because they uniquely possess macroscopic quantum coherence.

In particular we have seen that both charge and flux based qubit devices have been demonstrated to show quantum coherence. Two qubit devices suitable for implementing quantum logic gates have been designed. At the present time the next major challenge in the field is to demonstrate quantum entanglement in such coupled qubit devices.

In terms of an eventual implementation of quantum computation, the DiVincenzo checklist [22] is a helpful guide. Superconducting devices clearly satisfy the criteria of state preparation and readout, although the flux based qubits still have not achieved time-domain quantum state control similar to that obtained for charge qubits [3]. Scalability and logic gates should not be a problem, although there has still not been an explicit demonstration of entanglement. Probably the most difficult target to achieve among the DiVincenzo requirements is that of coherence times of over $10^4$ gate operations. A number of estimates of theoretical limits imply that this is an achievable target, as we have discussed above. However the current generation of devices are still well below this figure of merit. Devices based upon manipulation of macroscopic quantum states have the advantage that they can be individually addressed, however they will also necessarily be more strongly coupled to their environments than microscopic systems, such as individual ions in a trap. Much more work is required in order to fully understand the dominant sources of decoherence in real systems and to attempt to reduce the effects of decoherence on gate operations. Certainly this will provide many exciting challenges for future research.

## 6. Acknowledgments

We are grateful to R. Scovell, for providing Fig. 1.11. We have benefited from many helpful discussions with colleagues, too numerous to list fully here.

## References

[1] E. Schrödinger, Naturwiss. **23** 807 (1935).
[2] A. J. Leggett, Suppl. Prog. Theo. Phys. **69** 80 (1980).
[3] Y. Nakamura, Yu. A. Pashkin, J. S. Tsai, Nature **398** 786 (1999).
[4] J. R. Friedman *et al.* Nature **406**, 43 (2000).
[5] C. H. van der Wal *et al.*, Science **290**, 773 (2000).
[6] P. W. Shor, p. 124 in *Proceedings of the 35th Annual Symposium on the Foundations of Computer Science*, ed. S. Goldwasser (IEEE Computer Society Press, Los Alamitos, CA, 1994); SIAM J. Computing **26**, 1484 (1997), quant-ph/9508027.
[7] L. K. Grover, p. 212 in *Proceedings, 28th Annual ACM Symposium on the Theory of Computing (STOC)*, (May 1996), quant-ph/9605043.
[8] Y. Makhlin, G. Schön and A. Shnirman, Rev. Mod Phys. **73**, 357 (2001).
[9] A. M. Turing, Proc. Lond. Math. Soc. Ser.2 **442**, 230 (1936); see also J. Church, Am. J. Math. **58**, 435 (1936).

[10] Strictly speaking *floating point* arithmetic is not within a Turing machine's capability. But the Turing machine, just like any modern computer, can implement approximate floating point calculations to any given (fixed) precision desired. In contrast, analogue computers which directly manipulate some real valued physical quantity, rather than bits of information, are not equivalent to Turing machines.

[11] *Statistical Physics II*, R. Kubo, M. Toda and N. Hashitsume, Springer Series in Solid State Sciences 31 (Springer-Verlag, Berlin, 1978).

[12] Malthus realized that the human population had also grown exponentially from pre-history to the present. He argued that the food supply only grew arithmetically, and hence world starvation was inevitable!

[13] R. Feynman, Int. J. Theor. Phys. **21**, 467 (1982).

[14] M. V. Berry, Proc. R. Soc. Lond. **A392**, 45 (1984).

[15] Y. Lecerf, Compt. Rend. **257** 2597 (1963); C. H. Bennett, IBM J. Res. Dev. **17**, 525 (1973); E. Fredkin and T. Toffoli, Int. J. Theor. Phys. **21**, 219, (1982); C. H. Bennett, Int. J. Theor. Phys. **21**, 905 (1982); R. Landauer, Physics Today **44**, 23 (May 1991).

[16] A. Ekert and R. Josza, Rev. Mod. Phys. **68**, 733 (1996).

[17] D. P. DiVincenzo, Phys. Rev. A **51**, 1015 (1995); S. Lloyd, Phys. Rev. Lett. **75**, 346 (1995); A. Barenco, C. H. Bennett, R. Cleve, D. P. DiVincenzo, N. Margolis, T. Sleator, J. Smolin and H. Weinfurter, Phys. Rev. A **52**, 3457 (1995).

[18] It should be noted that in fact it has not been *proven* that factoring is exponentially hard, but to date nobody has found an efficient algorithm. In principle a breakthrough in computational number theory could change this.

[19] D. Deutsch, Proc. R. Soc. London Ser. A **400**, 97 (1985).

[20] It should be noted that Grover's search algorithm [7] does not offer exponential speed up. The quantum algorithm reduces the classical search time by a square root—certainly not as dramatic as exponential speed up, but still potentially very useful in practice.

[21] A comprehensive but by no means exhaustive recent set of papers is Fortschritte der Physik **48**, Number 9-11 (2000).

[22] D. P. DiVincenzo, "Topics in Quantum Computers", in *Mesoscopic Electron Transport*, ed. L. Kowenhove, G. Schön and L. Sohn, NATO ASI Series E, (Kulwer Ac. Publ. Dordrecht, 1997), cond-mat/9612126.

[23] P. W. Shor, Phys. Rev. **A52**, 2493 (1995); A. M. Steane, Phys. Rev. Lett. **77**, 793 (1996); A. M. Steane, Proc. Roy. Soc. Lond. **A452**, 2551 (1996).

[24] P. W. Shor, "Fault-tolerant quantum computation", p. 56 in *Proceedings of the 37th Symposium on the Foundations of Computer Science*, (IEEE Computer Society Press, Los Alamitos, CA, 1996), quant-ph/9605011; A. M. Steane, Phys. Rev. Lett. **78**, 2252 (1997).

[25] A. Shnirman, G. Schön and Z. Hermon, Phys. Rev. Lett. **79**, 2371 (1997).

[26] D. V. Averin, Solid State Comm. **105**, 659 (1998).

[27] Yu. Makhlin, G. Schön and A. Shnirman, Nature **398**, 305 (1999).

[28] M. F. Bocko, A. M. Herr and M. J. Feldman, IEEE Trans. on Appl. Superconductivity **7**, 3638 (1997).

[29] J. E. Mooij, T. P. Orlando, L. Levitov, L. Tian, C. H. Van der Wal and S. Lloyd, Science **285**, 1036 (1999).

[30] G. Burkard G, H. A. Engel and D. Loss, Fortschr. Phys. **48** 965 (2000).

[31] M. N. Leuenberger and D. Loss, Nature **410**, 789 (2001).

[32] B. E. Kane, Nature **393**, 133 (1998).

[33] N. V. Prokof'ev and P. C. E. Stamp, Rep. Prog. Phys. **63** 669 (2000).

[34] J. A. Jones, Prog. Nucl. Mag. Res. Sp **38** (2001).

[35] B. D. Josephson, Phys. Lett. **1**, 251 (1962).

[36] P. W. Anderson, in *Lectures on the Many-body Problem*, Ravello 1963, (E. R. Caianiello, Ed.), Vol. 2 , 113 (Academic, 1964).

[37] A. Barone and G. Paterno, *Physics and Applications of the Josephson Effect*, (Wiley, 1982).

[38] H. Prance, T. D. Clark, R. J. Prance, T. P. Spiller, J. Diggins and J. F. Ralph, Nuc. Phys. (Proc. Suppl.) **33C**, 35 (1993).

[39] S. Han, J. Lapointe and J. E. Lukens, p. 241 in *Activated Barrier Crossing*, Eds. G. R. Fleming and P. Hanggi, (World Scientific, 1993).

[40] R. J. Prance, T. D. Clark, R. Whiteman, J. Diggins, J. F. Ralph, H. Prance, T. P. Spiller, A. Widom and Y. Srivastava, Physica **B203**, 381 (1994).

[41] R. Rouse, S. Han and J. E. Lukens, Phys. Rev. Lett. **75**, 1614 (1995).

[42] S. Han, R. Rouse and J. E. Lukens, Phys. Rev. Lett. **76**, 3404 (1996).

[43] A derivation is given in B. Jankó, "Number parity effects in superconductors", PhD. thesis, Cornell University (1996).

[44] Briefly, excess Cooper pair charge $n$ on a superconducting box is due to charge tunnelling across the Josephson junction $J$. The excess is given by the difference between the capacitor charges, $2ne = Q_J - Q_b$ where $Q_J$ is the charge on the combination of $C_J$ and $C_g$ ($V_g = 0$) and $Q_b$ is the charge on $C_b$. Kirchoff's law gives $V_b = Q_b/C_b + Q_J/(C_J + C_g)$ so the capacitor charges are determined in terms of the quantum variable $n$ and the external source $V_b$. For a given $n$ the total charging energy is given by the sum of the stored energies on the capacitors plus the work done by the voltage source $V_b$ in re-establishing the equilibrium values after each of $n$ tunnelling events, $E_{tot} = Q_b^2/2C_b + Q_J^2/2(C_J + C_g) - 2neC_bV_b/C$. (1980).

[45] Y. Nakamura and J. S. Tsai, J. Low Temp. Phys. **118**, 765 (2000).

[46] A.J. Leggett, S. Chakravarty, A.T. Dorsey, M.P.A. fisher, A. Garg and W. Zwerger, Rev. Mod. Phys. **59**, 1 (1987).

[47] N. Linden and S. Popescu, Phys. Rev. Lett. 87, 047901 (2001).

[48] P.W. Anderson, J. Phys. Chem. Solids **11**, 26 (1959).

[49] G. Falci, R. Fazio, G. Massimo Palma, J. Siewert and V. Vedral, Nature **407**, 355 (2000).

[50] Single qubit gates sandwiched around a conditional phase gate can, for example, turn it into C-NOT. The important issue is the generation of entanglement, which cannot come from (local) single qubit gates. Since any maximally entangled two qubit state can be transformed into any other by local operations, any entangling two qubit gate is universal.

[51] Y. Makhlin, G. Schön and A. Shnirman, J. Low Temp. Phys. **118**, 751 (2000).

[52] T. P. Orlando, J. E. Mooij, L. Tian, C. H. van der Wal, L. Levitov, S. Lloyd and J. J. Mazo, "A Superconducting Persistent Current Qubit", cond-mat/9908283.

[53] S. Lloyd, Science **261**, 1569 (1993).

[54] S. C. Benjamin, Phys. Rev. **A61**, 301 (2000).

[55] R. W. Scovell, B. L. Gyorffy, J. F. Annett and T. P. Spiller, "Quantum states of small superconductors", to appear in *IEE Proceedings: Science, Measurement and Technology*.

[56] R. Fazio, G. Massimo Palma and J. Siewert, Phys. Rev. Lett. **83**, 5385 (1999).

[57] L. Tian, L. S. Levitov, C. H. van der Wal. J. E. Mooij, T. P. Orlando, S. Lloyd, C. J. P. M. Harmans and J. J. Mazo, "Decoherence of the superconducting persistent current qubit", cond-mat/9910062.

[58] A. Shnirman and G. Schön, Phys. Rev. **B50**, 15400 (1998).

# Chapter 6

# GIANT MAGNETORESISTANCE AND LAYERED MAGNETIC STRUCTURES

D.M. Edwards
*Dept of Mathematics, Imperial College, London SW7 2BZ, UK*

## 1. INTRODUCTION

For more than fifty years electronic devices, such as the transistor and quantum well laser, have relied on the electron as a mobile charge in semiconductor heterostructures. However, as well as charge, the electron has a spin with an associated magnetic moment. This has been used, in a static way, for hundreds of years as the source of magnetization in permanent magnets. More recently it has been used, again statically, for storing information in magnetic recording media. During the last decade there has been an amazing development of *spin electronics,* or *spintronics,* in which the electron is exploited as a mobile particle with both charge and spin. This has already led to the introduction of a new magnetoresistive read head in commercial disc drives. In another very recent development, sometimes called *spin engineering,* magnetic materials are designed to control the direction of magnetization for a specific application. An example of this is in the development of possible perpendicular recording media. The natural direction of magnetization in a thin film or flat disc, in order to minimize magnetic field energy, is in the plane of the film. However the direction of magnetization can be reversed more rapidly across the film, with consequent higher data storage density, if the magnetization is perpendicular to the film. This can be achieved in a thin cobalt film by depositing an overlayer of palladium or platinum. The rotation of magnetization from in-plane to out-of-plane occurs due to strong coupling between the spin of an electron and its orbital motion around atoms in a region of reduced symmetry at the interface between two suitably-chosen metals.

## 1.1 Magnetoresistance

Magnetoresistance (MR), a change in electrical resistance produced by an applied magnetic field, is an effect associated with at least four different phenomena:

(i) Normal MR in pure non-magnetic metals

The resistance increases due to deflection of the charged electrons by the magnetic field. The effect is only large at low temperature, when $\omega_c \tau > 1$ so that the electrons execute helical motion. Hence $\omega_c$ is the cyclotion frequency and $\tau$ is the electron relaxation time. Spin is not involved in this normal MR.

(ii) Anisotropic MR (AMR) in ferromagnetic alloys e.g. permalloy $Ni_{0.8}Fe_{0.2}$.

There can be a 4% change in resistance at room temperature and this effect was used as a first choice in read heads until being superceded by GMR. Permalloy is chosen because of its small magnetic anisotropy, so that a weak magnetic field can change the direction of magnetization. This in turn changes the electron states, due to spin-orbit coupling, and hence the matrix elements for scattering which determine the resistivity.

(iii)    Giant MR (GMR) and tunnelling MR (TMR)

These are larger effects than AMR in which the resistance changes due to a field-induced change of magnetic configuration in a magnetic multilayer or tunnel junction. This is true spin electronics in heterostructures.

(iv)    Colossal MR (CMR) in manganites

This is a dramatic, but not potentially useful, effect in which a large field $(\sim 1T)$ drives the system through a metal-insulator transition.

In these lectures we shall discuss GMR and TMR.

## 1.2 Giant magnetoresistance (GMR)

A magnetic multilayer, consisting of alternate magnetic and nonmagnetic layers, is formed by layer-by-layer deposition using molecular beam epitaxy (MBE) or sputtering. Interfaces between layers of different metals can be sharp on the atomic scale. The magnetic layer is a ferromagnetic transition metal, usually iron or cobalt, and the non-magnetic spacer layers are normally a non-magnetic transition metal or noble metal. The magnetic moments of neighbouring magnetic layers are usually aligned either parallel or antiparallel, depending on the thickness of the spacer layers. In fact it is found that the exchange coupling between magnetic layers oscillates as a function of the spacer thickness. It is also found that the resistance of the multilayer structure to a current flowing parallel or perpendicular to the layers depends on the magnetic configuration.

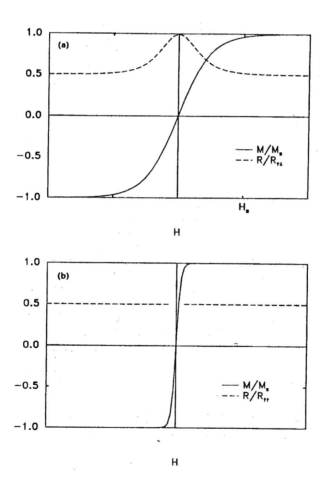

Fig. 1: Schematic plots of reduced resistance $R/R_{\uparrow\downarrow}$ and magnetization $M/M_s$, where $M_s$ is the saturation magnetization, as functions of applied field $H$ for (a) antiferromagnetic coupling (b) ferromagnetic coupling.

It is generally lower in the ferromagnetic configuration, with all magnetic moments parallel, than in the antiferromagnetic one with neighbouring moments antiparallel. An antiferromagnetic configuration can be changed by an applied magnetic field, attaining a ferromagnetic configuration in a saturation field $H_s$. The resistance may be reduced by as much as a factor of 2 and this sensitivity of resistance to applied field (GMR) is exploited in modern magnetoresistive read heads. In practical applications the current flows parallel to the layers (current-in-plane or CIP) since the much lower resistance involved in the current-perpendicular-to-plane (CPP) configuration can only be measured with special techniques.

Fig. 1 shows schematic plots of the resistance $R$ and magnetization $M$, as functions of applied field $H$, for (a) antiferromagnetic coupling (b) ferromagnetic coupling. In case (b) the multilayer saturates in a low field and no significant changes in resistance occur. It is conventional to characterize the magnetoresistance by the quantity

$$\frac{\Delta R}{R} = \frac{R_{\uparrow\downarrow} - R_{\uparrow\uparrow}}{R_{\uparrow\uparrow}} \qquad (1.1)$$

where $R_{\uparrow\downarrow}, R_{\uparrow\uparrow}$ are the resistances in the original antiferromagnetic configuration and the saturated ferromagnetic configuration respectively.

Early work [1,2,3] on Fe/Cr multilayers (superlattices) or Fe/Cr/Fe sandwiches indicated antiferromagnetic coupling between the iron layers. Subsequently Parkin et al [4] reported oscillations in the exchange coupling and CIP magnetoresistance as a function of the non-magnetic spacer layer in Co/Ru, Co/Cr and Fe/Cr sputtered multilayers. Their data for Fe/Cr is reproduced in fig.2 where the peaks in $\Delta R/R$ and saturation field $H_s$ indicate antiferromagnetic coupling and the troughs indicate ferromagnetic coupling. In the ranges of Cr thickness with antiferromagnetic coupling the observed $\Delta R/R$ is to be interpreted according to eq. (1) and $H_s$ is proportional to the exchange coupling. Other methods must be used to measure the strength of the exchange coupling when it is ferromagnetic. In this early work only a long oscillation period of about 18 Å (10-12 monolayers (ML)) was seen. Later work [5,6], on samples with sharper interfaces, found oscillation with a 2ML period superposed on the long period oscillations.

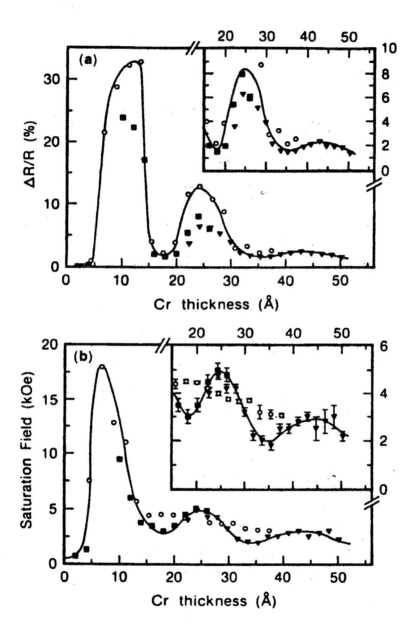

Fig.2: CIP magnetoresistance and saturation field of Fe/Cr multilayers as a function of thickness of the Cr layers [4].

We have shown the data on the Fe/Cr system for its historical importance. It is not typical of magnetic/non-magnetic multilayers since bulk Cr orders as an incommensurate spin density wave below 311K. Cr is also known to be ordered as a commensurate antiferromagnet in Fe/Cr multilayers with quite thin Cr layers [7]. The more typical case of Co/Cu is discussed later.

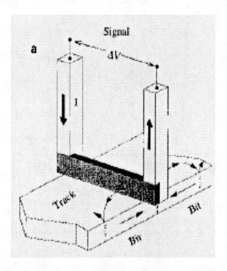

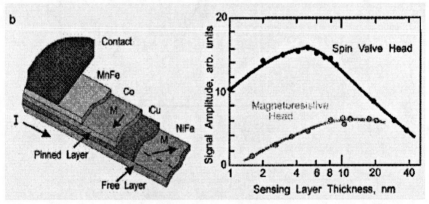

Fig. 3: Schematic diagrams of (a) AMR and (b) GMR read heads with (c) a comparison of their response [8,9].

## 1.3 Magnetic recording heads

To read a magnetic disc it is necessary to sense the magnetic field emanating from the magnetic 'bits' which constitute the information storage. The conventional read head is a small induction coil. Since the induced current depends on rate of change of field, and hence on the local disc velocity, one needs larger magnetic bits near the centre of the disc to obtain a uniform level of response across the disc. This implies non-uniform storage density. This disadvantage is overcome in a magnetoresistive head, using either AMR or GMR, which responds to the magnetic field itself, not its time derivative, through a charge in resistance. Schematic diagrams of AMR and GMR read heads are shown in figs. 3(a) and 3(b) respectively. In fig. 3(a) one sees part of a track on the disc which contains the two magnetic bits with opposite directions of in-plane magnetization. Curved lines indicate the magnetic field which passes through the darkly-shaded permalloy sensor. The GMR head in fig.3(b) is based on a Co/Cu/ permalloy trilayer. The magnetization direction of the Co layer is pinned by strong exchange coupling at its interface with an antiferromagnetic MnFe layer. Since the antiferromagnet has zero net magnetic moment it does not respond to a magnetic field and neither does the magnetization of the adjacent pinned Co layer. However the magnetization of the permalloy layer, with its low magnetic anisotropy responds readily so that fields of less than 1mT can be sensed by the change in resistivity along the length of the layered structure. This type of trilayer, with one of the magnetic layers pinned, is often called a *spin valve*. Fig.3(c) shows a comparison between the responses of a magnetoresistive head based on AMR and a spin valve head based on GMR. The GMR head is clearly superior with optimum response for a structure only 4nm thick. The small dimension is important for reading discs with ever-increasing storage density.

## 1.4 Models of GMR

For simplicity we shall discuss the resistivity of a multilayers in which the magnetization is everywhere collinear i.e. the magnetizations of different magnetic layers are parallel or antiparallel. It is important to discuss the various length scales which are relevant [10]. Electrons are scattered by non-spin-flip and spin-flip processes. Within a given layer the non-spin-flip scattering processes may be characterized by a mean free path $\lambda$ and a corresponding diffusion constant $D \sim \lambda v_F$, where $v_F$ is the Fermi velocity. Spin-flip processes lead to a finite spin diffusion length $\ell_{sf} \sim \left(D\tau_{sf}\right)^{1/2}$, where $\tau_{sf}$ is the relaxation time for non-equilibrium magnetization due to spin-flip. On a length scale less than $\ell_{sf}$ spin is conserved. Normally layer thicknesses are much smaller than $\ell_{sf}$ which, at low temperatures may exceed 1000 Å. Then, in a system with collinear magnetization, current is carried in two independent spin channels. We shall assume this is the case in our following discussion. Clearly, in the opposite limit where layer thicknesses are much larger then $\ell_{sf}$, all spin information is lost in transit between layers and the magnetoresistance $\Delta R/R = 0$.

It was suggested by Baibich et al [3] that spin dependent scattering at the interfaces between layers is responsible for GMR. A detailed theory of the spin dependent interfacial scattering based on the Boltzmann equation was worked out by Camley and Barnas [11], and was later extended to include spin dependent scattering in the bulk of the ferromagnetic layers [12]. As emphasized by Edwards et al [13], this bulk spin dependent scattering can arise from spin-independent disorder in combination with spin-dependent band structure. The mean free paths for majority and minority spin electrons may be very different because, according to Mott [14], the conduction electrons scatter predominantly into the d band with very different densities of final states for the two spins. For example in cobalt the majority spin d band is full whereas the Fermi level lies near a peak in the minority spin density of states. Hence for cobalt majority spin electrons have a much longer mean free path than minority spin electrons. The corresponding local resistivities are inversely proportional to the mean free paths. Recent calculations by Tsymbal and Pettifor [15] show that normally this type of bulk scattering is more important than interfacial scattering.

In these lectures we describe two simplified theoretical approaches to GMR:

(i) Boltzmann equation approach (CIP and CPP) [11-13, 10]

Electrons are considered to move in a conduction band which is common to both components of the multilayer. However the mean free paths in the magnetic and non-magnetic layers are different, and in a magnetic layer they are spin-dependent. In the simplest case $\ell_{sf} \to \infty$ and scattering at interfaces is neglected. This model contains the essence of the physics of GMR and in this introductory section we shall discuss certain limits in which it reduces to a simple resistor model.

(ii) Ballistic approach (CPP) [16-19]

In the ballistic limit, $\lambda_{sf} \to \infty$ and all mean free paths become infinite. In this case, for CPP, the Landauer formula relates conductance to the transmission coefficient across the sample. In section 3 we show how this may be calculated quantum-mechanically using realistic band structure for the multilayer components. Unfortunately it is difficult to realize the ballistic limit experimentally.

## 1.5 The resistor model for GMR

The Boltzmann equation approach, in the limit of long spin diffusion length $\ell_{sf}$, reduces to a simple resistor model in certain limiting cases. In the appropriate limits we can derive the resistor model simply without recourse to the Boltzmann equation. The limiting case results, whose derivation is sketched below, are as follows:

(i) mean free paths >> layer thicknesses

In this limit $\Delta R / R$ is given by the same resistor model for CIP and CPP.

(ii) mean free paths << layer thicknesses

$(\Delta R / R)_{CIP} = 0$

$(\Delta R / R)_{CPP} \neq 0$, and given by the resistor model.

The first result in (ii) is rather obvious since an electron rarely samples more than one magnetic layer as it drifts parallel to the layers. From the above limiting cases we might expect that in general $(\Delta R / R)_{CPP} > (\Delta R / R)_{CIP}$ and this is normally found to be true.

Since $\ell_{sf}$ is assumed to be large we can consider two independent spin channels in parallel. Suppose in a general layered structure the i$^{th}$ layer is of thickness $t_i$ and that the resistivity in a given spin channel is $\rho_i$. In the limit (i), whether in the CIP or CPP configuration, electrons traverse many different layers between scattering events. Hence the average scattering rate is the average of the scattering rates within layers and the same is true for the resistivity $\rho$.

Thus, in both CIP and CPP,

$$\rho = \sum_i t_i \rho_i \bigg/ \sum_i t_i .$$

The resistance in the given spin channel is proportional to $\rho$, for fixed dimensions of the multilayer, so that we can take the resistance as $\sum t_i \rho_i$, omitting a constant factor which cancels in $\Delta R / R$. Thus in the limit (i), for both CIP and CPP, we may consider the resistances of the layers as acting in series for a given spin channel. The $\uparrow$ and $\downarrow$ spin channels act in parallel as discussed above. Clearly in the limit (ii), in the CPP geometry, the layers act as standard resistances in series for each spin channel. Hence, as pointed out in ref [13], we can derive a single formula for $\Delta R / R$ which applies to CIP and CPP in limit (i) and to CPP in limit (ii). This idea has been used extensively by the Michigan group [20] to interpret their CPP measurements.

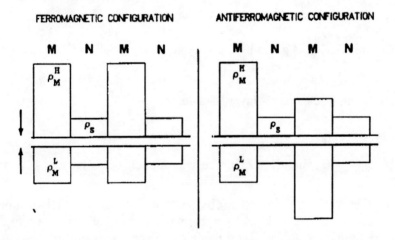

Fig. 4: Schematic representation of local resistivities in four adjacent layers of a multilayer: $\uparrow$ and $\downarrow$ spin channels are shown in the ferromagnetic and antiferromagnetic configuration.

To derive the formula for $\Delta R / R$ we consider the schematic representation of local resistivities shown in fig.4. This shows a superlattice cell of two magnetic (M) and two non-magnetic layers (N) with corresponding resistivities in the spin up ($\uparrow$) and down ($\downarrow$) channels for the ferromagnetic and antiferromagnetic configuration of magnetic layers. Here $\rho_s$ is the resistivity of the N layer, which is the same for both spin orientations, and $\rho_M^H, \rho_M^L$ are the high and low resistivities for the two different spin channels in the M layer. The corresponding effective resistor networks are shown in fig.5, where $t_M$ and $t_N$ are the thicknesses of M and N layers, respectively. It is straightforward to

# Giant Magnetoresistance and Layered Magnetic Structures

calculate the resistances $R_{\uparrow\uparrow}, R_{\uparrow\downarrow}$ of the networks representing the ferromagnetic and antiferromagnetic configuration. Hence, from eq.(1.1), we find

$$\frac{\Delta R}{R} = \beta^2 \frac{(t_M \rho_f)^2}{(t_M \rho_f + t_N \rho_s)^2 - (t_M \rho_f \beta)^2} \quad (1.2)$$

where we have written

$$\rho_M^H = \rho_f (1+\beta), \rho_M^L = \rho_f (1-\beta). \quad (1.3)$$

Clearly to make $\Delta R/R$ large we need $\beta$ to be as close to 1 as possible and $t_N$ should be small. Thus we need a big difference between the resistivity of majority and minority spin channels in the ferromagnet, as in Co. A lower limit on the spacer thickness $t_N$ is set in practice by the requirement of a fairly small antiferromagnetic exchange coupling in order to have a good magnetoresistive response in small fields.

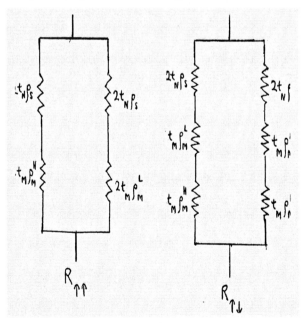

Fig.5: Effective resistor networks for calculating the resistances $R_{\uparrow\uparrow}$ and $R_{\uparrow\downarrow}$ of the ferromagnetic and antiferromagnetic configuration of a multilayer.

## 2. QUANTUM WELL STATES AND EXCHANGE COUPLING

In section 1.2 we noted the experimental observation [4] of oscillations in the exchange coupling between magnetic layers as a function of spacer thickness. Edwards et al [21] showed that these oscillations arise from quantum well (QW) states in the spacer passing through the Fermi level as the spacer thickness varies. Such QW states in a Cu overlayer on Co were observed directly in photoemission and inverse photoemission experiments by Ortega et al [22]. Spin polarisation of these QW states was observed by Garrison et al [23]. In section 2.1 we give an elementary introduction to QW states and compare with the experimental data. Then in section 2.2 we discuss two methods of calculating exchange coupling between magnetic layers and show how oscillatory exchange coupling is related to QW states. Finally in this chapter we describe a quantitative calculation of exchange coupling in a Co/Cu/Co (001) trilayer [24,25] and compare with experiment.

### 2.1 QW states in a trilayer

In a simple free-electron model of an A/B/A trilayer the potential energy $V(z)$ is sketched in

fig. 6(a), where $z$ is the electron coordinate perpendicular to the layers. The thickness of the B layer is $\ell$ and the depth of the well is $V$. Electron states $\psi$ satisfy the Schrödinger equation

$$\left(-\hbar^2/2m\right)\nabla^2\psi + V(z)\psi = E\psi \tag{2.1}$$

and, owing to in-plane translational symmetry, are classified by a Bloch wave-vector $k_{11} = (k_x, k_y, 0)$. Thus

$$\psi = e^{ik_{11}\cdot r}u(z) \tag{2.2}$$

where the electron position $r = (x,y,z)$. On substituting eq. (2.2) in (2.1) we find that $u(z)$ satisfies the one-dimensional (1D) Schrödinger equation

$$Hu \equiv -\frac{\hbar^2}{2m}\frac{d^2u}{dz^2} + V(z)u = E'u \tag{2.3}$$

with $E' = E - \hbar^2 k_{||}^2 /(2m)$. For a given $k_{||}$ we take the zero of energy to be such that $E' = 0$ corresponds to the bottom of the well. Thus if $E_F$ is the actual Fermi energy, measured from the bottom of the well, electrons fill states in the 1D well up to $E'_F = E_f - \hbar^2 k_{||}^2 /(2m)$. If $E'_F \leq 0$ there are no occupied states with the given $k_{||}$. Fig 6 (b) shows the potential V(z) for the system A/B/VAC, where an overlayer B of thickness $\ell$ separates the semi-infinite layer A from the vacuum. This is the potential corresponding to angle-resolved photoemission from an overlayer in which electrons are emitted from states of definite $k_{||}$ e.g. $k_{||} = 0$. The measurement probes the density of states $\rho(E,\ell)$ of the 1D well. For the present simple discussion it is sufficient to consider the symmetric well of fig. 6(a).

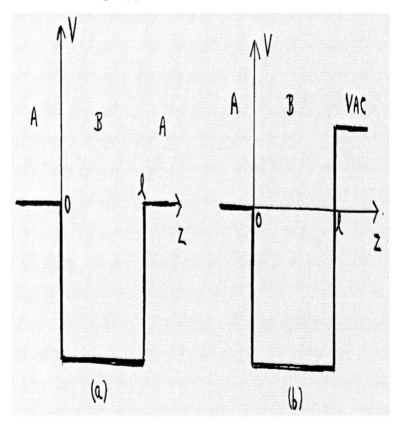

Fig. 6: Potential energy $V(z)$ of a an electron in (a) on A/B/A trilayer (b) an A/B/VAC system corresponding to a B overlayer on an A substrate.

In general the eigenstates of the 1D well consist of a number of discrete bound states with $E' < V$ and a continuum of states with $E' > V$. As $k_{11}$ is varied we may have a change in the nature of states near the effective Fermi level from continuum for $k_{11} = 0$ to bound states for some finite $k_{11}$. The density of states in the region of the well is given by

$$\rho(E',\ell) = \int_0^\ell dz \sum_i |u_i(z)|^2 \delta(E' - E_i) \qquad (2.4)$$

$$= -\pi^{-1} \operatorname{Im} \int_0^\ell G(z,z,E'+) dz \qquad (2.5)$$

Here $u_i(z), E_i$ are the eigenfunctions and eigenvalues of the Hamiltonian H in eq. (2.3) and $G(z,z',E+)$ is the one-particle Green function $\langle z|(E'-H)^{-1}|z'\rangle$. $E'+$ signifies that $E'$ has an infinitesimal positive imaginary part. The Green function is straightforward to calculate and we find that, in the continuum region $E' > V$,

$$\pi V \rho(E',\ell) = d(2\varepsilon - 1)(\varepsilon - 1)^{\frac{1}{2}} \Big/ \Big[ \sin^2(d\sqrt{\varepsilon}) + 4\varepsilon(\varepsilon - 1) \Big] \qquad (2.6)$$

where $\varepsilon = E'/V, d = \ell(2mV/\hbar^2)^{1/2}$ are dimensionless energy and well-width variables. Now

$$d\sqrt{\varepsilon} = \ell(2E'm/\hbar^2)^{1/2} = k_z \ell \qquad (2.7)$$

where $E = (\hbar^2/2m)(k_{11}^2 + k_z^2)$. Thus $k_z$ is the $k_z$ coordinate of a point on the constant energy surface, energy E, with a given $k_{11}$. Clearly, from eq.(2.6), $\ell^{-1}\rho(E',\ell)$ is a periodic function of $\ell$ with period $\pi/k_z$. In the later development this periodicity property of the density of states turns out to be very useful, and even for non-spherical energy surfaces the period is found to be determined by the same dimension $k_z$ of the constant energy surface.

## Giant Magnetoresistance and Layered Magnetic Structures 227

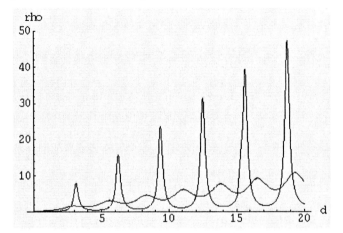

Fig. 7: Density of states in the well region (right-hand side of eq. (2.6)), as a function of dimensionless well-width $d$, for reduced energies $\varepsilon = 1.01$ (the peaky curve) and $\varepsilon = 1.3$ (the less peaky curve).

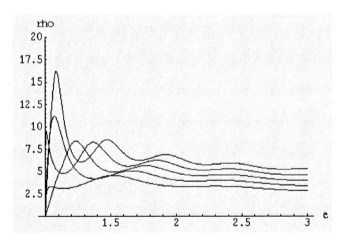

Fig. 8: Same density of states as in fig. 7, now as a function of reduced energy $\varepsilon$ for d=10,12,14,16,18. (The density of states is a monotonically increasing function of d at large energy).

In fig. 7 the right-hand side of eq. (2.6) is plotted as a function of $d$, the dimensionless width of the well, for two energies $\varepsilon = 1.01$ and 1.3. The first case corresponds to an energy $E'$ just above the top of the well and sharp resonances, corresponding to nearly bound states, pass through this energy as $d$ is varied. For $\varepsilon = 1.3$ the resonances are much broader. For $E'$ below the top of the well $(\varepsilon < 1)$ the resonances would become δ-functions, corresponding to bound states. In fig.8 the right-hand side of eq. (2.6) is plotted as a function of $\varepsilon(>1)$ for various values of $d$. This shows how the resonances seen in the density of states shift in energy with changing well-width.

Figs. 7 and 8 may be compared qualitatively with the spectral density observed [22] in angle-resolved photoemission and inverse photoemission with $k_{11} = 0$ from Cu overlayers on Co(001) or Fe(001). The results of Ortega al [22] shown in fig.9, corresponding to electrons emitted from the Fermi level, are similar to the plots in fig.7. This shows that quantum well resonances exist in the Cu overlayer and the relation of the oscillation period to the Cu Fermi surface is discussed later. The simple linear increase in amplitude is not observed since not all the photo-emitted electrons originate in the overlayer, and in any case emission is not uniform throughout the overlayer. The spectra shown in fig.10 may be compared with fig. 8 with a similar upward shift in resonances as the Cu thickness increases.

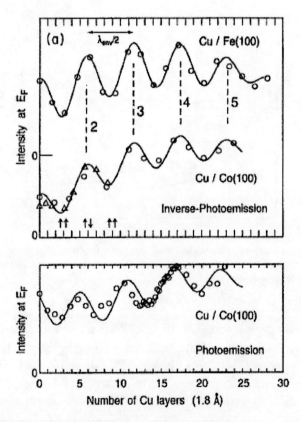

Fig. 9:   Quantum well resonances in Cu overlayers on Co(001) observed by Ortega et al [22] (cf. simple theoretical picture of fig.7).

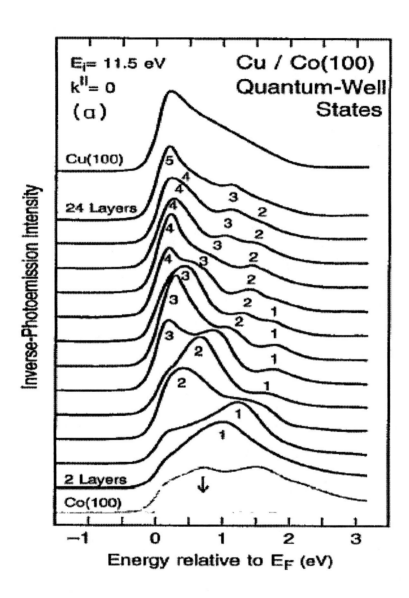

Fig.10: Photoemission spectra for Cu overlayers of different thickness on Co(001) [22] (cf. simple theoretical picture of fig. 8).

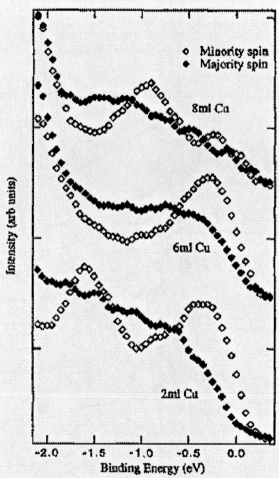

Fig.11: Spin-resolved photoemission spectra [23] for Cu overlayers on Co(001) showing distinct QW resonances for minority spin only.

Electrons in a ferromagnet such as Co and Fe have a spin-dependent potential energy, the difference in energy between that for minority and majority spin corresponding to an exchange splitting of 1 or 2 eV. Consequently the QW in a non-magnetic overlayer has a different depth for the two spin orientations. The effect of this is seen in the spin-resolved photoemission results of Garrison et al [23] shown in fig.11 where QW resonances are seen clearly only for minority spins. In general oscillations in the spectral density of large amplitude are associated with QW states strongly confined in the nonmagnetic layer.

## 2.2 Relation of QW states to exchange coupling

There are two starting-points for a quantitative calculation of exchange coupling. The first proceeds directly from the definition of exchange coupling between two semi-infinite magnetic layers across a non-magnetic spacer layer of thickness $\ell = Nd$. Here $d$ is the interplane spacing and $N$ is the number of atomic planes in the spacer. Thus we define the exchange coupling

$$J(\ell) = \Omega_{\uparrow\uparrow} - \Omega_{\uparrow\downarrow}, \tag{2.8}$$

where $\Omega_{\uparrow\uparrow}$, $\Omega_{\uparrow\downarrow}$ are thermodynamic potentials of the trilayer, per unit cross-sectional area, for the parallel and antiparallel configuration of the magnetic moments of the magnetic layers. The first method of calculating $J(\ell)$, which we use in this section, is to calculate this energy difference directly.

A second method, which we discuss later, is the torque method in which we consider the non-equilibrium situation where the magnetic moments of the two magnetic layers make an arbitrary angle $\theta$ with each other. If the corresponding thermodynamic potential is $\Omega(\theta)$ we define

$$J(\ell, \theta) = \Omega(0) - \Omega(\theta) \tag{2.9}$$

$$= \int_0^\theta (-d\Omega/d\theta) d\theta. \tag{2.10}$$

The integrand $-d\Omega/d\theta$ is the torque exerted on one magnetic moment by the other and is related to the spin current flowing across the spacer in the non-equilibrium configuration considered. We discuss this in section 3. Clearly, from eqs. (2.8) and (2.9), $J(\ell) = J(\ell, \pi)$ so that $J(\ell)$ can be calculated by integrating torque instead of taking energy difference. Both methods have been used and have been shown to be equivalent, at least in in a one-band model [35, 25].

To evaluate eq. (28) for $J(\ell)$ we make two approximations. The first is the so-called force theorem [24], which states that a good approximation to a total energy difference between different structures (in this case magnetic) is obtained by comparing sums of one-electron energies, using atomic potentials which are independent of the magnetic configurations. The second approximation is to neglect the dependence of the local densities of states within the magnetic layers on the magnetic configuration. The second approximation is to neglect the dependence of the local densities of states within the magnetic layers on the magnetic configuration. This approximation can be checked and is found to be a good one, at least for spacer thicknesses which are not too small. Thus, at temperature T=0,

$$J(\ell) = \sum_{k_{11}} \int^{\mu} (E-\mu)\{[\rho_\uparrow(k_{11},E,\ell) + \rho_\downarrow(k_{11},E,\ell)]_{FM} \quad (2.11)$$
$$-[\rho_\uparrow(k_{11},E,\ell) + \rho_\downarrow(k_{11},E,\ell)]_{AFM}\}dE$$

where $\mu$ is the chemical potential. Here $\rho_\sigma(k_{11},E,\ell)$ is the density of states within the spacer for given wave-vector $k_{11}$ and spin $\sigma$; these densities of states are calculated for the ferromagnetic (FM) and antiferromagnetic (AFM) configurations as indicated. At finite temperature $T$ the factor $E - \mu$ is replaced by

$$-k_B T \ln\{1 + \exp([\mu - E]/k_B T)\}. \quad (2.12)$$

Now, as pointed out in section 2.1, $\ell^{-1}\rho_\sigma(k_{11},E,\ell)$ is a periodic function of $\ell$ with period $\pi/k_z$ where $k_z = k_z(k_{11},E)$ defines the bulk spacer constant energy surface with energy E. This periodicity property has been proved rigorously by Umerski [27]. Thus we may introduce the Fourier series

$$\ell^{-1}\rho_\sigma(k_{11},E,\ell) = \sum_n c_{n\sigma}(k_{11},E)e^{2in\ell k_z(k_{11},E)} \quad (2.13)$$

and substitute into eq.(2.11). This is very convenient because for large $\ell$ we may use the stationary phase approximation to evaluate the integrals over $k_{11}$ and $E$ in eq. (2.11). The dominant contribution arises from $E \sim \mu$ and from points $k_{11} = k_{11}^0$ where the function $k_z(\mu,k_{11})$ Is stationary. Thus oscillations in $\rho_\sigma(k_{11}^0,\mu,\ell)$, which arise from QW states passing through the Fermi level μ as spacer thickness varies, give rise to corresponding oscillations in the exchange coupling $J(\ell)$. Oscillations of large amplitude are associated with strongly-confined QW states, as pointed out in section 2.1, so that one of the terms in eq. (2.11), for example $\left[\rho_\downarrow\right]_{FM}$ in the case of Co/Cu/Co(001), may be the most significant. We discuss the Co/Cu/Co example further in the next section.

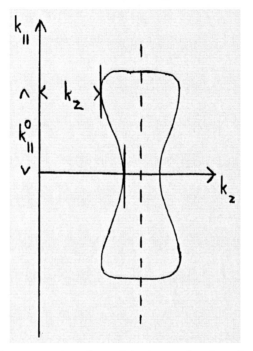

Fig.12: Sketch of a cross-section of part of the Cu Fermi surface, showing two extremal values of $k_z$.

### 2.3 Co/Cu/Co (001) trilayers

Fig.12 shows a sketch of a cross-section of part of the Cu Fermi surface. $k_z$ is in the (001) direction and $k_\parallel$ is the (110) direction. The dashed line is the zone boundary $k_z = \pi/d$ The two stationary points of $k_z(\mu, k_\parallel)$ are indicated. One occurs at $k_\parallel = 0$ and the other at a finite value $k_\parallel^0$. The corresponding oscillation periods are influenced by the discrete nature of the lattice, which we have not considered before. Thus $\ell = Nd$ where $N$ is an integer, so that

$$\exp(2in\ell k_z) = \exp(2in\ell[k_z - \pi/d]). \qquad (2.14)$$

When $\pi/d - k_z < k_z$ the observed period, considering the function of $\ell$ only at discrete points $Nd$, is $\pi/(\pi/d - k_z)$ rather than $\pi/k_z$. The stationary point at $k_{||} = 0$ therefore corresponds to a long period, in fact 5.9 ML, which is the one seen in fig.9. The other stationary point corresponds to a shorter period of 2.6ML (~4.7Å) and contributions with both periods should occur in the exchange coupling. Recently the shorter period has been observed [28] in angle-resolved photoemission from Cu overlayers on Co(001) with $k_{||} \sim k_{||}^0$.

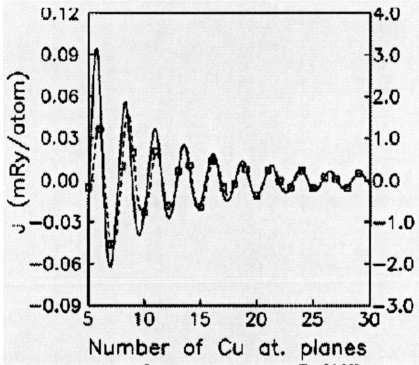

Fig.13: Exchange coupling $J$ in a Co/Cu/Co(001) trilayer at $T = 316K$ calculated as a function of Cu thickness fully numerically (dashed line) and using the stationary phase approximation (full line). [25].

We calculated [24, 25] the exchange coupling in Co/Cu/Co (001) trilayers using a fully numerical method, based on the torque (or "cleavage") method, as well as the semi-analytic stationary phase approximation described above. Both methods use a tight-binding model, with 9 orbitals (3d, 4s, 4p) and parameters fitted to first-principles calculations for the bulk metals. Given this model, the fully numerical method relies further only on the force theorem; the stationary phase method should give the same results for large $\ell$. Here we just discuss the results since full details are given in ref [25] and the torque method is discussed again in section 3.

In fig.13 we show a comparison between the numerical results for exchange coupling J and the stationary phase approximation. As expected the latter is very accurate when the number $N$ of Cu planes is large. Clearly the oscillatory exchange coupling is dominated by the short period contribution. The stationary phase approximation, which has the advantage of being able to separate the two contributions, show s that the long period contribution is less than 1% of the total. The reason for this is easy to understand in terms of confinement of QW states.

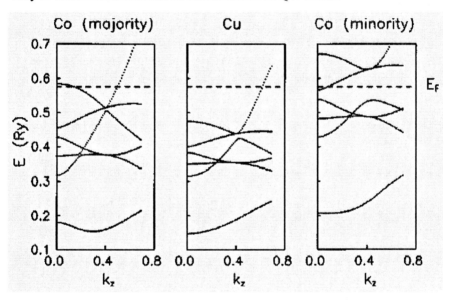

Fig.14: Energy bands for Cu and ferromagnetic FCC Co along a line in the $k_z$ direction with $k_{\parallel} = k_{\parallel}^0$ corresponding to a short-period Fermi surface extremum (see fig.12) [25].

Fig.14 shows the energy bands for bulk Co (majority and minority spin), and for Cu, along a line in $k$ space in the $k_z$ direction with $k_{\parallel} = k_{\parallel}^0$ corresponding to a short-period Fermi surface extremum (see fig.12). There is a good match between the sp-like bands for Cu and Co majority spin where they cross the Fermi level $\mu = E_F$. However for Co minority spin the Fermi level falls in an almost complete hybridization gap. Thus in the ferromagnetic (FM) configuration of the Co/Cu/Co trilayer minority spin electrons at the Fermi level are strongly confined in the Cu layer, which therefore acts like a deep QW in the simple free electron picture we used in section 2.1. However the majority spin electrons are hardly confined at all and will make only a small contribution to QW oscillations in photoemission and exchange coupling. In the AFM configuration electrons of both spin are free to escape into one of the Co layers and therefore contribute weakly to QW oscillations. Thus the short-period oscillations of fig.13 arise largely from the minority spin spectral density $T$ in eq.(2.11). The situation is quite different for $k_{||} = 0$, corresponding to the long-period Fermi surface extremum. The sp-like band crosses $E_F$ at very similar $k_z$ values in both spin bands of Co and in Cu, as shown in fig.15. Thus there are no strongly-confined QW states of either spin for $k_{||} = 0$. Consequently, the long-period oscillations in $J$ are much weaker than the short-period ones, as found in the calculations. Nevertheless the long-period oscillations can be seen in photoemission, at least for minority spin electrons, as described earlier.

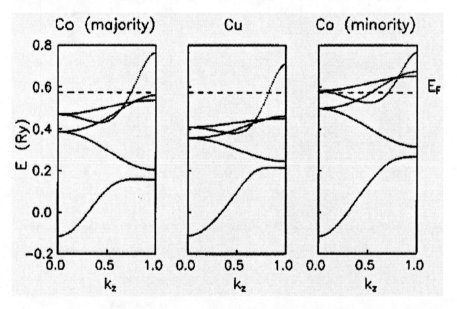

Fig.15: Energy bands as in fig.14 but with $k_{||} = 0$ corresponding to the long-period Fermi surface extremum [25].

The experimental situation is complicated by interfacial roughness. For an oscillation period close to 2ML the sign of the exchange coupling is changed by adding a spacer atomic layer. Thus the presence of steps will tend to give a zero exchange coupling averaged over the sample. The short-period oscillation in Cu is clearly much more susceptible to this effect than the long period one. In fact roughness has the effect of mixing states with different $k_{11}$, so that the very good band-matching found for $k_{11} = 0$ may be destroyed, with a consequent increase in QW confinement. Johnson et al [29] find in their experiments that the amplitudes of the short-and long-period oscillations are comparable. Their value of $J$ at the first antiferromagnetic peak (0.4mJ/m$^2$) is about 3 times smaller than we find, which may well be attributed to roughness. Weber et al [30], using a scanning electron microscope with polarization analysis (SEMPA), estimate that in their 'best' samples the amplitude of the short-period oscillation is at least a factor 7 larger than that of the long-period oscillation. This is more in accord with our predictions. Unfortunately SEMPA does not give the absolute values of the exchange coupling.

## 3. BALLISTIC CPP MAGNETORESISTANCE AND EXCHANGE COUPLING

In this section we present a unified treatment of charge and spin currents in the CPP configuration of a trilayer. Charge current is involved in giant magnetoresistance (GMR) and tunnelling magnetoresistance (TMR) and spin current is involved in exchange coupling. The only difference between CPP GMR and TMR is that in GMR the spacer between the two ferromagnets is a metal and in TMR it is an insulator. The insulator most commonly used is amorphous alumina (Al$_2$O$_3$)and systems like Co/Al$_2$O$_3$/Co are being considered for applications such as magnetic random access memory (MRAM) and the next generation of read heads. The large tunnelling resistance makes TMR a much more practical proposition than CPP GMR. The theoretical problem is to calculate the current across the spacer in the non-equilibrium situation when a voltage difference is applied between the two ferromagnets. The ferromagnets will usually be considered to have their magnetization parallel or antiparallel.

Another non-equilibrium situation exists even with zero applied voltage, when the magnetic moments $M_1, M_2$ of the left and right hand ferromagnet make an angle $\theta(\neq 0, \pi)$ with each other. Owing to exchange coupling the left hand (LH) ferromagnet exerts a torque on the spins of the right hand (RH) ferromagnet which causes them to precess. Their total rate of change of angular momentum is equal to the torque $-d\Omega/d\theta$, using the notation of section 2.2. Hence if $M_1 = (0, 0, M_1)$ is in the z direction and $M_2 = M_2(\sin\theta, 0, \cos\theta)$,

$$-d\Omega/d\theta = \hbar \left\langle \sum_i \dot{S}_{iy} \right\rangle \tag{3.1}$$

where the right hand side is the rate of change of the y component of spin, summed over all the spins in the RH ferromagnet. By continuity this is equal to the rate of flow of $y$ component of spin across the spacer (spin current). Hence

$$-d\Omega/d\theta = \hbar I_{sy} \tag{3.2}$$

where $I_s = (I_{sx}, I_{sy}, I_{sz})$ is the spin current. The exchange coupling between the ferromagnets can be calculated from the spin current using eq. (2.10).

## 3.1 Current and spin current in a tight-binding model

The Hamitonian of the tight-binding model can be written as

$$H = \sum_{k_{11}\sigma} \sum_{m,n} T_{mn}(k_{11}) c^\dagger_{k_{11}m\sigma} c_{k_{11}n\sigma} + H_{int} \tag{3.3}$$

where $c^\dagger_{k_{11}m\sigma}$ creates an electron in a Bloch state, wave-vector $k_{11}$, formed from a given atomic orbital in a given plane. $m$ and $n$ include plane and orbital indices. $H_{int}$ is an effective Coulomb interaction between electrons which we neglect in the spacer but which is necessary to give the exchange splitting in the ferromagnets, within the Hartree-Fock approximation. It commutes with spin and charge density operators and therefore does not contribute to current and spin current, these being related to the time derivatives of charge and spin density by equations of continuity. The current flowing across the spacer can be evaluated by considering the current flowing between two adjacent spacer planes L and R. It is given by (eg. [18])

$$I = \frac{ie}{\hbar} \sum_{k_{11}\sigma} \sum_{m \in L} \sum_{n \in R} \left\langle T_{mn}(k_{11}) c^\dagger_{k_{11}m\sigma} c_{k_{11}n\sigma} - T_{nm}(k_{11}) c^\dagger_{k_{11}n\sigma} c_{k_{11}m\sigma} \right\rangle. \tag{3.4}$$

Instead of evaluating the current using the Kubo formula, which involves a two-particle Green function as in ref. [18], we use here the Keldysh formalism [31,32]. This involves a generalisation of the one-particle Green function to deal with non-equilibrium systems. We can then consider a finite applied voltage V, whereas the Kubo formula gives only the linear response for infinitesimal V. The thermal average $\langle ... \rangle$

does not refer to an equilibrium state but to a steady non-equilibrium state. We consider an initial state at time $t = -\infty$ in which the hopping integrals $T_{mn}$ Between planes L and R are switched off. Then both sides of the system are in equilibrium but with different chemical potentials, $\mu + eV$ on the left and $\mu$ on the right. The hopping is then turned on slowly and the system evolves to a steady state.

Following Keldysh [31] we define a matrix

$$G_{RL}^+(t,t') = i \langle c_L^\dagger(t') c_R(t) \rangle \tag{3.5}$$

where $R \equiv (n\sigma')$ and $L \equiv (m\sigma)$ and we suppress the $k_{\parallel}$ label. Then eq. (3.4) becomes

$$I = \frac{e}{\hbar} \sum_{k_{\parallel}} Tr\left\{ \left[ G_{LR}^+(t,t) - G_{RL}^+(t,t) \right] T \right\} \tag{3.6}$$

where the trace is carried out over orbital and spin indices. The matrices $G^+$ and $T$ have dimensions 2m x 2m, where m is the number of orbitals on each atomic site, and are written so that the m x m upper diagonal block contains matrix elements between ↑ spin orbitals and the m x m lower diagonal block relates to ↓ spin. Only these diagonal blocks are non-zero in the hopping matrix $T$ and this is also true of the $G^+$ matrices if the ferromagnets have collinear moments. Introducing the Fourier transform $G_{LR}^+(\omega)$ of $G_{LR}^+(t,t')$, which is a function of $t-t'$, we have

$$I = \frac{e}{\hbar} \sum_{k_{\parallel}} \int \frac{d\omega}{2\pi} Tr\left\{ \left[ G_{LR}^+(\omega) - G_{RL}^+(\omega) \right] T \right\}. \tag{3.7}$$

Similarly the spin angular momentum current (torque) is given by

$$\hbar \mathbf{I}_s = \frac{1}{2} \sum_{k_{\parallel}} \int \frac{d\omega}{2\pi} Tr\left\{ \left[ G_{LR}^+(\omega) - G_{RL}^+(\omega) \right] T \boldsymbol{\sigma} \right\}. \tag{3.8}$$

Here the 2m x 2m matrix $\boldsymbol{\sigma}$ is the direct product of the 2 x 2 Pauli matrices $(\sigma_x, \sigma_y, \sigma_z)$ and the m x m unit matrix. Now

$$G_{LR}^+(\omega) = F_{LR} + G_{LR}^a - G_{LR}^r \tag{3.9}$$

where $F_{LR}(\omega)$ is the Fourier transform of

$$F_{LR}(t,t') = -i \langle \left[ c_L(t), c_R^\dagger(t') \right]_- \rangle \tag{3.10}$$

and $G^a, G^r$ are the usual advanced and retarded Green functions, eg [33] (note that the definitions of $G^a$ and $G^r$ are interchanged in Keldysh's paper [31]).

Charge current $I$ and spin current $I_s$ are related by eqs (3.7) - (3.9) to the quantities $G_{LR}$ and $F_{LR}$. The latter are calculated for the coupled system by starting with decoupled left and right systems (each in equilibrium) and turning on the hopping between planes L, R as a perturbation. Hence we express $G_{LR}$ and $F_{LR}$ in terms of surface Green functions $g_{LL}, g_{RR}$ for the decoupled equilibrium systems. As an example we consider $G_{LR}$, which may be $G_{LR}^r$ or $G_{LR}^a$ (depending on the sign of the infinitesimal imaginary part of $\omega$), in a one-electron approximation where $H_{int}$ is treated in the Hartree-Fock approximation to produce exchange-split on-site potentials in the ferromagnets. The matrix G is given by

$$G = (\omega - H)^{-1} = (\omega - H_0 - T)^{-1}$$
$$= (\omega - H_0)^{-1} + (\omega - H_0)^{-1} T (\omega - H)^{-1}$$
(3.11)

where $H_0$ is the one-electron Hamiltonian for the decoupled system and $T$ provides the hopping between planes L and R, as before. Eq. (3.11) takes the form of the Dyson equation

$$G = g + gTG,$$ (3.12)

where $g = (\omega - H_0)^{-1}$ is the Green function for the decoupled system. The only non-zero matrix elements of $T$ are those between L and R, $(T_{LR} = T, T_{RL} = T^\dagger)$ and clearly $g_{LR} = g_{RL} = 0$. Hence from eq. (3.12)

$$G_{LR} = g_{LL} T G_{RR}$$ (3.13)

and

$$G_{RR} = g_{RR} + g_{RR} T^\dagger G_{LR}.$$ (3.14)

Hence we find

$$G_{LR} = \left(1 - g_{LL} T g_{RR} T^\dagger\right)^{-1} g_{LL} T g_{RR}. \tag{3.15}$$

Eqs. (3.7) and (3.9), together with eq. (3.15) and a similar equation for $F_{LR}$, enable us to write down general formulae for $I$ and $I_s$. It is first instructive to give results for the special case of a one-band model. For collinear ferromagnets at $T = 0$, writing $I = \sum I_\sigma = I_\uparrow + I_\downarrow$, we find

$$I_\sigma = \frac{4e}{h} \sum_{k_{11}} T(k_{11})^2 \int_\mu^{\mu+eV} \frac{\operatorname{Im} g_{L\sigma}(\omega) \operatorname{Im} g_{R\sigma}(\omega)}{\left|1 - T(k_{11})^2 g_{L\sigma}(\omega) g_{R\sigma}(\omega)\right|^2} d\omega \tag{3.16}$$

where $g_{L\sigma} = g^r_{LL,k_{11}\sigma}$ etc.

We note that, as $V \to 0$, the conductance for spin $\sigma$, given by $I_\sigma/V$, is proportional to a product of the left-hand and right-hand densities of states at the Fermi level, but with an important denominator. We shall return to this point later. An alternative derivation of this conductance, using the Kubo formula, may be found in ref [34].

For the spin current, or torque, we consider the magnetic moments $M_1, M_2$ of the two ferromagnets to make an angle $\theta$ with each other and quantize spin initially along the $M_1$ direction. However we rewrite the local Green functions $g_R$ for the coupled right-hand side of the trilayer in terms of $g_{R\uparrow}, g_{R\downarrow}$ with spin quantized along the $M_2$ direction. Thus in both $g_L$ and $g_R$ spin directions refer to the local magnetization direction of the decoupled system. Then from eqs (3.2), (3.8) and (3.9), together with eq. (3.15) and a similar equation for $F_{LR}$, we find that for applied voltage $V = 0$ the torque exerted by the LH ferromagnet on the RH ferromagnet is given by

$$-\frac{d\Omega}{d\theta} = \frac{1}{\pi} \sum_{k_{11}} T \int d\omega f(\omega) \operatorname{Im} P(\omega, \theta), \tag{3.17}$$

where

$$P(\omega,\theta) = \frac{1}{2}T^2 \sin\theta \left(g_{L\uparrow} - g_{L\downarrow}\right)\left(g_{R\uparrow} - g_{R\downarrow}\right)$$

$$\times \left\{ 1 - \frac{1}{2}T^2\left(g_{L\uparrow} + g_{L\downarrow}\right)\left(g_{R\uparrow} + g_{R\downarrow}\right) + T^4 g_{L\uparrow}g_{L\downarrow}g_{R\uparrow}g_{R\downarrow} \right. \quad (3.18)$$

$$\left. -\frac{1}{2}T^2\cos\theta\left(g_{L\uparrow} - g_{L\downarrow}\right)\left(g_{R\uparrow} - g_{R\downarrow}\right)\right\}^{-1}$$

and $T = T(k_{\parallel})$. $f(\omega)$ is the Fermi function. Eq. (3.17) can be integrated with respect to $\theta$ to get a closed-form expression for the exchange coupling $J(\theta) = \Omega(0) - \Omega(\theta)$ (c.f. eq. (2.9)) [35]. $J(\theta)$ can be expanded in powers of $\cos\theta$ so that we have couplings like Heisenberg exchange $\hat{M}_1 \cdot \hat{M}_2$, biquadratic exchange $(\hat{M}_1 \cdot \hat{M}_2)^2$ etc, where $M_i = M_i / |M_i|$.

Finally in this section we give results for general tight-binding bands, which are used in applications to real materials in section 4. A typical trilayer is Co/Cu/Co and to model this we need to use s, p and d orbitals, nine in all. The on-site energies and hopping parameters are fitted to "first principles" band structure calculations which include the exchange splitting between ↑ and ↓ spin on-site energies in the ferromagnetic layers. All Green functions are then 9 x 9 matrices, or 18 x 18 if we include spin as in eqs. (3.7) –(3.15). The matrices double in size again if we include 2$^{nd}$ nearest –neighbour hopping and use "principal layers" consisting of two atomic planes. The hopping matrix $T_{LR} = T(k_{\parallel})$ between L and R is of the same size as the Green functions.

For the case of charge current $I$ it is convenient to use the representation including spin, as in eq. (37), and the result for an applied voltage $V$ at $T = 0$ is

$$I = \frac{e}{h}\sum_{k_{\parallel}} \int_{\mu}^{\mu+eV} Tr\left[\tau\left(g_R - g_R^\dagger\right)\tau^\dagger\left(g_L - g_L^\dagger\right)\right]d\omega \quad (3.19)$$

where

$$\tau = T\left(1 - g_R T^\dagger g_L T\right)^{-1}. \quad (3.20)$$

$g_{L,R} = g_{L,R}(\omega, k_{11})$ are the surface Green functions for the LH and RH parts of the decoupled (cleaved) systems. These LH and RH systems can in general be arbitrary layered structures and the magnetization need not be collinear throughout the multilayer. Clearly the single-orbital formula (3.12) is a special case of eq (3.19). We can obtain a similar general formula for the spin current associated with the charge current; the corresponding torques can produce current –induced switching of the magnetization direction of one of the magnetic layers, as proposed by Slonczewski [36].

We can find a general formula for the exchange coupling $J = \Omega(0) - \Omega(\pi)$, as defined by eqs. (2.8) –(2.9), using eqs. (3.2) and (3.8). However in the multi-orbital case it does not seem possible to integrate $d\Omega/d\theta$ analytically as in the single-orbital case (cf. eq. (3.18)). Nevertheless an explicit integral for $J$, involving the surface Green function matrices $g_{L\sigma}, g_{R\sigma}$ (now m x m rather than 2m x 2m with spin included) can be obtained by the energy-difference method [25]. We give this formula below; it is not clear at present why the torque and energy-difference methods are non-equivalent in the multi-orbital case. The energy-difference result for exchange coupling is

$$J = \sum_{k_{11}} \int d\omega f(\omega) Q(k_{11}, \omega) \qquad (3.21)$$

where

$$Q = -\pi^{-1} \operatorname{Im} Tr \ln \left[ 1 + S^{\uparrow} \left( g_{L\uparrow} - g_{L\downarrow} \right) S^{\downarrow} \left( g_{R\uparrow} - g_{R\downarrow} \right) \right] \qquad (3.22)$$

with

$$S^{\uparrow} = T^{\dagger} \left[ 1 - g_{L\uparrow} T g_{R\uparrow} T^{\dagger} \right]^{-1} \qquad (3.23)$$

$$S^{\downarrow} = T \left[ 1 - g_{R\downarrow} T^{\dagger} g_{L\downarrow} T \right]^{-1}. \qquad (3.24)$$

To use the expressions (3.19) and (3.21) for $I$ and $J$ we need to be able to calculate surface Green functions for arbitrary layered structures, often on a semi-infinite substrate. The best way to do this, using the matrix Möbius transformation, is described in the next section.

## 3.2 The matrix Möbius transformation and applications

The bilinear, or Möbius, transformation $\zeta = (az+b)(cz+d)^{-1}$ is a well-known conformal transformation which transforms circles in the complex $z$ plane into circles in the $\zeta$ plane. Umerski [27] showed how a generalisation of this to the case where a,b,c,d,z, are $N \times N$ matrices, in general complex, is very useful in the tight-binding theory of layered structures. We follow his discussion in this section.

We define

$$A \bullet z = (az+b)(cz+d)^{-1} \qquad (3.25)$$

where $A$ is the $2N \times 2N$ partitioned matrix

$$A = \begin{pmatrix} a & b \\ c & d \end{pmatrix}. \qquad (3.26)$$

It is straight-forward to show that

$$A \bullet (B \bullet z) = (AB) \bullet z. \qquad (3.27)$$

Suppose that a number of identical atomic layers are deposited on a substrate whose local surface Green function matrix $(N \times N)$ is $g_0$. Typically $N = 9$ or 18 as discussed in the previous section. Suppose that after $n$ layers have been deposited the local Green function for the last layer is $g_n$. We now show that $g_n$ is related recursively to $g_{n-1}$ by a transformation $g_n = X \bullet g_{n-1}$. The hopping matrix $T \equiv T(k_{11})$ and the on-site energy matrix $u$ for the deposited layers are both $N \times N$. Using the Dyson equation (3.11) we find that

$$g_n = \left(\omega - u - T^\dagger g_{n-1} T\right)^{-1}$$
$$= T^{-1}\left[(\omega - u)T^{-1} - T^\dagger g_{n-1}\right]^{-1} = X \bullet g_{n-1} \qquad (3.28)$$

with

$$X = \begin{pmatrix} 0 & T^{-1} \\ -T^\dagger & (\omega - u)T^{-1} \end{pmatrix}. \tag{3.29}$$

We note that in general $T(k_{11})$ is non-singular and that $X$ is a $2N \times 2N$ matrix. By iterating eq (3.28), and using eq. (3.27), we obtain the useful result

$$g_n = X^n g_0. \tag{3.30}$$

We proceed to find the eigenvalues $\lambda$ of $X$ in order to diagonalize $X$ and calculate $X^n$. In fact some of the eigenvalues are related to the bulk band structure of the overlayer. The eigenvalues $\lambda$ satisfy $\det(X - \lambda I) = 0$, where $I$ is the unit matrix, and

$$\begin{pmatrix} I & 0 \\ -\lambda T^\dagger & I \end{pmatrix}(X - \lambda I) = \begin{pmatrix} -\lambda I & T^{-1} \\ 0 & -\lambda^{-1}T^\dagger T^{-1} \\ & +(\omega - u)T^{-1} - \lambda I \end{pmatrix}. \tag{3.31}$$

Hence

$$\det(X - \lambda I) = (-1)^N \det(\lambda T^{-1}) \det(\omega - u - \lambda T - \lambda^{-1}T^\dagger) \tag{3.32}$$

so that the eigenvalues satisfy

$$\det(\omega - u - \lambda T - \lambda^{-1}T^\dagger) = 0 \tag{3.33}$$

where the determinant is $N \times N$ and $\omega$ stands for $\omega I$, $I$ being the $N \times N$ unit matrix. Furthermore as $\lambda \to 0$ we see from eq. (3.32) that $|\det X| = 1$, so there are no zero eigenvalues. We also note, from eq. (3.33), that if $\lambda$ is an eigenvalue so is $(\lambda^*)^{-1}$. In general there are $2(N-P)$ eigenvalues with $|\lambda| \neq 1$ and $2P$ eigenvalues with $|\lambda| = 1$, for some integer $P(0 \leq P \leq N)$. If $|\lambda| = 1$, we write

$\lambda = \exp(ikd), -\pi < kd \le \pi$, where $d$ is the interplane spacing in the overlayer. Eq. (3.33) is then the equation determining the bulk overlayer band structure $\omega = E(\mathbf{k}_{\shortparallel}, k_z)$ with $k_z = k$. Hence, for a given $\mathbf{k}_{\shortparallel}$, $k_z = \pm k$ are points on the constant energy surface of the bulk spacer with energy $\omega$. If there is only one such pair of points $P = 1$ but for a constant energy surface with more than one sheet we may have $P > 1$. Thus eigenvalues $\lambda$ on the unit circle are determined by the overlayer bulk band structure. Those with $|\lambda| \ne 1$ relate to surface states which are exponentially attenuated within the bulk. In section 2.2 we saw how the dimension $k_z = k$ of the spacer Fermi surface determined the oscillatory behaviour of the exchange coupling as a function of spacer thickness. We now pursue this connection more generally, noting that both charge current I, given by eq. (3.19), and exchange coupling $J$, given by eq. (3.24), depend on the number of spacer layers only through $g_L$ or $g_R$. (We may choose the cleavage plane dividing the LH and RH systems at the extreme right or left of the spacer). Thus $I$ and $J$ have the same periodicity, as a function of number of atomic layers $n$, as $g_n$. We now investigate this periodicity for large $n$ using eq. (3.30).

The eigenvalues and eigenvectors of $X$ must be determined numerically, so that $X$ may be diagonalized by a similarity transformation:

$$O^{-1}XO = \Lambda = \begin{pmatrix} \Lambda_1 & 0 \\ 0 & \Lambda_2 \end{pmatrix} \qquad (3.34)$$

where the $N \times N$ eigenvalues in the diagonal matrices $\Lambda_1, \Lambda_2$ are ordered in ascending order of magnitude down the diagonal ie $|\lambda_1| \le |\lambda_2| \ldots \le |\lambda_{2N}|$. Eigenvalues off the unit circle occur in pairs, $\lambda = re^{i\theta}, (\lambda^*)^{-1} = r^{-1}e^{i\theta}$, one of the pair (with modulus less than 1) appearing in $\Lambda_1$ and the other (with modulus greater than 1) in $\Lambda_2$. The eigenvalues on the unit circle also occur in pairs $e^{\pm ikd}$ $(k > 0)$, and we may put $e^{ikd}$ in $\Lambda_1$ and $e^{-ikd}$ in $\Lambda_2$; the precise order is not important at this stage.

It follows from eqs. (3.30) and (3.34) that

$$g_n = (O\Lambda^n O^{-1}) \square g_0. \qquad (3.35)$$

We define $f_n = O^{-1} \bullet g_n$ so that

$$f_n = (\Lambda^n O^{-1}) \square g_0 = \Lambda^n \square f_0. \qquad (3.36)$$

Hence, using the definition (3.25), $f_n = \Lambda_1^n f_0 \Lambda_2^{-n}$ and

$$g_n = O \square f_n = O \square (\Lambda_1^n f_0 \Lambda_2^{-n}). \qquad (3.37)$$

For large $n$ only eigenvalues $e^{\pm i k_i d}$ $(i = 1, 2..P)$ on the unit circle contribute; the other diagonal elements of $\Lambda_1^n$ and $\Lambda_2^{-n}$ tend to zero as $n \to \infty$. Thus $g_n$ depends on $n$ through quantities like

$$\begin{aligned} e^{in(k_i + k_j)d} &= e^{in_i k_i d} \, e^{in_j k_j d} \\ e^{2ink_i d} &= e^{2in_i k_i d} \end{aligned} \qquad (3.38)$$

with $n_1 = .. = n_p = n$. Hence $g_n$, and any function of it, is periodic in the variables $n_1, ... n_p$ (which can be considered as continuous) with periods $2\pi / (k_1 d), ... 2\pi / (k_p d)$ respectively. We can therefore expand the integrands in eqs. (3.19) and (3.21) as multivariate Fourier series and finally put $n_1 = .... n_p = n$ which yields a quasi-periodic function of $n$. In the case considered in section 2.2 $P = 1$, with only one wave-vector $k_1$, so the period is $\pi / (k_1 d)$ rather than $2\pi / (k_1 d)$. This can also happen with $P > 1$, as discussed in an example in section 4.

To calculate $g_n$ from eq. (3.37) we need $f_0 = O^{-1} \bullet g_o$, and therefore have to calculate the surface Green function $g_0$, for a semi-infinite substrate. To do this it is necessary to be more precise about how we order the eigenvalues on the unit circle. The trick is to include the small positive imaginary part $\delta$ in $\omega(\omega \to \omega + i\delta)$ and to let $n \to \infty$ before

$\delta \to 0$. This shifts the pairs of eigenvalues on the unit circle in such a way that one moves just inside the circle and the other just outside. The former are in $\Lambda_1$ and the latter in $\Lambda_2$, so that $\Lambda_1^n$ and $\Lambda_2^{-n}$ tend to zero as $n \to \infty$. Hence if we write

$$O = \begin{pmatrix} o_1 & o_2 \\ o_3 & o_4 \end{pmatrix} \quad (3.39)$$

it follows from eq. (3.37) that

$$g_\infty = O \Box 0 = o_2 o_4^{-1}. \quad (3.40)$$

Thus $g_\infty$ is determined by half the eigenvectors of $X$, which form the columns of the right-hand half of $O$.

It is interesting to reformulate the calculation of $g = g_\infty$. It follows from eq. (3.28) that $g$ satisfies an equation of the form $g = A \Box g$ with $A = X$, This general form of equation, with $A$ given by eq. (3.26), becomes $g = (ag+b)(cg+d)^{-1}$ or

$$gcg + gd - ag - b = 0. \quad (3.41)$$

Thus we have a simple method of solving this quadratic equation involving $N \times N$ matrices. We assemble the matrix coefficients to form the $2N \times 2N$ matrix of eq (3.26) and determine half its eigenvectors to obtain the solution (3.40). This method of solving matrix quadratic equations has been used previously in control theory for obtaining the steady state solution of the matrix Ricatti equation [37]

$$dg/dt = g(t)c(t)g(t) + g(t)d(t) - a(t)g(t) - b(t). \quad (3.42)$$

However the matrix Möbius transformation is used much more powerfully in the present context of layered structures.

### 3.3. The stationary phase approximation

Quantities associated with a trilayer like conductance $\Gamma = \lim_{V \to 0} (I/V)$ and exchange coupling $J$ can always be calculated as a function of integer spacer thickness $n$ purely numerically. The advantage of the Möbius transformation method is that even for complex spacer Fermi surfaces, with several sheets, dominant contributions and oscillation periods can be ascribed definitively to features of the Fermi surface. Integrands in eqs. (3.19) and (3.21) are functions of $g_{n\sigma}$ and $g_{o\sigma}$ if we take the "cleavage plane" to be between the RH ferromagnetic layer and the last spacer atomic plane. Thus, for example, $Q(\mathbf{k}_{11},\omega)$ in eq. (3.21), for a given $\mathbf{k}_{11}$ and $\omega$, is a function of $g_n$ whose periodicity as a function of large $n$ we discussed in section 3.2. Using the notation of that section we may expand $Q$ in the Fourier series

$$Q = \sum_{s_1...s_p} c_{s_1...s_p} \exp\left[i\left(s_1 n_1 k_1 + .. + s_p n_p k_p\right)\right] \tag{3.43}$$

and then put $n_1 = .... = n_p = n$. The integers $s_1,...s_p$ define a vector $\mathbf{s}$ and, writing $\phi_s = s_1 k_1 + ...s_p k_p$, we have

$$Q = \sum_s |c_s| e^{i\psi_s} e^{in\phi_s}. \tag{3.44}$$

The Fourier representation here, valid for an arbitrary number of Fermi surface sheets, is more general than the one introduced in eq. (2.13) with a single $k_z$. The dominant contributions to the integrals over $\omega$ and $\mathbf{k}_{11}$ in eq. (3.21) come from $\omega \sim \mu$, the Fermi cut-off, and from stationary points $\mathbf{k}^0_{11s}$ of the phase $\phi_s(\mathbf{k}_{11})$. In the case of the conductance additional oscillations, not related to specific dimensions of the spacer Fermi surface, arise from a cut-off in the $\mathbf{k}_{11}$ integral [39]. This cut-off occurs since the only states contributing to the spin $\sigma$ current are those with values of $\mathbf{k}_{11}$ which are common to the spacer Fermi surface and the ferromagnet spin $\sigma$ Fermi surface.

It may be shown [40] from eqs (3.21) and (3.44), using the stationary phase approximation, that for large spacer thickness ($n$ atomic layers) the exchange coupling at temperature $T$ is given by

$$J(n) = -\frac{(2\pi)^2 k_B T}{2n A_{BZ}} \sum_s \sum_{k^0_{11s}} \mathrm{Re} \left( \frac{e^{i\frac{\pi}{4}(N_+ - N_-)} c_s e^{in\phi_s}}{\left|\det(\partial^2 \phi_s)\right|^{1/2} \sinh(\pi k_B T(n\phi'_s + \psi'_s))} \right)_{\mu, k^0_{11s}}$$

$$\tag{3.45}$$

where $A_{BZ}$ is the area of the two-dimensional $k_{11}$ Brillouin zone. Here the 2x2 matrix $\partial^2 \phi_s$ is defined by

$$\left(\partial^2 \phi_s\right)_{\alpha\beta} = \left(\frac{\partial^2 \phi_s}{\partial k_{11\alpha} \partial k_{11\beta}}\right)_{k_{11}^0}$$

(3.46)

and $N_+, N_-$ are the number of positive and negative eigenvalues of $\partial^2 \phi_s$. Also $\phi'_s$ and $\psi'_s$ are derivatives with respect to $\omega$, evaluated at the chemical potential $\mu$ like the other quantities in the large bracket of eq. (3.45).

Clearly $J(n) \propto 1/n^2$ at $T = 0$ if $\psi'_s$ is negligible, but this is not always the case. The most important factors in determining the magnitude of $J$ are usually the Fourier coefficients $c_s$ and $\left|\det\left(\partial^2 \phi_s\right)\right|$ which is related to spacer Fermi surface curvature. Small curvature at the relevant extremal points gives large $J$. Large $c_s$, giving large $J$, occurs when states of one spin are strongly confined in the spacer in the ferromagnetic configuration of the trilayer, as discussed in section 2.2. In the next section a detailed application of eq. (3.45) to a Fe/Mo/Fe (001) trilayer, where the spacer Mo has a complicated Fermi surface, is described.

## 4 EXAMPLES OF EXCHANGE COUPLING, GMR AND TMR

### 4.1.1 Exchange coupling in Fe/Mo/Fe(001)

Mo is an interesting spacer because of its similarity to Cr, but with no antiferromagnetism. Fe and Mo have BCC lattices and the two-dimensional $k_{11}$ Brillouin zone (BZ) for a BCC lattice grown in the (001) direction is a square with corners at $(\pm\pi/a, \pm\pi/a)$ where $a$ is the cube side of the BCC lattice. Fig 16 shows the triangular irreducible part of this BZ with corners at $\Gamma(0,0), \Delta(\pi/a, 0)$ and $N(\pi/a, \pi/a)$. Umerski and Mathon [38] use the stationary phase method to analyse the exchange coupling $J(n)$, where $n$ is the number of Mo atomic layers, and find contributions from the five $k_{11}$ points marked with an asterisk. We shall denote the point on $\Gamma\Delta$ by A and the point on $\Gamma N$ by B. The bulk Mo band-structure may be displayed in a three-dimensional BZ which is a cuboid defined by $|k_x| < \pi/a, |k_y| < \pi/a, |k_z| < 2\pi/a$; its projection on the $x, y$ plane is the two dimensional BZ defined above. Figs. 17 and 18 show sections of the bulk Mo Fermi surface cut by the $k_y = 0$ and $k_x = k_y$ planes respectively. The partially occupied d band of Mo gives rise to the complicated Fermi surface with several sheets. The apparent distortion from cubic symmetry in these figures is due to use of a different scale for $k_z$ and $k_{11}$.

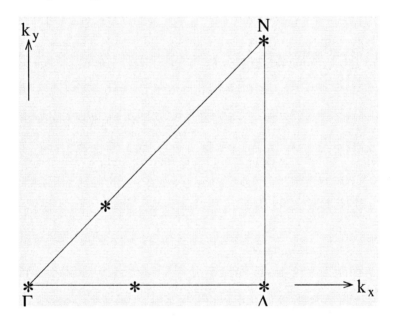

Fig.16: Irreducible part of the two-dimensional Brillouin zone for a BCC lattice grown in the (001) direction.

At the $\Gamma$ point of $k_{11}$ there are four intersections with the Fermi surface for $k_z > 0$, with $k_z = \gamma_1, \gamma_2, \gamma_3 = \gamma_4$. It turns out that no interband mixing occurs (ie only quantities like the second ones in eq (3.38) appear) so that only terms in eq (3.43) with $s_i$ even $(i = 1, ... 4)$ are non-zero. Thus there are three periods $\pi/(\gamma_1 d), \pi/(\gamma_2 d), \pi/(\gamma_3 d)$ and it is found that the first, equal to 3.30ML, dominates the contribution to $J$ arising from the stationary points at $\Gamma$. This contribution, seen in fig 19(a), is fairly small. The case of the point N is qualitatively similar, with two Fermi surface intersections at $k_z = v_1, v_2$ (cf. fig.18). There are two periods $\pi/(v_1,d), \pi/(v_2 d)$ but the contribution to $J$ from the stationary points at N is two orders of magnitude smaller than the total $J$ (cf. fig. 19 (b)). The situation at $\Delta$ is quite complicated, as one can see from the contribution to $J$ shown in fig. 19 (c). As may be seen in fig. 17 the four intersections with the Fermi surface are not stationary points of $k_z$. It is necessary to look for stationary points of $\phi_s = \sum s_i k_i$ for different sets of integers $s_i$. The amplitude of the long period oscillation (7.67ML) is significant and arises from $s = (0,1,-1,0)$, the period being given by $2\pi/(d_1 d)$, where $d_1 = \delta_3 - \delta_2$ is indicated in fig. 17. The fairly large amplitude arises from a large Fourier coefficient $c_s$ and the low curvature of the $\delta_2, \delta_3$ parts of the Fermi surface. The largest contributions to $J$ arise from the $k_{11}$ points A and B on the $\Gamma\Delta, \Gamma N$ lines, respectively. They correspond to stationary points of $\phi_s$ with $s = (0,1,0,1)$ the stationary values of $k_2 + k_4$ being indicated as $v_1$ and $v_2$ in figs 17 and 18 respectively. The corresponding periods are 2.06 ML and 2.00 ML and the huge contributions to $J$ are shown in fig. 19 (d) and (e). The very small curvature of the relevant Fermi surfaces, with almost perfect nesting over a wide area (which gives rise to anitferromagnetism in the similar metal Cr ) is the predominant cause of the large exchange coupling.

The total exchange coupling $J(n)$ at $T = 315K$ as a function of the number $n$ of the Mo planes, is shown in fig. 20 . The thick line is the result of a full numerical calculation and the thin line is the result of the stationary phase analysis. The agreement is quite good for large $n$ and suggests that all the important stationary points have been located. However the agreement with experiment is strikingly bad. Qui et al [39,40] find a small coupling of about 0.009 mRy/atom at $n \square 5$ML and a period of about 3ML. This contrasts strongly with the result of Umerski and Mathon [38] (fig. 20) which is about two orders of magnitude larger with a period of about 2ML. Mirbt et al [41, 42] find similar numerical results but are not able to carry out a stationary phase analysis of the various contributions to $J$.

The origin of the huge discrepancy between theory and experiment is unclear. Certainly interface roughness will tend to cancel out the 2ML period contribution, as discussed in section 2.3 in connection with a factor

3 discrepancy between theory and experiment in Co/Cu/Co. But the discrepancy for Fe/Mo/Fe seems too large to ascribe to this effect. Exchange coupling can also be strongly attenuated by scattering due to defects, or possibly spin fluctuations, in the spacer but it is not clear why the 2ML period should be completely suppressed. Further analysis of this problem is required.

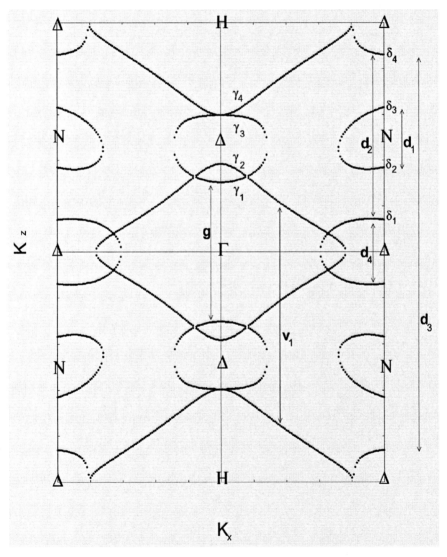

Fig.17: Section of the bulk Mo Fermi surface cut by the plane $k_y = 0$ [38].

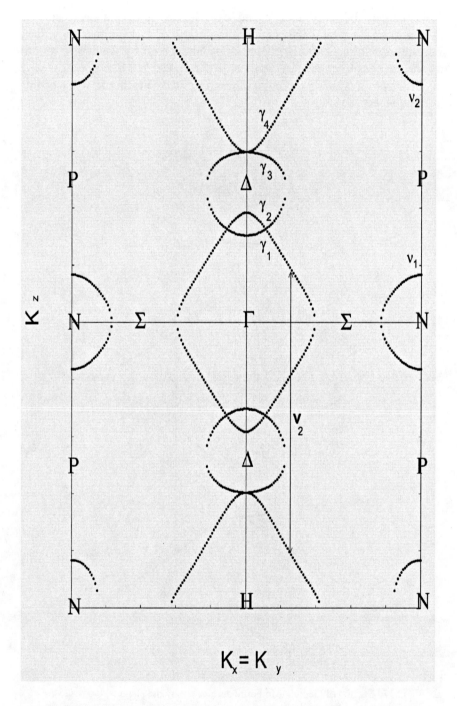

Fig.18: Section of the bulk Mo Fermi surface cut by the plane $k_x = k_y$ [38].

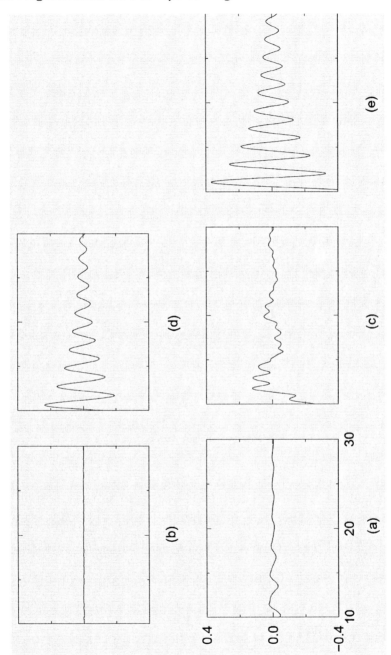

Fig.19: Contributions to exchange coupling $J(n)$ in BCC Fe/Mo/Fe(001) trilayers, as a function of number of atomic layers $(n)$ of Mo, arising from different extremal points in the stationary phase analysis [38].

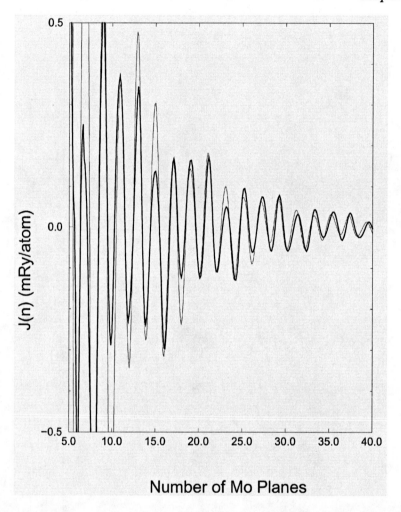

Fig.20: The total exchange coupling $J(n)$ in a Fe/Mo/Fe(001) trilayer at $T = 315K$ calculated fully numerically (thick line) and using stationary phase (thin line) [38].

### 4.2 Ballistic CPP GMR in Co/Cu/Co (001)

Mathon et al [18] investigated oscillations of CPP GMR in a trilayer consisting of $n$ atomic planes of Cu between Co layers with semi-infinite Cu leads attached. They assumed ballistic transport with perfect interfaces and no scattering within the layers. In terms of conductance $\Gamma = R^{-1}$ eq. (1.1) for the magnetoresistance becomes

$$\frac{\Delta R}{R} = \frac{\Gamma_{\uparrow\uparrow} - \Gamma_{\uparrow\downarrow}}{\Gamma_{\uparrow\downarrow}} = \frac{\Gamma_{\uparrow\uparrow}^{\uparrow} + \Gamma_{\uparrow\uparrow}^{\downarrow} - 2\Gamma_{\uparrow\downarrow}^{\uparrow,\downarrow}}{2\Gamma_{\uparrow\downarrow}^{\uparrow,\downarrow}}, \qquad (4.1)$$

where $\Gamma_{\uparrow\uparrow}^{\sigma}, \Gamma_{\uparrow\downarrow}^{\sigma}$ are the conductance's for spin $\sigma$ in the ferromagnetic and antiferromagnetic configuration respectively (clearly $\Gamma_{\uparrow\downarrow}^{\uparrow} = \Gamma_{\uparrow\downarrow}^{\downarrow}$ ). A numerical evaluation of the conductance, using the equivalent of eq. (3.19) in the limit $V \to 0$, gave the results for $100 \Delta R / R$ shown in fig. 21 for 7 atomic planes of Co in each magnetic layer. Oscillations as a function of Cu thickness occur about a mean GMR of about 88%. The oscillatory part of the GMR is found to have little dependence on Co thickness, as long as it is larger than 6ML, so the origin of the oscillations was investigated for a trilayer Co/Cu(n)/Co with semi-infinite Co layers.

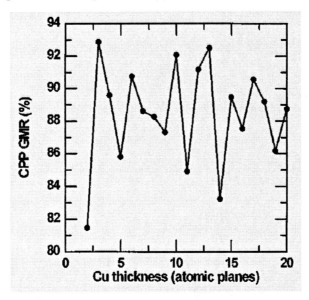

Fig.21: CPP GMR versus Cu thickness for a trilayer with seven atomic planes of Co in each magnetic layer.

The oscillatory part of the conductances $\Gamma_{\uparrow\uparrow}^{\uparrow}, \Gamma_{\uparrow\uparrow}^{\downarrow}$ and $\Gamma_{\uparrow\downarrow}^{\uparrow}, \Gamma_{\uparrow\downarrow}^{\downarrow}$ are shown in figs. 22, 23, 24 respectively. Numerical results given by full lines are compared with stationary phase results (dashed lines) associated with the two stationary points of the Cu Fermi surface shown in fig. 12. The oscillations in $\Gamma_{\uparrow\uparrow}^{\uparrow}$ clearly arise from the Fermi surface extrema but those in $\Gamma_{\uparrow\uparrow}^{\downarrow}$ and $\Gamma_{\uparrow\downarrow}^{\uparrow,\downarrow}$ must have a different origin. They arise from the cut-offs in the $k_{11}$ integral mentioned in section 3.3.

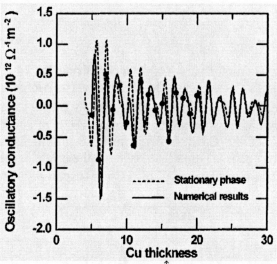

Fig.22: Oscillatory part of the conductance $\Gamma_{\uparrow\uparrow}^{\uparrow}$ in the Co(∞)/Cu(n)/Co(∞) trilayer obtained fully numerically (full line) and by stationary phase analysis (dashed line).

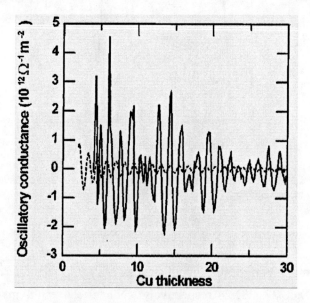

Fig.23: Similar to fig.22 but showing $\Gamma_{\uparrow\uparrow}^{\downarrow}$. The oscillations of the full line arise from Fermi cut-offs rather than Fermi surface extrema (see text).

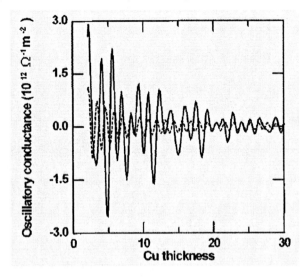

Fig.24: Similar to fig.23 but showing $\Gamma^\uparrow_{\uparrow\downarrow} = \Gamma^\downarrow_{\uparrow\downarrow}$.

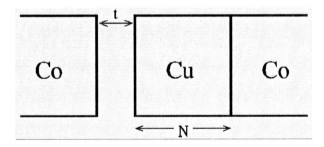

Fig.25: The system used [43] to study TMR between Co electrodes, one capped with Cu.

## 4.3 TMR between Co electrodes, one capped with Cu

As a last example we discuss some theoretical results of Mathon and Umerski [43] on the system shown in fig. 25. The right-hand Co electrode is capped with $N$ atomic layers of Cu. Tunnelling between the left-hand Co layer and the Cu layer, through vacuum or an insulator, is modelled by a hopping matrix with zero hopping involving d orbitals and with s-p hopping matrix elements reduced to 20% of their bulk values. The TMR is almost independent of further reduction, although the individual conductances decrease rapidly.

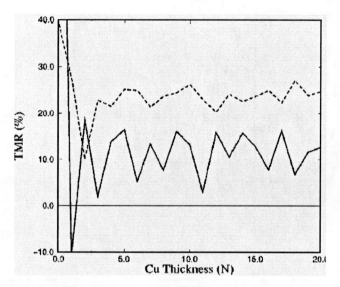

Fig.26: The calculated TMR (full line) and the result (dashed line) of a simplified theory (see text) [43].

The main point of this example is to show the limitation of Jullière's well-known tunnelling formula [44]. This proposes that conductance $\Gamma \propto N_L(E_F) N_R(E_F)$, where $N_L$ and $N_R$ are densities of states to the left and right of the tunnelling barrier, evaluated at the Fermi level $E_F$. Then, according to eq. (4.1),

$$\frac{\Delta R}{R} = \frac{N_L^\uparrow R_L^\uparrow + N_L^\downarrow N_R^\downarrow - N_L^\uparrow N_R^\downarrow - N_L^\downarrow N_R^\uparrow}{N_L^\uparrow R_R^\downarrow + N_L^\downarrow N_R^\uparrow} = \frac{2 P_L P_R}{1 - P_L P_R} \qquad (4.2)$$

where spin polarisations $P_L, P_R$ are defined in terms of $N_L^\sigma, N_R^\sigma$ respectively by

$$P = \left[ N^\uparrow(E_F) - N^\downarrow(E_F) \right] / \left[ N^\uparrow(E_F) + N^\downarrow(E_F) \right]. \qquad (4.3)$$

For the system of fig. 25, $P_R = P_{Cu} = 0$ so that Jullière's formula predicts $\Delta R / R = 0$. However experimentally a non-zero magnetoresistance is observed [45]. Mathon and Umerski [43], using eq. (3.19) in the limit $V \to 0$ find the non-zero TMR shown in fig. 26. The important factors in eq (3.19) to achieve this are $\tau$ and $\tau^\dagger$. These provide the important denominator of the integrand which is shown more explicitly for the one-band case in eq. (3.16). Without this denominator the integrand is a product of left-hand and right-hand densities of states, as in the Jullière formula. The effect of the denominator is to suppress the contribution of bound states which do not carry current. In the present example this applies to minority spin states confined in the Cu layer, as discussed in section 2.3, which contribute to $N_R$ but not to conductance. Mathon and Umerski define an effective transport density of states, in which confined states are removed, which can be used in a Jullière-type formula in a qualitative way. The result is the dashed line of fig. 26.

## ACKNOWLEDGMENTS:

For the last decade I have enjoyed a very fruitful collaboration with J.Mathon (City University) which has been funded continuously by the EPSRC. The material of these lecture notes has been developed in this context and important contributions have been made by Andrey Umerski, Murielle Villeret, R Bechara Muniz and José d'Albuquerque e Castro. Reference is made to some work by other authors but no attempt has been made at a complete bibliography in this popular field.

## REFERENCES

1. P. Grünberg, R. Schreiber, Y. Pang, M.N. Brodsky and H. Sowers, Phys. Rev. Lett. **57** 2442 (1986)
2. C. Carbone and S.F. Alvorado, Phys. Rev. B **36** 2433 (1987)
3. M.N. Baibich, J.M. Borot, A, Fert, F. Nguyen Van Dau, F. Petroff, P. Etienne, G. Creuzet, A. Friederch and J. Chazelas, Phys. Rev. Lett. **61** 2472 (1988)
4. S.S.P. Parkin, N. More and K.P. Roche, Phys. Rev. Lett. **64** 2304 (1990)
5. P. Grünberg, S. Demokritov, A. Fuss, M.Vohl and J.A. Wolf, J. Appl. Phys. **69** 4789 (1991)
6. J. Unguris, R.J. Celotta and D.T. Pierce, Phys. Rev.Lett. **67** 140 (1991)
7. A. Schreyer, C.F. Majkrzak, Th. Zeidler, T. Schmitte, P. Bödeker, K. Theis-Bröht, A. Abromeit, J.A. Dura and T. Watanabe, Phys. Rev. Lett. **79** 4914 (1997)
8. F.J. Himpsel, J.E. Ortega, G.J. Mankey and R.F. Willis Adv. Phys. **47** 511 (1998)
9. E. Grochowski and D.A. Thompson, IEEE Trans. Magn. **30,** 3797 (1994)
10. T. Valet and A. Fert, Phys. Rev. B **48** 7099 (1993)
11. R.E. Camley and J Barnás, Phys. Rev.Lett. **63** 644 (1989)
12. J.Barnás, A. Fuss, R.E. Camley, P. Grünberg and W. Zinn, Phys. Rev. B **42** 8110 (1990)
13. D.M. Edwards, J. Mathon and R.B. Muniz, IEEE Trans. Magn. **27** 3548 (1991)
14. N.F. Mott, Adv. Phys. **13** 325 (1964)
15. E. Yu. Tsymbal and D.G. Pettifor, Phys. Rev. B **54** 15314 (1996)
16. J. Mathon, M. Villeret and H. Itoh, Phys. Rev. B **52** R6983 (1995)
17. K.M. Schep, P.J. Kelly and G.E.W Bauer, Phys. Rev. Lett. **74** 586 (1995)

18. J. Mathon, A. Umerski and M. Villeret, Phys. Rev. B **55** 14378 (1997)
19. S. Sanvito, C.J. Lambert, J.H. Jefferson and A.M. Bratkovsky, Phys. Rev B **59** 11936 (1999)
20. S.F. Lee, W.P. Pratt, Q. Yang, P.Holody, R, Loloee, P.A. Schroeder and J. Bass, J. Magn. Magn. Mater. **118** 1 (1993)
21. D.M. Edwards, J. Mathon, R.B. Muniz and M.S. Phan, Phys. Rev. Lett **67** 493 (1991); J. Phys.: Condens. Matter **3** 4941 (1991)
22. J.E. Ortega, F.J. Himpsel, G.J. Mankey and R.F. Willis, Phys. Rev. B **47** 1540 (1993)
23. K. Garrison, Y. Chang and P.D. Johnson, Phys. Rev. Lett. **71** 2801 (1993)
24. J. Mathon, M. Villeret, R.B. Muniz, J.d'Albuquerque e Castro and D.M. Edwards, Phys. Rev Lett. **74** 3696 (1995)
25. J. Mathon, M. Villeret, A. Umerski, R.B. Muniz, J.d'Albuquerque e Castro, Phys. Rev. B **56** 11797 (1995)
26. D.G. Pettifor and C. Varma, J. Phys. C **12** L253 (1979)
27. A. Umerski, Phys. Rev. B **55** 5266 (1997)
28. R. Kläsges, D. Schmitz, C. Carbone, W. Eberhardt, P. Lang, R. Zeller and P.H. Dederichs, Phys. Rev. B **57** R696 (1998)
29. M.T. Johnson, P.J.H Bloemen, R. Coehoorn, J. J. de Vries, N.W.E. McGee, R. Jumgblut, A. Reinders and J. aan de Stegge, MRS Symposia Proceedings No. 313 (Materials Research Society, Pittsburgh, 1993) p.93
30. W. Weber, R. Allenspach and A. Bischof, Europhys. Lett **31,** 431 (1995)
31. L. V. Keldysh, Soviet Physics JETP **20** 1018 (1965)
32. C. Caroli, R. Cambescot, P.Nozières and D. Saint-James, J. Phys. C **4** 916 (1971)
33. G.D. Mahan, "Many Particle Physics" $2^{nd}$ Ed. 1990, Plenum Press
34. J. Mathon, Phys. Rev. B **56** 11810 (1997)
35. D.M. Edwards, A.M. Robinson and J.Mathon, J. Magn. Magn. Mat. **140-144** 517 (1995)
36. J. Slonczewski, J. Magn. Magn. Mat. **159** L1 (1996)
37. S. Barnett, "Matrices in Control Theory" , Van Nostrand Reinhold Company 1971
38. A. Umerski and J.Mathon, preprint
39. Z.Q. Qui, J. Pearson, A. Berger and S.D. Bader, Phys. Rev. Lett. **68** 1398 (1992)
40. Z.Q. Qui, J, Pearson and S.D. Bader, J. Appl. Phys. **73** 5765 (1993)
41. S. Mirbt, A.M.N. Niklasson, H.L. Skriver and B. Johansson, J. Magn. Magn. Mater. **148** 209 (1995)
42. S. Mirbt, A.M.N. Niklasson, B. Johansson and H.L. Skriver, Phys. Rev. B **54** 6382 (1996)
43. J. Mathon and A. Umerski, Phys. Rev. B **60** 1117 (1999)
44. M. Jullière, Phys. Lett. A **54** 225 (1975)
45. S.S.P. Parkin (private communication)

# Chapter 7

# FROM THE FERMI LIQUID TOWARDS THE WIGNER SOLID IN TWO DIMENSIONS

Jean-Louis Pichard [a], Giuliano Benenti, Georgios Katomeris
Franck Selva and Xavier Waintal
*Service de Physique de l'Etat Condensé, CEA-Saclay, 91191 Gif sur Yvette cedex, France*

**Abstract**

The quantum-classical crossover from the Fermi liquid towards the Wigner solid is numerically revisited, considering small square lattice models where electrons interact via a Coulomb $U/r$ potential. We review a series of exact numerical results obtained in the presence of weak site disorder for fully polarized electrons (spinless fermions) and when the spin degrees of freedom are included. A novel intermediate regime between the Fermi system of weakly interacting localized particles and the correlated Wigner solid is obtained. A detailed analysis of the non disordered case shows that the intermediate ground state is a solid entangled with an excited liquid. For electrons in two dimensions, this raises the question of the existence of an unnoticed intermediate liquid-solid phase. Using the Coulomb energy to kinetic energy ratio $r_s \propto U \propto n_s^{-1/2}$, we discuss certain analogies between the numerical results obtained as a function of $U$ for a few particles and the low temperature behaviors obtained as a function of the carrier density $n_s$ in two dimensional electron gases. Notably, the new "exotic state of matter" numerically observed at low energies in small clusters occurs at the same intermediate ratios $r_s$ than the unexpected low temperature metallic behavior characterizing dilute electron gases. The finite size effects in the limit of strong disorder are eventually studied in the last section, providing two numerical evidences that the weak coupling Fermi limit is delimited by a second order quantum phase transition when one increases $U$.

**Keywords:**

- Numerical studies of lattice models with Coulomb repulsions.
- Intermediate liquid-solid quantum phase.
- Metal-insulator transition in two dimensions.
- Finite size scaling with Coulomb repulsions.

"The very simplest form of the theory of the energy bands in metals gave for many problems such accurate explanations of often very intricate properties of metals and alloys that it may well appear superfluous to consider extensions of the simple form of the theory". Those words written by Wigner [1] in 1938 come again as an objection against the need to develop a more rigourous theory, since the Fermi liquid theory (FLT) was improved by Landau [2] in the sense of a perturbation theory based on renormalized single-particle excitations and adapted to include the effects of elastic scattering by the impurities [3]. The need to go outside conventional FLT for explaining the unexpected two dimensional metallic phase [4] discovered by Kravchenko and Pudalov is a subject of controversy. On one hand, certain characteristic FLT behaviors [5] seem to remain in the vicinity of the metal-insulator transition (MIT), suggesting that an "apparent" metallic behavior could be the consequence of "classical" effects (interband scattering [6], temperature dependent screening [7], temperature dependent scattering [8, 9] or classical percolation [10]). On the other hand, the observation of an unexpected MIT is first the result (see for instance Refs. [11, 12]) of new possibilities of studying controlled many body systems which are closer to the strong coupling limit than the previously studied systems in two dimensions. This gave us the motivation to numerically revisit the classic problem of the crossover from the weak coupling Fermi limit towards the strong coupling Wigner limit for electrons in two dimensions. In this chapter, we review our main numerical results. The interest of the information given from exact diagonalization of small systems can be questionned, but it may have the merit to raise questions which may be relevant for explaining the behaviors observed around the "two dimensional MIT".

## 1. WEAK AND STRONG COUPLING LIMITS

A convenient measure of the electron gas density $n_s$ is the dimensionless parameter $r_s$

$$r_s = \frac{1}{\sqrt{\pi n_s} a_B^*} \tag{1}$$

defined as the radius $1/\sqrt{\pi n_s}$ of the unit disk divided by the Bohr radius $a_B = \hbar^2/(me^2)$. The unit disk encloses an area equal to the area per electron of the gas. For a real two dimensional electron gas (2DEG) created in a field effect device, one uses an effective Bohr radius $a_B^* = \hbar^2 \epsilon/(m^* e^2)$ which includes the dielectric constant $\epsilon$ of the medium in which the 2DEG is created and the effective mass $m^*$ of the carriers. Wigner was the first to consider the dilute limit where $r_s$ becomes large, the Coulomb interactions dominate the kinetic energy in determining the wave function and the electrons tend to arrange themselves in a regular lattice. It may be argued that a lattice configuration is not consistent with the translational symmetry characterizing the 2DEG Hamiltonian in the absence of a random substrate. This objection can be removed by forming a new wave function which is a linear combination of all translations of the original lattice: the resultant wave function will have a uniform electronic charge density, as symmetry demands, with an unchanged energy. Originally, Wigner assumed a bcc electron lattice. It was later shown [13] that the hexagonal lattice has a lowest electrostatic energy in two dimensions.

As explained in Refs. [14, 15, 16], when one considers the Hamiltonian of $N$ electrons in two dimensions,

$$H = \frac{1}{r_s^2} \sum_i^N \nabla_i^2 + \frac{2}{r_s} \sum_{\substack{i,j \\ i \neq j}}^N \frac{1}{|r_i - r_j|} \qquad (2)$$

where the lengths are given in units of $1/\sqrt{\pi n_s}$, the question has been from the early days to obtain the asymptotic behaviors of the ground state energy $E_0$ around the weak coupling limit ($r_s \ll 1$):

$$E_0 = \frac{h_0}{r_s^2} + \frac{h_1}{r_s} + 0(\ln r_s) \qquad (3)$$

and around the strong coupling limit ($r_s \gg 1$):

$$E_0 = \frac{f_0}{r_s} + \frac{f_1}{r_s^{3/2}} + \frac{f_2}{r_s^2} + 0(r_s^{-5/2}), \qquad (4)$$

to calculate the coefficients $h$ and $f$, and to discuss the expected range of validity for those asymptotic expansions. Then, one can try to numerically determine the value of $r_s$ where the weak coupling energy exceeds the large coupling energy. This can be done at the price of certain approximations which are controlled in the two limits and remain more uncertain in the middle, and after extrapolating finite size studies towards the thermodynamic limit.

The most advanced works in this field are Quantum Monte Carlo studies [16, 17] of systems involving many electrons (typically more than $10^2$) and assuming two trial wave functions $\Psi_T(R)$ adapted to describe either the weak coupling limit, or the strong coupling limit. $\Psi_T(R)$ are of the Slater-Jastrow form

$$\Psi_T(R) = D(R) \exp(-\sum_{i<j}^N u(|r_i - r_j|)), \qquad (5)$$

where $D(R)$ is a Slater determinant of extended plane waves for weak coupling, of localized single-particle orbitals for strong coupling. The liquid and crystal pseudopotentials $u(r)$ are repulsive and include in an approximate way the effects of electronic correlations. Then a simple variational approach, or a more involved fixed node Green's-function approach, are used to obtain the energies $E_0(r_s)$ dictated by the chosen $\Psi_T(R)$ or by its nodal structure. Both the variational energies and the fixed node energies give an upper bound to the exact energy. Comparing for intermediate $r_s$ the energies given by the $\Psi_T(R)$ adapted to describe the liquid and the crystal, one concludes [17] that there is a first order quantum liquid-solid transition at $r_s \approx 37$, with a possible division [16, 18] of the liquid phase into a non polarized liquid at small $r_s$ and a polarized liquid for larger $r_s$. However, the nature of the quantum mechanism of melting is still debated, and the possibility of a continuous transition has been very recently proposed[19]. The solid is assumed to be a frustrated antiferromagnet[20] before becoming ferromagnetic [21] at very large $r_s$. The same Monte Carlo method has been used[22] in the presence of impurities. The conclusion was that disorder can stabilize the solid to weaker values of $r_s$.

Andreev and Lifshitz have discussed [23] in 1969 the possibility to have a more complex intermediate state between the solid and the liquid, which should be neither a solid nor a liquid. Two kinds of motion should be possible in it; one possesses the properties of motion in an elastic solid, the second possesses the properties of motion in a liquid. This idea comes from a theory of defects in quantum solids. The nature of the relevant defects is a complicated issue. Let us give a possible example: a vacancy yielded by one electron hopping from the Wigner lattice towards some interstitial site. In a classical solid, this defect has a certain electrostatic cost and remains localized. In a quantum solid, we have in addition the tunneling effect, and if this defect can be created, it will be delocalized since the system is invariant under translation. Therefore, when $r_s$ decreases from the strong coupling limit, the increasing band width of the zero point defects of this type may exceed their decreasing electrostatic energy cost, leading to two possibilities for intermediate $r_s$: Either the total melting of the solid to directly give a liquid, as implicitly assumed for instance in Refs. [19], or a quantum floppy solid coexisting with a liquid of delocalized defects, as conjectured by Andreev and Lifshitz. A phenomenological FLT theory à la Landau of such gapless "delocalized excitations" of a floppy quantum solid has been later proposed in Ref. [24]. The discussion of this second possibility is one of the central points of this chapter.

## 2. DILUTE 2DEG IN FIELD EFFECT DEVICES

A pure 2DEG can be realized by trapping electrons on the surface of liquid helium, but it is difficult to reach a sufficient density to study the quantum regime. Another possible realization is given by the new classes of superconducting cuprates where the electronic motion is essentially two dimensional. The charge density can be varied by chemical doping, and a complex phase diagram is obtained, with insulating, superconducting and metallic behaviors.

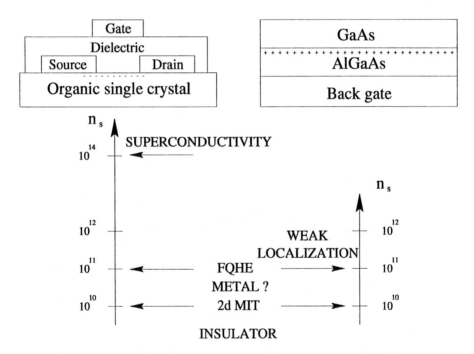

*Figure 1.* Schematic picture of an organic field effect transistor (upper left) and of a GaAs heterostructure (upper right) where the location of the 2DEG or 2DHG (symbol plus for the holes) is indicated. The characteristic low temperature behaviors are summarized below as a function of the (typical) carrier density $n_s$ ($cm^{-2}$): 2dMIT, 2$d$ metal, FQHE for a sufficient magnetic field, weak localization correction to the Boltzmann conductivity, superconductivity (see Ref.[25]). The densities give a typical order of magnitude, the observed behaviors depending also on the effective mass of the carriers.

Eventually, one can create a two dimensional gas of charges (electrons or holes) at the interface between two doped semiconductors (GaAs-AlGaAs heterostructures), between a semi-conductor and an insulator (Si-Mosfet), or very recently [12] between an organic crystal (pentacene, tetracene and anthracene) and an insulator. The carrier density can be varied by a gate from a very dilute limit towards larger densities. In Fig. 1, we mention the remarkable phenomena observed in organic (left) or doped semi-conductor (right) field effect devices. A clean interface may give a high carrier mobility, may allow the observation [26] of the fractional quantized Hall effect (FQHE) and may give a measurable conductivity in a very dilute limit (typically $n_s \approx 10^9 - 10^{11}$ carriers per cm$^2$) for GaAs heterostructures and organic devices. If the effective mass of the carriers is large enough (a condition which is not satisfied by the electrons in Ga-As heterostructures) the effective factor $r_s$ can be in the vicinity of the values where the Fermi-Wigner crossover is expected. One of the surprises in those high quality field effect transistors has been the observation of a metallic low temperature behavior [4, 12, 27] in a certain intermediate range of carrier densities, where a large perpendicular magnetic field yields FQHE or magnetically induced Wigner crystals [28].

## 3. METAL-INSULATOR TRANSITION

A recent review of the 2D-MIT can be found in Ref. [4] with an extended list of references. We summarize by a few sketches some of the behaviors which have been observed, and which are useful for discussing our numerical results. The main surprise was given by the temperature dependence of the 2DEG resistivity $\rho(T)$ around a low critical density $n_{c1}$. As sketched in Fig. 1 (see Fig.1 of Ref. [29]), $\rho(T)$ decreases as a function of $T$ when $n_s < n_{c1}$, becomes temperature independent at $n_s = n_{c1}$ and increases when $n_s > n_{c1}$. A decay is the expected behavior for an insulator, while an increase usually characterizes a metal. These behaviors occur [30] in a low temperature range $35mk < T \leq T_F$, where $T_F \approx 0.8 - 5K$ are the typical Fermi temperatures of those dilute 2DEGs. The temperature increase of $\rho(T)$ can be large for a 2DEG created in a Si-Mosfet (typically one order of magnitude), but remains weak in a 2DHG created in a GaAs heterostructure. For the densities $n_s$ where $\rho(T)$ has a metallic behavior, a parallel field $B$ induces a large positive magnetoresistance which saturates above a certain field $B_{sat}$, as sketched in Fig.3 (see Fig. 1 of Ref. [31] and Fig. 3 of Ref. [32]). From small angle Shubnikov-de Haas measurements done in a Si-mosfet, it was concluded in Ref. [31] that $B_{sat}$ signals also the onset of full spin polarization. Close to the MIT, $B_{sat}$ is very small and increases as $n_s - n_{c1}$ above $n_{c1}$ [32]. This corresponds to the intermediate values of $r_s$ (typically $3 < r_s < 10$) where the metallic behavior is observed. When $B > B_{sat}$, the metallic increase of $\rho(T)$ disappears, but the $I - V$ characteristics

# From the Fermi liquid towards the Wigner solid in two dimensions  269

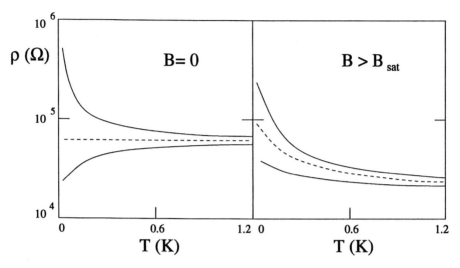

*Figure 2.* Resistivity versus temperature without (left) and with (right) a large parallel magnetic field $B > B_{sat}$. $n_s < n_{c1}$ (upper curves), $n = n_{c1}$ (middle curves) and $n > n_{c1}$ (lower curves).

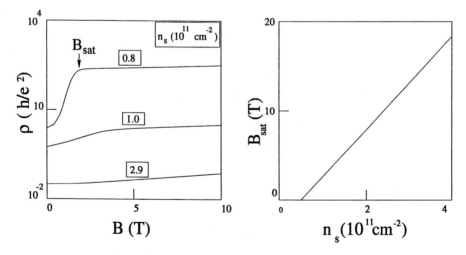

*Figure 3.* Resistivity versus parallel magnetic field (left) and saturation field $B_{sat}$ as a function of the carrier density (right).

sketched in Fig. 4 (Fig 2 of Ref. [29]) indicates the existence of a critical density $n_{c2}$ below which a non linearity is observed and above which it disappears. The density $n_{c1}$ and $n_{c2}$ are close to each others, if not identical when $B = 0$. The dependence of the characteristic $n_{c2}$ as a function of a parallel magnetic field $B$ is sketched in Fig. 4. (Fig. 4 of Ref. [29]).

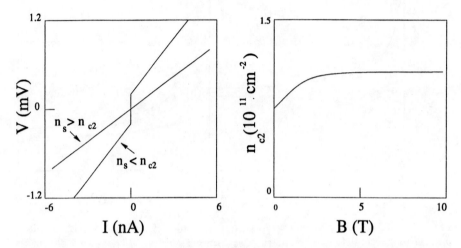

*Figure 4.* Current-voltage non linear characteristics of the insulating phase which are suppressed above the MIT (left). Critical density $n_{c2}$ above which the non linearity of the I-V characteristics disappears as a function of the parallel magnetic field.

The critical density $n_{c1}$ does not give a unique critical value for the factor $r_s$. Impurity scattering plays a role. For clean systems, one needs to have a much larger factor $r_s$ than in a dirty system, as sketched in Fig. 5. (see inset of Fig. 1 in Ref. [33]). In an (undoped) organic field effect transistor, the 2DEG is less scattered by impurities and the MIT is seen[27] at an even smaller density ($r_s \approx 50$).

The surprise caused by this unexpected metallic behavior was mainly due to its discrepancy with the scaling theory of localization, which does not take into account electron-electron interactions. When the conductance $g$ is larger than the conductance quantum $e^2/h$, a weak disorder perturbative expansion gives for the average conductance a universal logaritmic correction to the Drude conductivity which defavors transport, unless there is a sufficient spin orbit scattering. In the weak disorder limit, one can also take into account the interaction when $r_s < 1$ and one obtains additional corrections which reduce transport in a similar way. However, the extrapolation of the small $r_s$ interaction dependent correction to larger $r_s$ suggests a possible change of the sign of the corrections due to the interactions, indicating the possibility of a metallic phase in two dimensions, as mentioned by Finkelshtein [34]. Computer calculations without interaction and transport measurements at not too low densities made in the eighties have confirmed the absence of metallic

From the Fermi liquid towards the Wigner solid in two dimensions 271

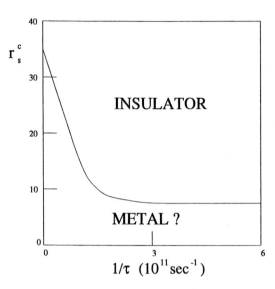

*Figure 5.* Critical factor $r_s^c$ at which the MIT is observed as a function of the inverse elastic scattering time.

behavior in two dimensions. The difference between the recent experiments giving a MIT and the former experiments confirming the absence of metallic behavior seems to be the quality of the interfaces at which the 2DEG is created. This feature makes possible to have a measurable conductivity at much lower carrier densities than previously. An intermediate range of density, where the factor $r_s$ is too large (too small) to allow expansion in powers of $r_s$ ($1/r_s$) and a weak elastic scattering seems to be necessary for observing the metallic behavior. This hypothesis was supported by Ref. [35] where a study of a 2DHG in a Ga-As heterostructure gives a range $r_s^F < r_s < r_s^W$ for having a weak metallic behavior in a disordered sample. When $r_s < r_s^F$, one would have weakly interacting quasi-particles dominated by Anderson localization when the temperature $T \to 0$. When $r_s > r_s^W$, one would have a highly correlated set of charges. Between $r_s^F \approx 6$ and $r_s^W \approx 9$ in the studied sample, a problematic small metallic behavior is observed between two insulating behaviors of different nature (Anderson insulator for large densities, pinned Wigner solid for low densities). The re-entrant MIT at low $r_s$ is not easy to observe, if it exists, since the localization length of a clean device can be very large, and the observation of a possible re-entrant insulating behavior at high densities can require very low temperatures. This is why many observations of a MIT have been reported for $r_s \approx 10$, while very few experiments give a possible re-entrant MIT at $r_s \approx 3$.

Moreover, more recent works [5, 7, 9]) put doubts about the reality of this intermediate metallic behavior when the temperature goes to zero, since the effect of a weak perpendicular magnetic field can be described by usual weak localization theories, even for values of $r_s$ as large as 15 (see Fig. 6 taken from Ref. [5]). The hypothesis of a certain temperature dependent screening was suggested for explaining the anomalous temperature dependence, and it was proposed that usual quantum interferences should drive at possibly very low temperatures the system to the formerly expected insulating behavior. Recent measures performed down to $5mK$ do not confirm[36] this hypothesis. The estimate of the phase breaking length $L_\phi$ which is traditionally done for estimating the low temperature dependence of the resistance from a zero temperature theory leads again to the famous problem of the saturation [37] of $L_\phi$ when $T \to 0$, problem leading also to many possible and controversial explanations.

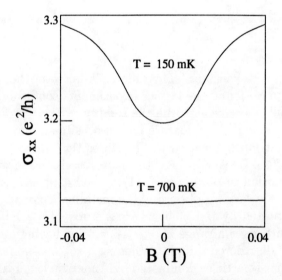

*Figure 6.* Usual weak localization behavior of the conductance induced by a perpendicular magnetic field in a 2DHG created in Ga-As heterostructure.

In summary, a significant metallic behavior can be seen using a 2DEG created in a Si-mosfet, while a weaker one occurs in a 2DHG created in a Si-Ge quantum well or a Ga-As hetrostructure. Nevertheless, in the latter system, the study of the compressibility gives complementary signatures [38, 39, 40] of a possible quantum phase transition. Local compressibility measurements show that the system is more homogenous in the intermediate metallic phase than in the low density insulating phase. Very recently, the possibility that the MIT would be accompanied by a magnetic transition has been suggested [32, 41].

## 4. LATTICE MODEL

The previous experimental observations lead us to numerically revisit the Fermi-Wigner crossover using a two dimensional model describing $N$ particles on $L \times L$ square lattice with periodic boundary conditions (BCs), i.e. with a torus topology. The most general Hamiltonian $\mathcal{H}$ of the lattice model we will focus on reads,

$$\mathcal{H} = \sum_{i,\sigma}(-t\sum_{i'} c^\dagger_{i',\sigma}c_{i,\sigma} + v_i n_{i,\sigma})$$
$$+\frac{U}{2}\sum_{\substack{i,i' \\ i \neq i'}} \frac{n_{i,\sigma}n_{i',\sigma'}}{|i-i'|} + 2U\sum_i n_{i,\uparrow}n_{i,\downarrow}, \qquad (6)$$

where the operators $c_{i,\sigma}$ ($c^\dagger_{i,\sigma}$) destroy (create) an electron of spin $\sigma$ at the site $i$ and $n_{i,\sigma} = c^\dagger_{i,\sigma}c_{i,\sigma}$. $\mathcal{H}$ consists of

- a hopping term $-t$ that couples nearest-neighbor sites, and accounts for the quantum kinetic energy,
- the pairwise electron-electron interaction, which itself consists of a $2U$ Hubbard repulsion when two electrons are at the same site $i$ with opposite spins and a $U/|i-i'|$ spin independent Coulomb repulsion when they are separated by a distance $|i-i'|$ (smallest distance between the sites $i$ and $i'$ on a square lattice with periodic BCs),
- on site random potentials $v_i$ which are uniformly distributed inside the interval $[-W/2, W/2]$.

The *clean* system is obtained when the disorder strength $W$ is set to zero. In our model with a lattice spacing $a$, $\hbar^2/(2m^*a^2) \to t$, $e^2/(\epsilon a) \to U$, such that the factor $r_s$ becomes:

$$r_s = \frac{1}{\sqrt{\pi n_s}a^*_B} = \frac{U}{2t\sqrt{\pi n_e}}, \qquad (7)$$

for a filling factor $n_e = N/L^2$. This dimensionless ratio $r_s$ will allow us to compare our results obtained as a function of $U$ for a fixed filling factor $n_e$ and the experimental results obtained as a function of $n_s$.

We denote $\mathcal{S}$ and $\mathcal{S}_z$ the total spin and its component along an arbitrary direction $z$. Since $[\mathcal{S}^2, \mathcal{H}] = [\mathcal{S}_z, \mathcal{H}] = 0$, $\mathcal{H}$ can be written in

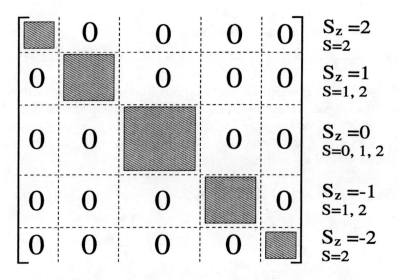

*Figure 7.* Structure of the Hamiltonian matrix for 4 electrons in a 6×6 square lattice. The size of the different diagonal non zero sub-blocks are $N(S_z) = 396900, 257040$ and $58905$ for $S_z = 0, \pm 1, \pm 2$ respectively.

a block-diagonal form, with $N+1$ blocks where $S_z = -N/2, \ldots, N/2$ respectively. When $B = 0$, there is no preferential direction and the groundstate energy $E_0$ does not depend on $S_z$. For a groundstate of total spin $S$, $\mathcal{H}$ has $2S+1$ blocks with the same lowest eigenenergy $E_0(S^2)$ since $E_0(S^2) = E_0(S^2, S_z)$; $S_z = -S, -S+1, \ldots, S-1, S$. Therefore, the number $N_b$ of blocks of different $S_z$ and of same lowest energy gives the total spin $S = (N_b - 1)/2$ of the groundstate.

If $N$ and $L$ are small enough, the ground state and the first excitations can be exactly calculated using Lanczos algorithm. Otherwise, certain approximations are unavoidable. Let us focus on the case $N = 4$ and $L = 6$ where the structure of $\mathcal{H}$ is given in Fig. 7 for different $S_z$. Without magnetic field, we have the symmetry $\pm S_z$, and we have only to diagonalize the three sub-blocks with $S_z \geq 0$. $\mathcal{H}(S_z = 2)$ corresponds to fully polarized electrons (spinless fermions) where the orbital part of the wave-functions is totally anti-symmetric.

## 5. STUDIED QUANTITIES

From exact diagonalization for small systems and using approximations for larger systems and weak coupling, we will study:

- the lowest eigenenergies $E_n(S_z)$ of the $S_z$ sub-blocks and the corresponding eigenvectors $|\Psi_n(S_z)>$. $n = 0, 1, 2, \ldots$ corresponds to the states ordered by increasing energies.

- The local persistent currents $\vec{J}(i)$ created at a site $i$ by an Aharonov Bohm flux $\phi$ which is enclosed along the longitunal $l$-direction as sketched in Fig. 8. The flux $\phi$ can be included by taking

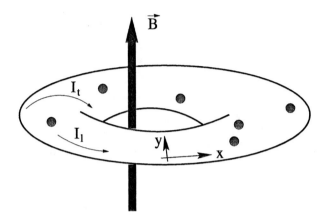

*Figure 8.* 2D Torus with $N$ electrons enclosing an Aharonov-Bohm flux $\phi = BL_x^2$

appropriate longitudinal BCs (antiperiodic BCs corresponding to $\phi = \pi$ in our convention). The BCs along the transverse $t$-direction remain periodic. The $\vec{J}(i)$ are vectors defined by their longitudinal and transverse components $(\vec{J}(i) = (J_{i,l}, J_{i,t}))$, or by their angles $\theta_i = \arctan(J_{i,t}/J_{i,l})$ and their absolute values $J_i = |\vec{J}_i|$. The longitudinal component $J_{i,l}$ of an eigenstate $|\Psi\rangle$ is defined as

$$J_{i,l} = 2\text{Im}\langle\Psi|c^\dagger_{i_x+1,i_y}c_{i_x,i_y}\exp(i\phi/L)|\Psi\rangle, \qquad (8)$$

and $J_{i,t}$ is given by

$$J_{i,t} = 2\text{Im}\langle\Psi|c^\dagger_{i_x,i_y+1}c_{i_x,i_y}|\Psi\rangle. \qquad (9)$$

The total current $I^{(n)}$ of the $n^{\text{th}}$ many-body wavefunction $|\Psi_n\rangle$ of energy $E_n$ has a total longitudinal component $I_l(n)$ given by

$$I_l(n)(\overline{\phi}) = -\left.\frac{\partial E_n}{\partial \phi}\right|_{\phi=\overline{\phi}} = \frac{\sum_i J_{i,l}(n)}{L}. \qquad (10)$$

which will be calculated for $\overline{\phi} = \pi/2$.

- The crystallization parameter $\gamma$ defined using the function $C(r) = N^{-1}\sum_i \rho_i \rho_{i-r}$, where $\rho_i = \langle\Psi|n_i|\Psi\rangle$ is the electronic density of the state $|\Psi\rangle$ at the site $i$. The crystallization parameter $\gamma$ is given by

$$\gamma = \max_r C(r) - \min_r C(r) \tag{11}$$

Note that $\gamma = 1$ when the $N$ particles are localized on $N$ lattice sites and form a rigid solid and 0 when they are extended on the $L^2$ sites and form an homogenous liquid.

- The participation ratio $\chi = N^2(\sum_i \rho_i^2)^{-1}$, which gives the typical number of lattice sites occupied by an eigenstate $|\Psi>$.

- The spectral parameter $\eta$ which characterizes the level repulsion. Uncorrelated spectra exhibit Poisson statistics. Correlated spectra can be described by Random Matrix Theory with Wigner-Dyson (W-D) statistics. For the one body spectra, the distribution $P(s)$ of the normalized energy spacings between consecutive levels has two different forms when $L \to \infty$: the Poisson distribution $P_P(s) = \exp(-s)$ if the wavefunctions are localized, the Wigner surmises $P_W^O(s) = (\pi s/2)\exp(-\pi s^2/4)$ with time reversal symmetry (TRS) and $P_W^U(s) = (32s^2/\pi^2)\exp(-4s^2/\pi)$ without TRS, if the wave functions are extended. For the normalized $N$-body energy spacings $s_n = (E_{n+1} - E_n)/ < E_{n+1} - E_n >$ (the brackets denote ensemble average), we define a spectral parameter:

$$\eta_{(O,U)} = \frac{\text{var}(P(s)) - \text{var}(P_W^{(O,U)}(s))}{\text{var}(P_P(s)) - \text{var}(P_W^{(O,U)}(s))}, \tag{12}$$

($\eta_O$ with TRS, $\eta_U$ in the absence of TRS, for instance when $\phi = \pi/2$). var$(P(s))$ denotes the variance of $P(s)$. The spectral parameter $\eta = 1$ when $P(s) = P_P(s)$ and $\eta_{(O,U)} = 0$ when $P(s) = P_W^{(O,U)}(s)$.

- The Zeeman energy necessary to polarize a non magnetized cluster. A *parallel* magnetic field $B$ does not induce orbital or Aharonov-Bohm effects, but defines the $z$-direction and removes the $S_z$ degeneracy by the Zeeman energy $-g\mu B S_z$. The ground state energy and its magnetization are given by the minimum of $E_0(S^2, S_z, B = 0) - g\mu B S_z$. For a $S = 0$ groundstate without field, the value $B^*$ for which $E_0(S_z) - g\mu B^* S_z = E_0(S_z = 0)$ defines the field necessary to polarize the system to $S \geq S_z$. If one studies $N = 4$ electrons, the total $Q_2 = E_0(S_z = 2) - E_0(S_z = 0)$ and partial $Q_1 = E_0(S_z = 1) - E_0(S_z = 0)$ polarization energies give the Zeeman energies necessary to yield $S = 2$ and $S = 1$ respectively for a cluster with $S = 0$.

## 6. INTERMEDIATE COUPLING REGIME FOR SPINLESS FERMIONS AND WEAK DISORDER

We first consider an ensemble of disordered clusters with $L = 6$ and $N = 4$. The ground state (GS) and the first excitations of the fully polarized sub-block ($S_z = 2$, spinless fermions) of the Hamiltonian matrix shown in Fig.7 have been obtained using the Lanczos algorithm. The statistical ensemble typically includes $10^3 - 10^4$ samples obtained from a disorder distribution with $W = 5$. This is a relatively weak disorder for which one has quantum diffusion when $r_s = 0$ inside the small clusters (no Anderson localization).

### 6.1. GROUND STATE

We summarize in this subsection the main results published in Ref. [42] and complementary unpublished results. As one switches on $U$, a first characteristic threshold $r_s^F$ can be identified by looking at the average total longitudinal persistent current $I_l$ of the GS at $\phi = \pi/2$, and comparing the exact quantity with the Hartree-Fock (HF) approximation (see appendix). Below $r_s^F \approx 5$, the mean field approximation reproduces the exact $I_l$, but strongly underestimates $I_l$ above $r_s^F$. This sharp breakdown of the HF approximation shown in Fig. 9 means that strong correlation effects occur above $r_s^F$, such that the shift the GS energy when the BCs are changed cannot be obtained assuming the best possible SD for the ground state.

A closer investigation of the persistent currents on a typical sample gives three regimes, as shown in Fig.10. See also Ref. [43]. For weak coupling, the local currents flow randomly inside the cluster, due to elastic scattering on the site potentials. For intermediate coupling, the pattern of the persistent currents becomes oriented along the shortest direction enclosing $\phi$. For large coupling, the oriented currents vanish. Ref. [44] gives a detailed study of the large coupling limit where one can use perturbation theory for having the sign and the magnitude of $I_l$.

If one looks at the distribution of the angles $\theta_i$ of the local currents, one can see in Fig.11 that the currents are randomly scattered without interaction, and that they become aligned when one goes to the strong coupling limit. The ensemble average value $< |\theta| >$ allows us to quantify the progressive change. If $p(\theta) = 1/(2\pi)$, $< |\theta| > = \pi/2$, a value obtained for the low ratios $r_s$. At large $r_s$, $< |\theta| > \to 0$. The ratio $r_s$ at which the local currents cease to be oriented at random is consistent with the critical ratio $r_s^F$ where the HF approximation breaks down.

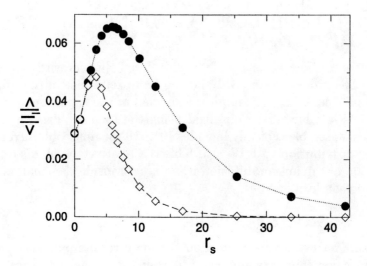

*Figure 9.* Ensemble average longitudinal GS current $< I_l >$ as a function of $r_s$ for $N = 4$, $L = 6$ and $W = 5$. Exact values (filled symbols) and HF values (empty symbols).

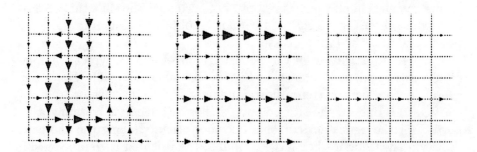

*Figure 10.* Map of the local persistent currents in a given sample for small (left), intermediate (center) and large (right) values of $r_s$.

By studying the average amplitude of the local currents, one can see in Fig. 12 that $< J_i >$ is essentially independent of $r_s$ up to a second threshold $r_s^W \approx 10$ which exceeds $r_s^F$. Moreover, comparing in Fig. 12 the GS average crystallization parameter $< \gamma >$ and $< J_i >$, on can see that the suppression of the persistent currents coincides with the formation of a solid Wigner molecule inside the disordered clusters.

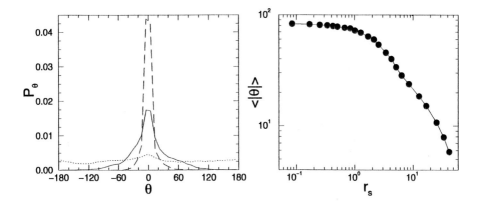

*Figure 11.* Left: Distribution $P_\theta$ of the GS local current angles for $r_s = 0$ (dotted line), $r_s = 6.3$ (full line) and $r_s = 42$ (dashed line). Right: Ensemble average angle $<|\theta|>$ as a function of $r_s$.

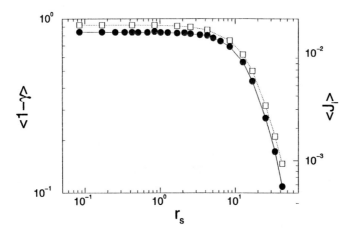

*Figure 12.* Averages of the GS crystallization parameter $<1 - \gamma>$ (left scale, empty symbols) and GS local current amplitude $<J_i>$ (right scale, filled symbols) as a function of $r_s$.

The response of the ground state to an enclosed Aharonov-Bohm flux shows us that an intermediate correlated regime takes place between the Fermi limit and the Wigner limit, when typically $5 < r_s < 10$.

## 6.2. LOW ENERGY EXCITATIONS

In Ref. [45], the low energy excitations of the same clusters have been studied, notably their statistics when the microscopic configurations of the random substrate are changed. For intermediate ratios $r_s$, the GS and the 8 first following low energy excitations are characterized by oriented non random persistent currents and do not exhibit Wigner-Dyson (W-D) spectral statistics. Above those states, when the excitation energy $\epsilon$ exceeds an energy of the order of the Fermi energy $\epsilon_F$, the local currents become randomly oriented and the levels obey W-D statistics. Incidentally, let us note that the metallic behavior observed for intermediate couplings disappears also when the temperature exceeds a temperature of the order of the Fermi temperature.

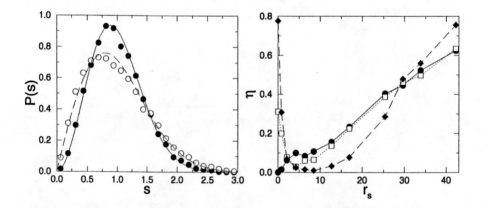

*Figure 13.* Left: Spacing distribution $P(s)$ for $r_s = 6.3$ when $\phi = 0$ (○) and $\phi = \pi/2$ (•) for energy levels above the 9 first levels (excitation energies $1.4 < \epsilon/\epsilon_F < 1.9$), compared to $P_O^W(s)$ (dashed line) and $P_U^W(s)$ (continuous line). Right: Spectral parameter $\eta_U$ as a function of $r_s$ for the first excitation $s_0$ (circles), $s_2 - s_4$ (squares) and $s_{10} - s_{20}$ (diamonds).

In Fig. 13 (left) one can see that the spacing distribution $P(s)$ calculated for $r_s = 6.3$ using the low energy levels except the 9 first levels ($n = 10, \ldots, 20$), is given by the Wigner surmise, with an orthogonal-unitary crossover when one turns on an Aharonov-Bohm flux $\phi = 0 \to \pi/2$. This corresponds to excitation energies $1.4 < \epsilon/\epsilon_F < 1.9$. Taking $\phi = \pi/2$, the variation of the spectral parameter $\eta_U$ as a function of $r_s$ is given for the successive level spacings in Fig. 13. The first excitation is described by the Wigner surmise ($\eta_U = 0$) without interaction but becomes more and more Poissonian when $r_s$ increases. The spacings characterizing the levels with $n = 10, \ldots, 20$ have an opposite behavior. For $r_s = 0$, those excitations being the sum of more than one single-electron excitation are essentially uncorrelated, but become correlated for intermediate $r_s$ ($\eta_U \approx 0$) before being again uncorrelated at larger $r_s$.

In summary, when one considers the low energy spectral statistics, a complementary signature of an intermediate regime is obtained, given by W-D statistics and randomly oriented local persistent currents outside the 9 first states for which the absence of W-D statistics for intermediate $r_s$ is accompanied by a non random orientation of the persistent current angles $\theta$ (see Ref. [45]). This behavior does not appear for weak and strong couplings, where the low energy spectral correlations decrease as the excitation energy increases.

## 6.3. INTERMEDIATE LIQUID-SOLID REGIME IN THE CLEAN LIMIT

The fact that the 9 first states do not display quantum ergodicity for intermediate coupling and weak disorder suggests the existence of 9 low energy collective excitations. A collective motion cannot be due to impurity scattering and should come from the corresponding clean limit. This limit has been investigated in Ref. [46].

When $W = 0$, one has a single system which remains invariant under rotation of angle $\pi/2$ and under translations and reflections along the longitudinal $x$ and transverse $y$ directions. Invariance under translations implies that the momentum $K$ is a good quantum number which remains unchanged when $U$ varies. The symmetries imply that the states are fourfold degenerate if $K \neq 0$ and can be non degenerate if $K = 0$.

When $U = 0$, the states are $N_H$ plane wave Slater determinants (SDs) $d^\dagger_{k(4)} d^\dagger_{k(3)} d^\dagger_{k(2)} d^\dagger_{k(1)} |0>$, where $d^\dagger_{k(p)}$ creates a particle in a state of momentum $k(p) = 2\pi(p_x, p_y)/L$ ($p_{x,y} = 1, \ldots, L$) and $|0>$ is the vacuum state. For $N = 4$ and $L = 6$, $N_H = 58905$. The low energy eigenstates are given by the following plane wave SDs:

- 4 degenerate ground states (GSs) $|K_0(\beta)>$ ($\beta = 1, \ldots, 4$) of energy $E_0(U = 0) = -13t$ and of momenta $K_0 = (0, \pm\pi/3)$ and $(\pm\pi/3, 0)$.

- 25 first excitations of energy $E_1(U = 0) = -12t$ out of which 4 plane wave SDs $|K_1(\beta)>$ will play a particular role for describing the intermediate GS. They correspond to a particle at an energy $-4t$ with $k(1) = (0,0)$, two particles at an energy $-3t$ and a fourth particle of energy $-2t$ with momenta such that $\sum_{j=2}^{4} k(j) = 0$. One has $k(2) = (0, \pm\pi/3)$, $k(3) = (\pm\pi/3, 0)$ and $k(4) = (\mp\pi/3, \mp\pi/3)$ or $k(2) = (0, \mp\pi/3)$, $k(3) = (\pm\pi/3, 0)$ and $k(4) = (\mp\pi/3, \pm\pi/3)$.

- 64 second excitations $|K_2(\alpha)>$ of energy $E_2(U = 0) = -11t$.

- 180 third excitations $|K_3(\alpha)>$ of energy $E_3(U=0) = -10t$.

- 384 fourth excitations of energy $E_4(U=0) = -9t$ out of which 16 plane waves SDs $|K_4(\delta)>$ will play a particular role for describing the intermediate GS. They are given by the condition that the total momentum is zero, which selects 20 SDs out of which 4 where the single particle state of energy $-3t$ is not occupied do not contribute. The $|K_4(\delta)>$ are 16 SDs of energy $-9t$, given by 8 SDs where the particles have energies $-4t, -3t, -2t, 0t$ respectively and by 8 other SDs where the particles have energies $-3t, -3t, -2t, -t$ respectively.

When $t = 0$, the states are $N_H$ Slater determinants $c_i^\dagger c_j^\dagger c_k^\dagger c_l^\dagger |0>$ built out from the site orbitals. The configurations $ijkl$ correspond to the $N_H$ different patterns characterizing 4 different sites of the $6 \times 6$ square lattice. The low energy part of the spectrum is made of the following site SDs:

- 9 squares $|S_0(I)>$ $(I=1,\ldots,9)$ of side $a=3$ and of energy $E_0(t=0) \approx 1.80U$.

- 36 parallelograms $|S_1(I)>$ of sides $(3, \sqrt{10})$ and of energy $\approx 1.85U$.

- 36 other parallelograms $|S_2(I)>$ of sides $(\sqrt{10}, \sqrt{10})$ and of energy $\approx 1.97U$.

- 144 deformed squares $|S_3(I)>$ obtained by moving a single site of a square $|S_I>$ by one lattice spacing and of energy $\approx 2U$.

For the first low energy states, the crossover from the $U = 0$ eigenbasis towards the $t = 0$ eigenbasis is shown in Fig. 14 when one increases the ratio $r_s$. If we follow the 4 GSs $E_0(r_s = 0)$ ($K_0 \neq 0$), one can see a first level crossing at $r_s^F \approx 9.3$ with a non degenerate state ($K_0 = 0$) which becomes the GS above $r_s^F$, followed by two other crossings with two other sets of 4 states with $K_I \neq 0$. When $r_s$ is large, 9 states coming from $E_1(r_s = 0)$ have a smaller energy than the 4 states coming from $E_0(r_s = 0)$. The degeneracies ordered by increasing energy become $(1, 4, 4, 4, \ldots)$ instead of $(4, 25, 64, \ldots)$ for $r_s = 0$. Since the degeneracies are $(9, 36, 36, \ldots)$ when $t = 0$, these 9 states give the 9 square molecules $|S_0(I)>$ when $r_s \to \infty$. When $r_s^{-1}$ is very small, the first 9 states correspond to a single massive molecule free to move on a restricted $3 \times 3$ lattice, the single non frozen degree of freedom in this limit being the location $R_I$ of the center of mass of the $|S_0(I)>$. One has an effective hopping term $T \propto t r_s^{-3}$ when $N = 4$ and the total momentum is quantized ($K_l(I) = 2\pi p_l/3$ and $K_t(I) = 2\pi p_t/3$ being its longitudinal

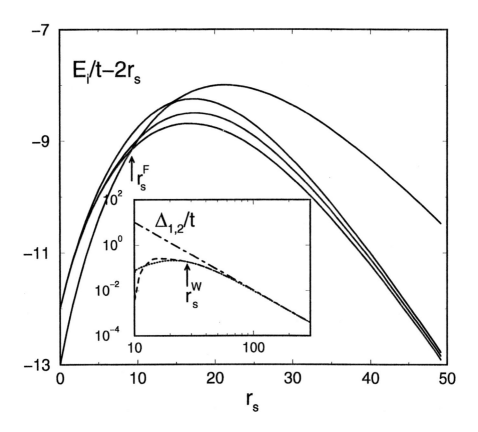

*Figure 14.* As a function of $r_s$, low energy part of the spectrum exhibiting a GS level crossing at $r_s^F$. Inset: two first level spacings $\Delta_1/t$ (dashed) and $\Delta_2/t$ (dotted) which become equal at $r_s^W$ and the perturbative result $\Delta_1/t = \Delta_2/t \approx 10392/r_s^3$ valid when $r_s \to \infty$ (dot-dashed).

and transverse components respectively with $p_{l,t} = 1, 2, 3$). For a square lattice at a filling factor $1/9$, the $R_I$ are indeed located on a periodic $3 \times 3$ square lattice. This is an important simplification of our model. This gives 9 states of kinetic energies given by $-2T(\cos K_l(I) + \cos K_t(I))$. The kinetic part of the low energy spectrum is then $-4T, -T, +2T$ with degeneracies $1, 4, 4$ respectively. This structure with two equal energy spacings $\Delta_1$ and $\Delta_2$ appears (inset of Fig. 14) when $r_s$ is larger than the crystallization threshold $r_s^W \approx 28$. Above $r_s^W$, to create a defect in the rigid molecule costs a high energy available in the $10^{th}$ excitation only. We have seen in the previous section that the 9 first levels do not obey Wigner-Dyson statistics at intermediate $r_s$ when a random potential is added, in contrast to the following levels. The study of the clean limit gives us the explanation. The two characteristic thresholds $r_s^F$ (level crossing) and $r_s^W$ (9 first states having the structure of the spectrum of a single massive molecule free to move on a 3 × 3 square lattice) can also be detected by other methods given in Ref. [46].

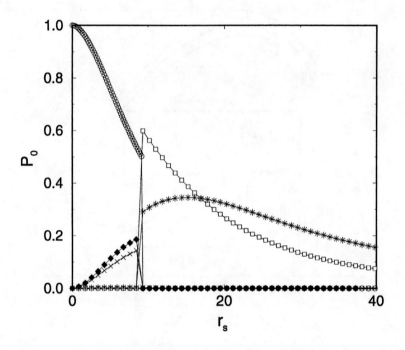

*Figure 15.* Ground state projections $P_0(r_s)$ onto a few plane wave SDs, given by the $4|K_0(\beta)>$ (empty circle), the $4\,|K_1(\beta)>$ (empty square), the $64\,|K_1(\alpha)>$ (filled diamond), the $180\,|K_2(\alpha)>$ (×), the $16\,|K_4(\delta)>$ (asterisk) respectively, as a function of $r_s$.

To understand further the nature of the intermediate GS, we have projected the GS wave functions $|\Psi_0(r_s)>$ over the low energy eigenvectors of the two eigenbases valid for $U/t = 0$ (Fig. 15) and for $t/U = 0$ (Fig. 17) respectively.

Let us begin by studying the GS structure in the $U = 0$ eigenbasis. Below $r_s^F$, each of the 4 GSs $|\Psi_0^\alpha(r_s)>$ with $K_0 \neq 0$ has still a large projection

$$P_0(r_s, 0) = \sum_{\beta=1}^{4} |<\Psi_0^\alpha(r_s)|K_0(\beta)>|^2 \qquad (13)$$

over the 4 non interacting GSs. There is no projection over the 25 first excitations and smaller projections $P_0(r_s, 2)$ and $P_0(r_s, 3)$ over the 64 second and 180 third excitations of the non interacting system. Above $r_s^F$, the non degenerate GS with $K_0 = 0$ has a large projection

$$P_0(r_s, 1) = \sum_{\beta=1}^{4} |<\Psi_0(r_s)|K_1(\beta)>|^2 \quad (14)$$

which is equally distributed over the 4 excitations $|K_1(\beta)>$ of momentum $K_1 = 0$ and a second significant contribution

$$P_0(r_s, 4) = \sum_{\delta=1}^{16} |<\Psi_0(r_s)|K_4(\delta)>|^2 \quad (15)$$

given by its projection onto the 16 plane wave SDs $|K_4(\delta)>$ belonging to the fourth excitation of the non interacting system. Above $r_s^F$, its projections onto the 4 $|K_0(\beta)>$, the 21 other first excitations and the second and third excitations of the non interacting system are zero or extremely negligible.

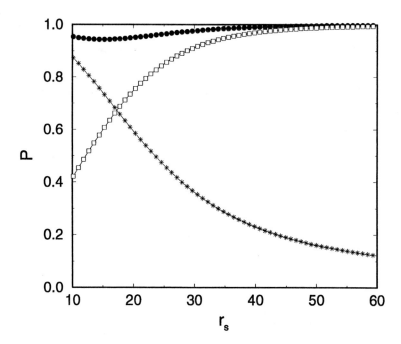

*Figure 16.* Ground state projection $P_0^t(r_s)$ (asterisk) and $P_\infty^t(r_s)$ (empty square) onto the subspace spanned by the low energy plane wave and site SDs respectively, and total GS projection $P$ (filled circle) onto the re-orthonormalized basis using the low energy eigenvectors of the two limiting bases.

The total GS projection

$$P_0^t(r_s) = P_0(r_s, 1) + P_0(r_s, 4) \qquad (16)$$

onto the 4 $|K_1(\beta)>$ and 16 $|K_4(\delta)>$ is given in Fig. 16 when $r_s > r_s^F$. This shows us that a large part of the system remains an excited liquid above $r_s^F$, given by a special rule of occupation of the one particle plane wave states. The occupation of the one body states is very different from the usual Pauli rule after the level crossing $r_s^F$. A necessary, but non sufficient condition for a plane wave SD to significantly contribute to the zero momentum GS is of course to have a zero total momentum. Those projections decrease as $r_s$ increases and become negligible in the strong coupling limit. A complete GS description in this limit will require more and more plane wave SDs.

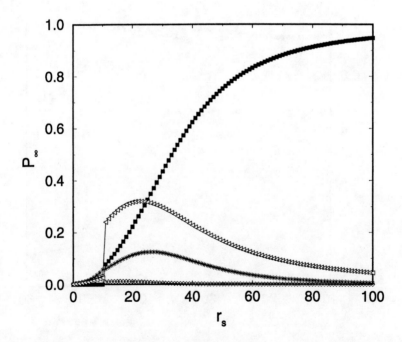

*Figure 17.* Ground state projection $P_\infty(r_s)$ onto a few site SDs, given by the 9 squares $|S_0(I)>$ (filled square), the 36 parallelograms $|S_1(I)>$ (asterisk), the 36 parallelograms $|S_2(I)>$ (diamond), and the 144 deformed squares $|S_3(I)>$ (left triangle) respectively, as a function of $r_s$.

We now study the GS projections $P_\infty$ onto the $t = 0$ eigenbasis. The GS projection

$$P_\infty(r_s, 0) = \sum_{I=1}^{9} |<\Psi_0^\alpha(r_s)|S_0(I)>|^2 \qquad (17)$$

onto the 9 square site SDs $|S_0(I)>$ is given in Fig. 17, together with the GS projection $P_\infty(r_s, J)$ onto the site SDs corresponding to the $J^{th}$ degenerate low energy excitations of the $t = 0$ system. The total GS projection

$$P_\infty^t(r_s) = \sum_{p=0}^{3} P_\infty(r_s, p) \qquad (18)$$

onto the 9 squares $|S_0(I)>$, the 36 parallelograms $|S_1(I)>$, the 36 other parallelograms $|S_2(I)>$ and the 144 deformed squares $|S_3(I)>$ is given in Fig. 16 when $r_s > r_s^F$. This shows us that the ground state begins to be a floppy solid also above $r_s^F$. When $r_s$ increases, $P_\infty(r_s, 0)$ goes to one, and the ground state is a simple rigid square Wigner molecule.

The site SDs and plane wave SDs are not orthonormal. After re-orthonormalization, the total projection $P$ of $|\Psi_0(r_s)>$ over the subspace spanned by the 4 $|K_1(\beta)>$ and 16 $|K_4(\delta)>$ and 225 site SDs of lower electrostatic energies (9 squares, 36 + 36 parallelograms, 144 deformed squares) are given in Fig. 16. One can see that $|\Psi_0(r_s)>$ is almost entirely located inside this very small part of a huge Hilbert space for intermediate $r_s$, spanned by no more than 245 low energy SDs of different nature, and adapted to describe a solid entangled with an excited liquid.

From the study of the GS projections emerges the conclusion that a minimal description of the intermediate GS requires to combine the low energy states of the two limiting eigenbases. In this sense, the GS is neither solid, nor liquid, but rather the quantum superposition of those two states of matter. This is strongly reminiscent of the conjecture proposed in Ref. [23] for the quantum melting of solid Helium in three dimensions. It suggests possible improvements of the trial GS to use for intermediate $r_s$ in variational or path integral quantum Monte Carlo approaches. Instead of using Jastrow wave functions improving the plane wave SDs for the liquid or the site SDs for the solid, or their nodal surfaces, it will be interesting to study if a combination of the two, with a more complex nodal structure, and describing a solid-liquid regime does not lower the GS energy for intermediate $r_s$. A positive answer would confirm that an unnoticed intermediate solid-liquid phase does exist in the thermodynamic limit for fermionic systems in two dimensions.

## 6.4. MAGNETIC SIGNATURE OF THE INTERMEDIATE REGIME FOR WEAK DISORDER

We return to the study of weakly disordered samples when the spin degrees of freedom are included. Their role and the consequences of an applied parallel magnetic field which align only the spins without inducing orbital effects, have been the subject of Ref.[47]. A statistical ensemble of matrices having the structure given in Fig. 7 have been studied for $W = 5$, $N = 4$ and $L = 6$ providing complementary signatures of a particular intermediate behavior.

When $r_s = 0$, the two one body states of lowest energy are doubly occupied and $S = 0$ ($S = 1/2$ if $N$ is odd). To polarize the $S = 0$ ground state to $S = 1$ corresponds to the transition of one electron at the Fermi energy and costs an energy equal to the one body level spacing. $p(Q_1)$ is then given by the spacing distribution $p(s)$ between consecutive one body levels, the Wigner surmise $P_W^O(s)$ in the diffusive regime. When $r_s$ is large, the 4 electrons occupy the four sites $c_j$ $j = 1, \ldots, 4$ of the square configuration $|S_0(I)>$ of side $a = 3$ with the lowest substrate energy $\sum_{j=1}^{4} v_{c_j}$. The ground state in this limit becomes $|\Psi_c> = \prod_{j=1}^{4} c_{c_j,\sigma_j}^{\dagger}|0>$ with a spin independent energy $E_c$. This square can support $2^N = 16$ spin configurations. We summarize the main results of a perturbative expansion of $E_c$ in powers of $t/U$. The spin degeneracy of $E_c$ is removed by terms of order $t(t/U)^{2a-1}$, which is the smallest order where the 16 spin configurations can be coupled via intermediate configurations allowing a double occupancy of the same site. Therefore, $2a - 1$ is the order where the perturbation begins to depend on $S_z$ and $Q_1$ as $Q_2 \propto t(t/U)^{2a-1} \to 0$ when $t/U \to 0$ (we have numerically checked this decay when $r_s > 100$). Moreover, the correction to $E_c$ depending on $S_z$ and $\propto t(t/U)^{2a-1}$ is given by an effective antiferromagnetic Heisenberg Hamiltonian. The $S = 0$ ground state for large $r_s$ correspond to 4 electrons forming an antiferromagnetic square Wigner molecule. However $Q_1$ and $Q_2$ are very small when $r_s$ is large, and the antiferromagnetic behavior can be an artefact due to the square lattice. Without impurities and in a continuous limit, a quasi-classical WKB expansion [21] valid for very large values of $r_s$ shows that 3 particle exchanges dominate, leading to ferromagnetism. Recent Monte-Carlo calculations [20] suggest that the crystal in the continuous limit becomes a frustrated antiferromagnet closer to the melting point.

The perturbative corrections $\propto t(t/U)^{2a-1}$ depend on the random variables $v_i$ via $\prod_{j=1}^{2a-1}(E_c - E_J)^{-1}$ where the $E_J$ are the classical energies of the intermediate configurations. $E_J$ is the sum of an electrostatic

From the Fermi liquid towards the Wigner solid in two dimensions 289

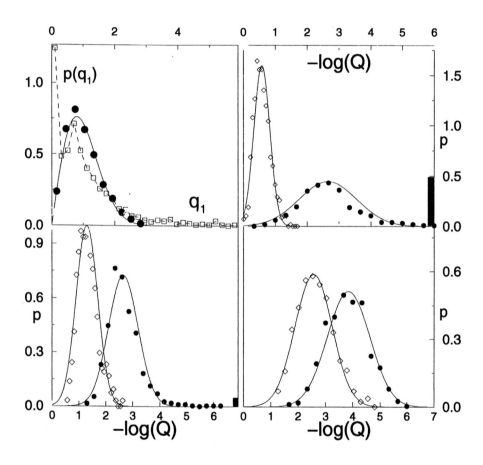

*Figure 18.* Distributions $p$ of the polarization energies $Q_1$ and $Q_2$ at different values of $r_s$. Upper left: $p(q_1)$ at $r_s = 0$ (circle) and 2.5 (square) where $q_1 = Q_1/<Q_1>$. The continuous line is the Wigner surmise. $p(-\log Q_1)$ (filled circle) and $p(-\log Q_2)$ (empty diamond) at $r_s = 2.5$ (upper right, right scale) 5.8 (lower left, left scale) and 16.8 (lower right, right scale) respectively. The thick bars (put at right edge of the figures) give the peaks $\delta(Q_1)$ of the bimodal $p(Q_1)$. The continuous lines are normal fits.

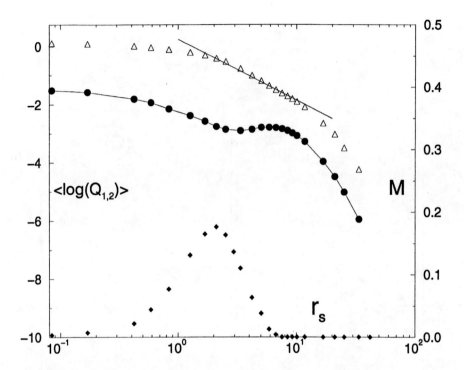

*Figure 19.* As a function of $r_s$, fraction $M$ of clusters with $S = 1$ at $B = 0$ (filled diamond, right scale), partial $<\log Q_1>$ (filled circle, left scale) and total $<\log Q_2>$ (empty triangle, left scale) energies required to polarize $S = 0$ clusters to $S = 1$ and $S = 2$ respectively. The straight line corresponds to $0.25 - 2\log r_s$.

energy and of a random substrate energy $E_s(J) = \sum_{k=1}^{4} v_{J(k)}$. Due to the high order $2a-1$ of the correction, a normal distribution for $E_J$ leads to a log-normal distribution for $\prod_{J=1}^{2a-1}(E_c - E_J)^{-1}$. Therefore $p(\Delta_1)$ and $p(\Delta_2)$ should be log-normal when $r_s$ is large.

$p(Q_1)$ is given in Fig. 18 for different $r_s$. The expected Wigner surmise takes place for $r_s = 0$. A small interaction quickly drives $p(Q_1)$ towards a bimodal distribution, with a delta peak at $Q_1 = 0$ and a main peak centered around a non zero value of $Q_1$. The delta peak gives the probability to have spontaneously magnetized clusters with $S = 1$. The main peak gives the field $B$ necessary to create $S = 1$ in a cluster with $S = 0$. The logarithmic scale used in Fig. 18 (upper right) underlines the bimodal character of the distribution and confirms that the main peak becomes log-normal when $r_s$ is large. The distribution of $Q_2$ is not bimodal: a fully polarized cluster has never been seen when $B = 0$. $Q_2$ becomes also log-normally distributed when $r_s$ is large.

In Fig. 19, the fraction $M$ of clusters with $S = 1$ at $B = 0$ is given as a function of $r_s$. One can see the mesoscopic Stoner instability (see

From the Fermi liquid towards the Wigner solid in two dimensions   291

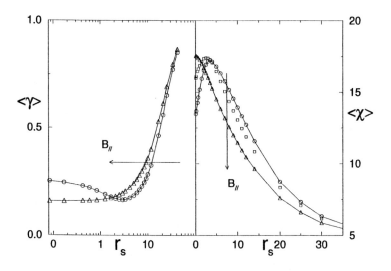

*Figure 20.* As a function of $r_s$, ensemble averages of the crystallization parameter $<\gamma(S_z)>$ (right) and of the numbers of occupied sites $<\chi(S_z)>$ (left). $S_z = 0$ (circle), $S_z = 1$ (square) and $S_z = 2$ (triangle). The arrows indicate the effect of a parallel magnetic field.

appendix) taking place at $r_s \approx 0.35$. The Stoner mechanism should eventually give fully polarized electrons. This is not the case, the increase of $M$ breaks down when $r_s = r_s^{FS} \approx 2.2$, a value where the Stoner mechanism and hence the HF approximation break down. In the same clusters, we have seen that the HF approximation fails to describe the persistent currents of the fully polarized sub-block of $\mathcal{H}$ (spinless fermions) when $r_s > r_s^{FP} \approx 5$. $r_s^{FP}$ takes a smaller value $r_s^{FS}$ when the spin degrees of freedom are included. Above $r_s^{FS}$, $M$ regularly decreases to reach a zero value for $r_s^{WS} \approx 9$ where an antiferromagnetic square molecule is formed. In the intermediate regime, there is a competition between the Stoner ferromagnetism and the Wigner antiferromagnetism. Since the $S = 0$ clusters are characterized by log-normal distributions, the ensemble averages $< \log Q_1 >$ and $< \log Q_2 >$ (without taking into account the $S = 1$ spontaneously magnetized clusters) define the typical fields $B$ necessary to yield $S = 1$ or $S = 2$ in a $S = 0$ cluster. Fig. 19 provides two magnetic signatures confirming the existence of a novel intermediate regime between the Fermi glass ($r_s < r_s^{FS}$) and the Wigner glass ($r_s > r_s^{WS}$): $< \log Q_1 >$ becomes roughly independent of $r_s$, while $Q_2 \propto r_s^{-2}$.

The number of occupied sites $\chi(S_z)$ depends on $S_z$ for small $r_s$ and becomes independent of $S_z$ for large $r_s$. At $r_s = 0$, $\Psi_0(S_z = 2, 1, 0)$ occupy respectively 4, 3, 2 one body states while the Wigner molecule occupies 4 sites only at large $r_s$. The ensemble average $<\chi(S_z=2)>$ shown in Fig. 20 is maximum when $r_s = 0$ and decays as $r_s$ increases, suggesting the absence of delocalization for the polarized system. The non polarized system behaves differently, since $<\chi(S_z=1,0)>$ first increase to reach a maximum $\approx<\chi(S_z=2,r_s=0)>$ before decreasing. In Fig.20 one can see also from the curves $\gamma(r_s)$ that charge crystallization is easier when the clusters are polarized than otherwise. The arrows indicated in Fig. 20 underline two consequences of a parallel field $B$: smaller number of occupied sites and smaller crystallization threshold.

## 6.5. NUMERICS VERSUS EXPERIMENTS

One can question whether the comparison between numerical simulations based on a small lattice model with a few particles and measures using $10^{11}$ electrons per $cm^2$ makes sense. It is nevertheless interesting to point out certains analogies in the obtained thresholds and behaviors.

- The ratios $r_s^W$ where the rigid Wigner molecule is formed in small clusters are of the order of the ratios $r_s^c$ where one observes the MIT. One finds a value $\approx 10$ for a weakly disordered cluster which increases to larger values when one goes to the clean limit ($r_s^W \approx 30-40$). The $r_s^W$ numerically obtained for larger disorder ($W = 5, 10, 15$) in Ref.[42] are roughly independent of $W$ when $W > 5$ and reproduce the flat part of the experimental curve sketched in Fig. 5.

- The large magnetoresistance sketched in Fig. 3 is consistent with the reduction of $\chi$ yielded by a parallel magnetic field as indicated in Fig. 20 (right).

- The shift of the critical threshold $r_s^W$ when one polarizes the electrons with a parallel magnetic field (Fig. 20 - left) is consistent with the shift of the critical densities sketched in Fig. 4.

- A polarization energy $Q_2 \propto r_s^{-2}$ is consistent with the $n_s$ dependence of $B_{sat}$ given in Fig. 3 for similar values $3 < r_s < 9$ [29].

- A threshold at $r_s^F \approx 3-5$ given by the study of the persistent currents and of the magnetization corresponds to the density at which the metallic behavior ceases to occur, according to Ref. [35].

- It can be argued that the usual small negative magnetoresistance yielded by a perpendicular magnetic field for intermediate $r_s$ (Fig. 6) would mean that transport is due to the gapless excitations of the intermediate floppy quantum solid. The propagation of such excitations could be reduced by the usual weak localization corrections when one has elastic scattering by impurities.

## 6.6. MESOSCOPIC 2D ELECTRON SYSTEMS CONFINED IN HARMONIC TRAPS

For the important issue of the melting of a macroscopic 2D Wigner glass (crystal without disorder) and a possibly associated 2D-MIT, the study of finite size systems can give hints, but requires to be complemented by a systematic study of the scale dependence of the finite size effects at a given electron density. Before reviewing some results where such a finite size scaling analysis has been done, let us underline that the formation of the mesoscopic Wigner molecule in a few electron system is by itself an important issue. One can use a few electrons [48] confined in a quantum dot or a few ions [49] trapped by electric and magnetic fields. Increasing the size of the trap, a crossover from independent-particle towards collective motion can be observed. Moreover, a localization-delocalization transition has been observed [50] in a quantum dot using single-electron capacitance spectroscopy and increasing the number of trapped electrons. Considering electrons confined in a harmonic trap, and not on a 2D torus, other numerical studies have reached the same crucial conclusion, namely that between the Fermi system and the "classical" rigid molecule, there is an intermediate regime of a floppy weakly formed Wigner molecule. This was shown for instance in an exact study of a two-electron artificial atom [51] and in a Monte Carlo study [52] of a few electron system. In Ref. [52], this new regime corresponds to particles having a radial ordering on shells without correlated intershell rotation. This was attributed to the special symmetry of the harmonic trap and to the imposed density gradient. Our studies show that mesoscopic Wigner crystallization takes also place in two stages when the particles are confined on a 2D torus, with a uniform electron density.

## 7. QUANTUM PHASE TRANSITION FOR WEAK COUPLING AND STRONG DISORDER

In this Section we consider the case of spinless fermions for strong disorder and low filling factors $N/L^2$. In contrast to the previously studied systems, the relevant length scale without interaction, the one-body localization length $L_1$, is not only smaller than the system size, but also smaller than the average distance between particles, $d \propto n_s^{-1/2}$. Truncated bases built out from single-particle wavefunctions or Hartree-Fock orbitals will be used to study larger systems. These approximations remain reliable when the interaction strenght $U$ does not exceed the hopping term $t$, but become uncontrolled in the strongly correlated regime $U \gg t$. They allow us to study the finite size effects in the limit of relatively weak coupling. They show the existence of a second order phase transition between the weak coupling Fermi limit and a new phase appearing above $U_c \approx t$ for $W = 10 - 15$. Two numerical evidences of a transition are presented. The first is given by the divergence of a characteristic length of the two dimensional system which allows us to map the finite size data onto a single scaling curve. The second is given by the existence of a size independent distribution of the first excitation at $U_c$.

In this limit, the use of the parameter $r_s$ becomes questionable, since it is no longer an appropriate measure of the Coulomb to Fermi energy ratio, the one body part of the Hamiltonian being deeply changed by the random substrate. The values $U_c$ obtained when $W$ is large give for the considered filling factors critical values $r_s^c = U_c/(2t\sqrt{\pi(N/L^2)})$ of the order of the previously obtained ratios $r_s^F$ for a weaker disorder. However, the use of a dimensionless ratio ignoring $W$ when $W$ is large can be misleading.

## 7.1. DIVERGENCE OF A CHARACTERISTIC LENGTH IN TWO DIMENSIONS

From a finite size study [53], one can obtain scaling laws consistent with the divergence of a characteristic length of the two dimensional system at a first lower threshold $U_c \approx t$, as in a second order phase transition. For defining a characteristic length of the $N$-body ground state, one considers the change $\delta \rho_j$ of the charge density induced by a small change $\delta v_i$ of the random potential $v_i$ located at a distance $r = |i - j|$. To improve the statistical convergence, one computes the change $\delta \rho(r) = \sum_{j_y} \delta \rho_{r,j_y}$ of the charge density on the $L$ sites of coordinate $j_x = r$ yielded by the change $v_{0,i_y} \to 1.01 v_{0,i_y}$ for the $L$ random potentials of coordinate $i_x = 0$. For a Slater determinant made with N occupied single particle eigenfunctions $(\psi_\alpha)$, first order perturbation theory gives:

$$\delta \rho_j = 2\delta v_i \sum_{\alpha=1}^{N} \sum_{\beta \neq \alpha} \frac{\psi_\alpha(i)\psi_\beta(i)\psi_\alpha(j)\psi_\beta(j)}{E_\beta - E_\alpha} \propto \exp{-\frac{2r}{L_1}}, \qquad (19)$$

the index $\beta$ varying over the one-body spectrum. Therefore, in the absence of interaction, the change $\delta \rho$ remains localized on a scale given by the one-body localization length ($\xi_L \approx L_1/2$).

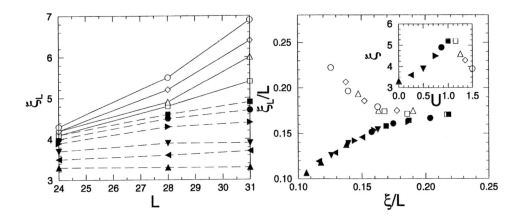

*Figure 21.* Right: Characteristic length $\xi_L$ as a function of $L$ for $U \leq 1$ (filled symbols): 0 (triangles up), 0.25 (triangles left), 0.5 (triangles down), 0.75 (triangles right), 0.85 (circles) and 1 (squares) and $U > 1$ (empty symbols): 1.15 (squares), 1.25 (triangles), 1.35 (diamonds) and 1.5 (circles). Left: Ratios $\xi_L/L$ mapped onto the scaling curve $f$ as a function of $\xi/L$. The two dimensional scaling length $\xi$ is given in the inset.

Let us study the dependence of $\xi_L$ on the interaction $U$ for an ensemble of $5 \times 10^3$ clusters, with $N = 3, 4, 5$ particles in square lattices of size $L = 24, 28, 31$ respectively, corresponding to a very low filling factor $n_e \approx 5 \times 10^{-3}$. To have Anderson localization inside these sizes we considered a large disorder to hopping ratio $W/t = 10$. Therefore the low energy tail of the one body spectrum is made of impurity states trapped at some site $i$ of exceptionally low $v_i$. As we are interested in studying the effect of Coulomb repulsion on genuine Anderson localized states we get rid of the band tail. Typically we ignore the $L^2/2$ first one-body levels (but results do not change, provided that the Fermi level is out of the band tail, $\epsilon_F > -4t$). From this restricted subset of one-body states we built a basis for the $N$-body problem, truncated to the $N_H = 10^3$ Slater determinants of lowest energy (convergency tests are discussed in Ref. [53]). We note that, for $W/t = 10$, the one-body localization length $L_1 \approx 4$ is smaller than the distance between particles $d \approx 15$.

We checked that $|\delta\rho(r)|$ is reasonably fitted by a log-normal distribution. Therefore it makes sense to characterize the typical strength of the fluctuations by

$$\delta\rho_{\text{typ}}(r) = \exp <\ln|\delta\rho(r)|> \qquad (20)$$

and extract the length $\xi_L$ over which the perturbation is effective from the exponential decay

$$\delta\rho_{\text{typ}}(r) \propto \exp(-r/\xi_L). \tag{21}$$

The size dependence of $\xi_L$ is presented in Fig. 21 left, for increasing Coulomb repulsions. One finds the behavior typical of a phase transition: $\xi_L$ converges towards a finite value when $U < U_c$ (localized phase), diverges linearly as a function of $L$ at $U = U_c \approx 1$ (critical point) and diverges faster than linearly when $U > U_c$ (delocalized phase). This is exactly the behavior [54, 55] which characterizes the one-body Anderson transition in three dimensions.

In Fig. 21 right we verify a usual scaling ansatz [54] inspired by the theory of second-order phase transitions:

$$\frac{\xi_L}{L} = f\left(\frac{L}{\xi}\right), \tag{22}$$

where we assume that it is possible to map the characteristic length $\xi_L$ at the system size $L$ onto a scaling curve $f(L/\xi)$, where $\xi$ is the scaling length characteristic of the infinite two dimensional system.

All the data of Fig. 21 left can be mapped onto the universal curve $f$ shown in Fig. 21 right, assuming the scaling length $\xi$ given in the inset. When $U < U_c$, this length characterizes the localization of the effect of a local perturbation of the substrate in the two dimensional thermodynamic limit. Our data are consistent with a divergence of $\xi$ at a threshold $U_c \approx t$.

Very often, additional corrections $\propto L^{-\alpha}$ to the scaling ansatz occurs for small sizes. We point out that our results can be fitted by a simple linear law $\xi_L = 0.17L$ for $U = U_c$. This is a further indication that the simple ansatz (22) describes scaling for $L \geq 24$, without noticeable additional $L^{-\alpha}$ corrections.

This tells us that the polarized electron system in a highly disordered $2d$ substrate becomes correlated when $U > U_c$ and that the effect of a local perturbation becomes delocalized.

## 7.2. SIZE INDEPENDENT DISTRIBUTION OF THE FIRST EXCITATION

In Ref. [56] we used the configuration interaction method, discussed in appendix, to study the statistics of the first many-body energy spacing. Questions related to the convergence of the method when the Hartree-Fock basis is truncated are discussed in Ref. [56].

Let us consider the first $N$-body energy levels $E_i$, $(i = 0, 1, 2, ...)$ for different sizes $L$, with a large disorder to hopping ratio $W/t = 15$

imposed to have Anderson localization and Poissonian spectral statistics for the one particle levels at $U = 0$ when $L \geq 8$. This corresponds to a strongly disordered limit where the one particle localization length $L_1$ is not only smaller than $L$ and the distance $d$ between the particles (as in the previous subsection), but becomes also of the order of the lattice spacing. We considered $N = 4, 9$, and 16 particles inside clusters of size $L = 8, 12$, and 16 respectively. This corresponds to a constant low filling factor $n_e = 1/16$. We studied an ensemble of $10^4$ disorder configurations.

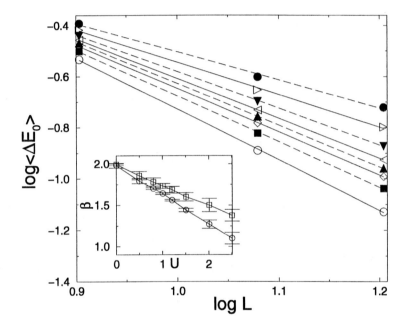

*Figure 22.* Size dependence of the average gap (first spacing $< \Delta E_0 > \propto L^{-\beta(U)}$), for $W = 15$, filling factor $n_e = 1/16$. From bottom to top: $U = 0, 0.5, 0.8, 1, 1.2, 1.5, 2, 2.5$. Inset: $\beta(U)$ (circles, characterizing $< \Delta E_0 >$ and squares, characterizing $< \Delta E_i >$, with an average over $i = 1 - 3$).

The first average spacing $< \Delta E_0 >$ is given in Fig. 22. It exhibits a power law decay as $L$ increases, with an exponent $\beta$ given in the inset. One finds for the first spacing that $\beta$ linearly decreases from $d = 2$ to 1 when $U$ increases from 0 to 2.5. The next mean spacings $< \Delta E_i > = < E_{i+1} - E_i >$ depend more weakly on $U$, as shown in Fig. 22.

For $U = 0$, the distribution of the first spacing $s = \Delta E_0 / < \Delta E_0 >$ becomes closer and closer to the Poisson distribution $P_P(s)$ when $L$ increases, as it should be for an Anderson insulator. For a larger $U$,

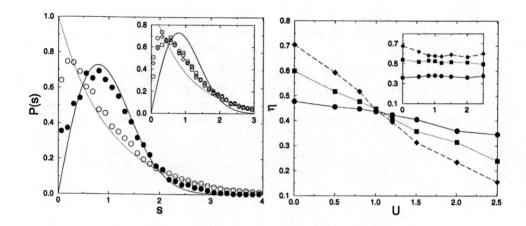

*Figure 23.* Left: gap distribution $P(s)$ for $U = 0$ (empty circles) and $U = 2.5$ (filled circles) when $N = 16$, $L = 16$, $W = 15$, compared to $P_P(s)$ and $P_W(s)$. Inset: size invariant $P(s)$ at $U_c \approx 1.1$ ; $L = 8$ (circles), 12 (squares), and 16 (diamonds). Right: parameter $\eta_O(U)$ corresponding to the first spacing $\Delta E_0$ at $L = 8$ (circles), 12 (squares), and 16 (diamonds). Inset: $\eta_O(U)$ for the second spacing $\Delta E_1$.

the distribution seems to become close to the Wigner surmise $P_W(s)$ characteristic of level repulsion in Random Matrix Theory, as shown in Fig. 23 left, for $U = 2.5$ and $L = 16$. To study how this $P(s)$ goes from Poisson to a Wigner-like distribution when $U$ increases, we use the spectral parameter $\eta_O$ which decreases from 1 to 0 when $P(s)$ goes from Poisson to Wigner. In Fig. 23 right, one can see that the three curves $\eta_O(U)$ characterizing the first spacing for $L = 8, 12, 16$ intersect at a critical value $U_c/t \approx 1.1$. Our data suggest that for $U < U_c$ the distribution tends to Poisson in the thermodynamic limit, while for $U > U_c$ it tends to a Wigner-like behavior. At the threshold $U_c$, there is a size-independent intermediate distribution shown in the inset of Fig. 23 left, exhibiting level repulsion at small $s$ followed by a $\exp(-as)$ decay at large $s$, with $a \approx 1.52$. Such a size independent distribution is known for characterizing a mobility edge in an one body spectrum. This Poisson-Wigner transition characterizes only the first spacing, the distributions of the next spacings being quite different. The inset of Fig. 23 right does not show any intersection for the parameter $\eta$ calculated for the second spacing.

The observed transition, and the difference between the first spacing and the following ones is mainly an effect of the H-F mean field. For the first spacings, the curves $\eta_O$ calculated with the HF data are qualitatively similar. At the mean field level the low energy excitations are particle-hole excitations starting from the ground state. The energy spacing between the first and the second excited states is given by the difference of two particle-hole excitations and a Poisson distribution follows if the low energy particle-hole excitations are uncorrelated.

We note that the energy of an electron-hole pair is given by $\epsilon_j - \epsilon_i - U/r_{ij}$ and the classical argument for the existence of a gap in the single particle density of states does not apply for the many-body spectrum[57]. Therefore the observed opening of a gap for the first energy excitation is a remarkable phenomenon beyond the predictions of the classical Coulomb gap model.

## 7.3. CHANGE OF THE INVERSE COMPRESSIBILITY

In Ref. [56] we studied also the inverse compressibility

$$\Delta_2(N) = E_0(N+1) - 2E_0(N) + E_0(N-1), \qquad (23)$$

i.e. the discretized second derivative of the ground state energy $E_0$ with respect to the number $N$ of particles.

We consider $N = 4, 9, 16$ particles on square lattices of size $L = 6, 9, 12$ respectively, corresponding to a constant filling factor $n_e = 1/9$. Here we focus on the strongly localized regime with disorder strength $W = 15$.

The following inverse compressibility data are obtained with the configuration interaction method. We have checked that the residual interaction does not change qualitatively Hartree-Fock results. Indeed HF approximation gives in this weak coupling regime a good estimate of the ground state energy (see Ref. [56]) and the inverse compressibility is a physical observable which only depends on the ground state energies at different number of particles. Neither a precise knowledge of the ground state wavefuction (as in the calculation of persistent currents) nor excited states energies (as in studies of spectral statistics) are required.

Fig. 24 left shows the $L$-dependence of the average inverse compressibility, which is well fitted by the power law $< \Delta_2(U) > \propto L^{-\alpha(U)}$, with $\alpha(U)$ going from 2 to 1 when $U$ goes from 0 to 3 approximately. The value $\alpha = 2$ is expected without interaction ($< \Delta_2 > = < \Delta > \propto 1/L^2$, with $< \Delta >$ single particle mean level spacing). The exponent $\alpha = 1$, in this strongly localized regime, can be related to the Coulomb gap in the single particle density of states [57]: According to Koopmans' theorem (see Refs. [58, 59, 60] for a thorough discussion about the limits of validity of the Koopmans' theorem for disordered quantum dots), one assumes that all the other charges are not reorganized by the addition

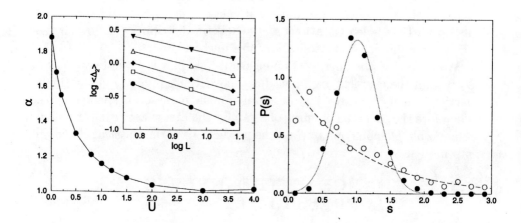

*Figure 24.* Left: Inset: size dependence of the average inverse compressibility $<\Delta_2>$, for $W = 15$, filling factor $n_e = 1/9$, $U = 0$ (circles), 0.5 (squares), 1 (diamonds), 2 (triangles up), and 4 (triangles down). Straight lines are power law fits $<\Delta_2(U)> \propto L^{-\alpha(U)}$. Main figure: exponent $\alpha(U)$. Right: distribution of the normalized inverse compressibilities for $N = 16$, $L = 12$, $W = 15$, $U = 0$ (empty circles) and $U = 3$ (filled circles), fitted by a Gaussian of standard deviation $\sigma = 0.30$ (dotted line). Dashed line gives Poisson distribution. Disorder average is over $10^3$ configurations.

of an extra charge,

$$\Delta_2 \approx \epsilon_{N+1} - \epsilon_N \propto \frac{1}{L} \qquad (24)$$

due to the Coulomb gap, where $\epsilon_k$ is the energy of the $k$-th HF orbital at a fixed number $N$ of particles.

The distribution of inverse compressibilities, for $L = 12$ and $W = 15$ is shown in Fig. 24 right. At $U = 0$, due to Anderson localization, $\Delta_2$ distributions are close to the Poisson distribution (deviations from the Poisson distribution at small $s$ values are due to the finite system size). On the contrary, the distribution at $U = 3$ shows a Gaussian shape. This can be understood within the Koopmans' theorem, since the HF energies are given by

$$\epsilon_k = \langle \psi_k | H_1 | \psi_k \rangle + \sum_{\alpha=1}^{N} \left( Q_{\alpha k}^{\alpha k} - Q_{\alpha k}^{k \alpha} \right), \qquad (25)$$

with $H_1$ one-body part of the spinless Hamiltonian and $Q_{\alpha\beta}^{\gamma\delta}$ interaction matrix elements given by Eq. (C.3) in appendix C. Due to the small correlations of eigenfunctions in a random potential, one can reasonably invoke the central limit theorem (see Ref. [60] for a more detailed discussion). We point out that a Gaussian-like distribution of inverse compressibilities has been observed in quantum dots experiments in the Coulomb blockade regime [61, 62, 63]. This implies that inverse compressibility fluctuations are dominated by interaction effects instead of single particle fluctuations.

We note that the Gaussian-like shape of the inverse compressibility distribution becomes more asymmetric when the disorder strength is increased. This asymmetry is due to the Coulomb gap and has been discussed in the classical limit ($t = 0$) in Ref. [64].

A simple model for the addition spectra of quantum dots has been recently proposed by B. Shapiro [65]. This model explains the change of the inverse compressibility distribution when $r_s$ increases assuming that the system can be decomposed into a stable N-electron Wigner lattice plus a lattice of interstitial sites where the added particle can move and for which a usual single particle description is assumed. This suggests possible relations between the problem of adding an extra particle to a Wigner solid and its melting when $r_s$ decreases.

## 8. CONCLUSION

We have numerically revisited the quantum-classical crossover from the weak coupling Fermi limit towards the strong coupling Wigner limit, using small lattice models. In the continuous limit, the assumed picture [17] is relatively simple: a liquid-solid first order transition for $r_s \approx 37$. For a clean lattice model at a filling factor 1/9, our results raise the question of the possible existence of an intermediate liquid-solid phase as first proposed by Andreev and Lifshitz. This may also raise questions about the differences between the continuous limit and lattice models. When the spin degrees of freedom are included, our lattice model is a Hubbard model with additional long range repulsions far from half filling. The physics of such models is not very well known, but at least suspected to be very rich and complex, as discussed in many works since the discovery of high $T_c$ superconductors. The role of the disorder is another source of complexity. We have divided this chapter in two main parts, the first where the disorder is weak, the second where the disorder is strong and makes the quantum kinetic effects less relevant. This raises also the question of the difference between the quantum Wigner glass and the classical Coulomb glass. For large disorder, a finite size scaling analysis leads us to conclude that the weak coupling Fermi phase is delimited by a second order quantum phase transition, when $r_s$ increases.

Eventually, our first motivation was the experimental discovery of a possible new metal for intermediate values of $r_s$. Our numerical studies have not directly addressed the transport properties for intermediate $r_s$.

One cannot claim that the observed intermediate regime gives a new metal in the thermodynamic limit. Nevertheless, if transport is mainly due to the presence of delocalized defects in a floppy and pinned electron solid for intermediate $r_s$, it might be interesting to extend the FLT theory à la Landau proposed in Ref. [24] for a clean system to the case where one has also elastic scattering by impurities. It will tell us if a modified FLT adapted to a very special Fermi system with a non conventional Fermi surface, still gives rise to usual quantum interferences leading to weak localization and eventually to Anderson localization for the conjectured gapless excitations of a floppy solid. This could help to determine whether one has a true new metal for intermediate $r_s$ and weak disorder, or only an "apparent" metallic behavior which should disappear at the true $T \to 0$ limit. This might help to explain why in measurements down to $5mK$ in a GaAs hetrostructure [36], the weak localization correction remains less than a few percent of the value predicted for a standard disordered $2d$ Fermi liquid.

## Appendix: Hartree-Fock approximation

This is the usual mean field approximation when one has electron-electron interactions, which we shortly review for the case of spinless fermions. The HF ground state (GS) is the Slater Determinant (SD) which minimizes the GS energy expectation value. In the HF approximation, one reduces the two-body part of the total Hamiltonian to an effective single particle Hamiltonian [66, 67, 68]

$$U(\sum_{i \neq j} \frac{1}{r_{ij}} n_i \langle n_j \rangle - \sum_{i \neq j} \frac{1}{r_{ij}} c_i^\dagger c_j \langle c_j^\dagger c_i \rangle), \qquad (A.1)$$

where $\langle ... \rangle$ stands for the expectation value with respect to the HF ground state, which has to be determined self-consistently. The first (Hartree) term describes the interaction of any electron with the charge distribution set up by all the other electrons, while the latter (Fock) term is a nonlocal exchange potential. The Hartree term comes out as an extra on-site disorder potential, while the Fock term introduces extra hopping amplitudes.

The main advantage of the Hartree-Fock approximation is that it reproduces the single-particle density of states $g(E)$ [68], particularly the Coulomb gap near the Fermi energy $E_F$: in the two-dimensional case, $g(E) \propto |E - E_F|$ [57]. The physical argument underlying this relation is that an empty site $j$ and an occupied site $i$ with a difference in energy $\epsilon_j - \epsilon_i$ smaller than $\delta$ must be at a distance larger than $U/\delta$ as the change $\Delta E_{ij} = \epsilon_j - \epsilon_i - U/r_{ij}$ of the system energy must be positive when we consider an excitation starting from the ground state. However, the HF approximation gives a Coulomb gap also in the delocalized regime at small disorder (single particle localization length larger than the system size), where complicated many-body effects beyond HF approximation should screen Coulomb interaction.

# Appendix: Stoner instability

Ferromagnetic instabilities come from the interplay between Coulomb repulsion and the Pauli principle. In the Pauli picture, electrons populate the orbital states of a system, such as a quantum dot or a metallic grain, in a sequence of spin up - spin down electrons. The resulting minimum spin state minimizes the kinetic energy: it costs energy to flip a spin since it must be promoted to a higher energy level. On the other hand, the maximum spin state requires a maximally antisymmetric coordinate wavefunction, thus reducing the effect of Coulomb repulsion. This is at the basis of Hund's rule for atoms: electrons occupy orbitals in an open shell so as to maximize their total spin.

Due to the locality of the Pauli principle, ferromagnetic instabilities can be studied within the Hubbard Hamiltonian,

$$H = -t \sum_{<i,j>\sigma} c^\dagger_{i\sigma} c_{j\sigma} + \sum_{i\sigma} v_i n_{i\sigma} + U \sum_i n_{i\uparrow} n_{i\downarrow}. \tag{B.1}$$

In the HF approximation the interaction part of the Hubbard Hamiltonian (B.1) is reduced to

$$U \sum_{i\sigma} n_{i\sigma} \langle n_{i-\sigma} \rangle. \tag{B.2}$$

The HF energies are given by

$$\epsilon_{\alpha\sigma} = \epsilon^0_\alpha + U \sum_{i\alpha\beta} |\psi_\alpha(i)|^2 |\psi_\beta(i)|^2 \langle n_{\beta-\sigma} \rangle, \tag{B.3}$$

with $\epsilon^0_\alpha$ one-body eigenenergies and $n_{\beta-\sigma} = d^\dagger_{\beta-\sigma} d_{\beta-\sigma}$ occupation numbers for the HF orbitals. The total ground state HF energy reads

$$E_0 = \sum_{\alpha\sigma} \epsilon^0_\alpha \langle n_{\alpha\sigma} \rangle + U \sum_{i\alpha\beta} |\psi_\alpha(i)|^2 |\psi_\beta(i)|^2 \langle n_{\alpha\uparrow} \rangle \langle n_{\beta\downarrow} \rangle. \tag{B.4}$$

In order to study the stability of the nonmagnetic solution,

$$\langle n_{\alpha\sigma} \rangle = \begin{cases} 1 & \text{if } \epsilon_{\alpha\sigma} < \epsilon_F \\ 0 & \text{if } \epsilon_{\alpha\sigma} > \epsilon_F, \end{cases} \tag{B.5}$$

with $\epsilon_F$ Fermi level, we take a layer (of thickness $\delta\epsilon$) of electrons with spin down below the Fermi level to put them in states with spin up [69]. This changes the spin state of $\delta n_s = \rho(\epsilon_F)\delta\epsilon$ electrons per unit volume, where $\rho(\epsilon_F)$ is the density of states per volume per spin. The change of the one-body energy density is given by

$$\delta w_0 = \rho(\epsilon_F)(\delta\epsilon)^2. \tag{B.6}$$

In the clean case, the change of the interaction energy (per volume) is

$$\delta w_{\text{int}} = U \left( \frac{n_s}{2} + \rho\delta\epsilon \right) \left( \frac{n_s}{2} - \rho\delta\epsilon \right) - \frac{U n_s^2}{4} = -U\rho^2(\epsilon_F)(\delta\epsilon)^2. \tag{B.7}$$

Therefore the nonmagnetic state becomes unstable when $\delta w_0 + \delta w_{\text{int}} < 0$, that is

$$U\rho(\epsilon_F) > 1, \tag{B.8}$$

which gives the Stoner criterion for ferromagnetic instability.

In the diffusive regime, the effective interaction strength is enhanced by the presence of disorder, leading to ferromagnetic instabilities already below the Stoner threshold, as pointed out in Ref. [70].

We also note that Stoner criterion predicts a ferromagnetic ground state for the Hubbard model even at finite temperatures in one and in two dimensions, thus violating the Mermin and Wagner's theorem. Therefore, when the Stoner criterion (B.8) is satisfied, we can only conclude that the nonmagnetic ground state is unstable.

The problem of a possible magnetization of the ground state is not only discussed in a dilute 2DEG, but is also central in the studies of mesoscopic quantum dots since their Ohmic resistances are measured as a function of a gate voltage in the Coulomb blockade regime. The possibility of a spontaneous magnetization $S$ of their ground state due to electron-electron interactions has been proposed to explain the observed conductance peak spacing distributions.

## Appendix: Configuration interaction method

Even though the approximations involved in the HF method are uncontrolled, the mean field HF results can be improved using a numerical method [71, 72, 56] familiar in quantum chemistry as the configuration interaction method (CIM) [73]. Once a complete orthonormal basis of HF orbitals has been calculated,

$$H_{HF}(|\psi_1\rangle, ..., |\psi_N\rangle)|\psi_\alpha\rangle = \epsilon_\alpha |\psi_\alpha\rangle, \tag{C.1}$$

with $\alpha = 1, 2, \ldots, L^2$, it is possible to build up a Slater determinants' basis for the many-body problem which can be truncated to the $N_H$ first Slater determinants, ordered by increasing energies. The two-body Hamiltonian can be written as

$$H_{\text{int}} = \frac{1}{2} \sum_{\alpha,\beta,\gamma,\delta} Q_{\alpha\beta}^{\gamma\delta} d_\alpha^\dagger d_\beta^\dagger d_\delta d_\gamma, \tag{C.2}$$

with

$$Q_{\alpha\beta}^{\gamma\delta} = U \sum_{i \neq j} \frac{\psi_\alpha(i)\psi_\beta(j)\psi_\gamma(i)\psi_\delta(j)}{r_{ij}} \tag{C.3}$$

and $d_\alpha^\dagger = \sum_j \psi_\alpha(j) c_j^\dagger$. One gets the residual interaction subtracting Eq. A.1 from Eq. C.2.

## Acknowledgments

The authors wish to acknowledge the participation of D. Shepelyansky in one of the reviewed works (Ref. [56]) and to thank B. Spivak for discussions about the work of Andreev and Lifshitz.

Present addresses:

G. Benenti: Università degli Studi dell'Insubria and Istituto Nazionale per la Fisica della Materia, via Valleggio 11, 22100 Como, Italy.

G. Katomeris: Department of Physics, University of Ioannina, 45 110, Greece.

X. Waintal: Laboratory of Atomic and Solid State Physics 530 Clark Hall, Cornell University, Ithaca, NY 14853-2501, USA.

## References

# REFERENCES

[1] E. P. Wigner, Trans. of the Faraday Soc. **34**, 678 (1938). See also Phys. Rev. **46**, 1002 (1934).
[2] L. D. Landau, JETP **30**, 1058 (1956).
[3] B. L. Altshuler and A. G. Aronov, in *Electron-Electron Interactions in Disordered Systems*, edited by A. L. Efros and M. Pollak (North Holand, Amsterdam, 1985).
[4] E. Abrahams, S. V. Kravchenko and M. P. Sarachik, cond-mat/0006055, to be published in Rev. Mod. Phys. **73**, 251 (2001) and Refs. therein.
[5] M. Y. Simmons, A. R. Hamilton, M. Pepper, E. H. Linfeld, P. D. Rose and D. A. Ritchie, Phys. Rev. Lett. **84**, 2489 (2000).
[6] Y. Yaish, O. Prus, E. Buchstab, S. Shapira, G. Ben Joseph, U. Sivan and A. D. Stern, Phys. Rev. Lett. **85**, 4954 (2000).
[7] V. Senz, T. Ihn, T. Heinzel, K. Ensslin, G. Dehlinger, D. Grutzmacher and U. Gennser, Phys. Rev. Lett. **85**, 4357 (2000).
[8] B. L. Altshuler and D. L. Maslov, Phys. Rev. Lett. **82**, 145 (1999).
[9] S. S. Safonov, S. H. Roshko, A. K. Savchenko, A. G. Pogosov, and Z. D. Kvon, Phys. Rev. Lett. **86**, 272 (2001).
[10] Y. Meir, Phys. Rev. Lett. **83**, 3506 (1999).
[11] A. P. Mills, A. P. Ramirez, L. N. Pfeiffer and K. W. West, Phys. Rev. Lett. **83**, 2805 (1999).
[12] J. H. Schön, S. Berg, Ch. Kloc and B. Batllog, Science **287**, 1022 (2000).
[13] Lynn Bonsall and A. A. Maradudin, Phys. Rev. B **15** (1977) 1959.
[14] D. Pines *The many body problem*, W. A. Benjamin, Inc. New York (1961).
[15] W. J. Carr, Phys. Rev. **122**, 1437 (1961).
[16] D. Ceperley, Phys. Rev. B **18**, 3126 (1978).
[17] B. Tanatar and D. Ceperley, Phys. Rev. B **39**, 5005 (1989).
[18] D. Varsano, S. Moroni and G. Senatore, Europhys. Lett. **53**, 348 (2000).
[19] Ladir Cândido, Philip Phillips and D. M. Ceperley, Phys. Rev. Lett. **86**, 492 (2001).
[20] B. Bernu, L. Cândido and D. M. Ceperley, cond-mat/9908062.
[21] M. Roger, Phys. Rev. B **30**, 6432 (1984).
[22] S. T. Chui and B. Tanatar, Phys. Rev. Lett. **74**, 458 (1989).
[23] A. F. Andreev and I. M. Lifshitz, JETP **29** (1969) 1107.
[24] I. E. Dzyaloshinskii, P. S. Kondatenko and V. S. Levchenko, JETP **35** (1972) 823 and 1213.
[25] J. H. Schön, Ch. Kloc and B. Batllog, Letters to Nature **406**, 702 (2000).
[26] J. H. Schön, Ch. Kloc and B. Batllog, Science **288**, 2338 (2000).
[27] B. Batllog, colloqium given in Orsay (october 2000).
[28] E. Y. Andrei, G. Deville, D. C. Glattli, F. I. B. Williams, E. Paris and B. Etienne, Phys. Rev. Lett. **60**, 2765 (1988).
[29] A. A. Shashkin, S. V. Kravchenko and T. M. Klapwijk, cond-mat/0009180.
[30] S. V. Kravchenko and T. M. Klapwijk, Phys. Rev. Lett. **84**, 2909 (2000).
[31] S. A. Vitkalov, H. Zheng, K. M. Mertes, M. P. Sarachik and T. M. Klapwijk, Phys. Rev. Lett. **85**, 2164 (2000).
[32] S. A. Vitkalov, H. Zheng, K. M. Mertes, M. P. Sarachik and T. M. Klapwijk, cond-mat/0009454.
[33] J. Yoon, C. C. Li, D. Shahar, D. C. Tsui and M. Shayegan, Phys. Rev. Lett. **82**, 1744 (1999).

[34] A. M. Finkelshtein, Sov. Phys. JETP **57**, 98 (1983).
[35] A. R. Hamilton, M. Y. Simmons, M. Pepper, E. H. Linfield, P. D. Rose and D. A. Ritchie, Phys. Rev. Lett. **82**, 1542 (1999).
[36] A. P. Mills, Jr., A. P. Ramirez, X. P. A. Gao, L. N. Pfeiffer, K. W. West and S. H. Simon, cond-mat/0101020.
[37] P. Mohanty, E. M. Q. Jariwala and R. A. Webb, Phys. Rev. Lett. **77**, 3366 (1997).
[38] C. Dultz and H. W. Jiang, Phys. Rev. Lett. **84**, 4689 (2000).
[39] S. Ilani, A. Yacoby, D. Mahalu and Hadas Shtrikman, Phys. Rev. Lett. **84**, 3133 (2000).
[40] S. Ilani, A. Yacoby, D. Mahalu and Hadas Shtrikman, Science **292**, 1354 (2001).
[41] M. Reznikov, conference talk given in Rencontres de Moriond, Les Arcs, 2001.
[42] G. Benenti, X. Waintal and J.-L. Pichard, Phys. Rev. Lett. **83**, 1826 (1999).
[43] R. Berkovits and Y. Avishai, Phys. Rev. B **57**, R15076 (1998).
[44] F. Selva and D. Weinmann, Eur. Phys. J. B **18**, 137 (2000).
[45] G. Benenti, X. Waintal and J.-L. Pichard, Europhys. Lett. **51**, 89 (2000).
[46] G. Katomeris and J.-L. Pichard, cond-mat/0012213.
[47] F. Selva and J.-L. Pichard, cond-mat/0012015.
[48] R. C. Ashoori, Nature **379**, 413 (1996).
[49] D. H. Dubin and T. M. O'Neil, Rev. Mod. Phys. **71**, 87 (1999).
[50] N. B. Zhitenev, M. Brodsky, R. C. Ashoori, L. N. Pfeiffer and K. W. West, Science **285**, 715 (1999).
[51] C. Yannouleas and U. Landman, *Phys. Rev. Lett.* **85**, 1726 (2000).
[52] A. V. Filinov, M. Bonitz and Yu. E. Lozovik, *Phys. Rev. Lett.* **86**, 3851 (2001).
[53] X. Waintal, G. Benenti, and J.-L. Pichard, Europhys. Lett. **49**, 466 (2000).
[54] J.-L. Pichard and G. Sarma, J. Phys. C **14**, L127 and L617 (1981).
[55] A. MacKinnon and B. Kramer, Phys. Rev. Lett. **47**, 1546 (1981); B. Kramer and A. MacKinnon, Rep. Prog. Phys. **56**, 1469 (1993).
[56] G. Benenti, X. Waintal, J.-L. Pichard, and D.L. Shepelyansky, Eur. Phys. J B **17**, 515 (2000).
[57] A.L. Efros and B.I. Shklovskii, J. Phys. C **8**, L49 (1975); M. Pollak, Phil. Mag. B **65**, 657 (1992); see also *Electron-Electron Interactions in Disordered Systems*, edited by A.L. Efros and M. Pollak, (North-Holland, Amsterdam, 1985).
[58] S. Levit and D. Orgad, Phys. Rev. B **60**, 5549 (1999).
[59] P.N. Walker, G. Montambaux, and Y. Gefen, Phys. Rev. B **60**, 2541 (1999).
[60] A. Cohen, K. Richter, and R. Berkovits, Phys. Rev. B **60**, 2536 (1999).
[61] U. Sivan, R. Berkovits, Y. Aloni, O. Prus, A. Auerbach, and G. Ben-Yoseph, Phys. Rev. Lett. **77**, 1123 (1996).
[62] S.R. Patel, S.M. Cronenwett, D.R. Stewart, A.G. Huibers, C.M. Marcus, C.I. Duruöz, J.S. Harris, Jr., K. Campman, and A.C. Gossard, Phys. Rev. Lett. **80**, 4522 (1998).
[63] F. Simmel, D. Abusch-Magder, D.A. Wharam, M.A. Kastner, and J.P. Kotthaus, Phys. Rev. B **59**, R10441 (1999).
[64] A.A. Koulakov, F.G. Pikus, and B.I. Shklovskii, Phys. Rev. B **55**, 9223 (1997).
[65] B. Shapiro, cond-mat/0008366, to appear in Phil. Mag. **82**, No 3.
[66] H. Kato and D. Yoshioka, Phys. Rev. B **50**, 4943 (1994).
[67] G. Bouzerar and D. Poilblanc, J. Phys. I France **7**, 877 (1997).

[68] F. Epperlein, M. Schreiber, and T. Vojta, Phys. Rev. B **56**, 5890 (1997); Phys. Status Solidi (b) **205**, 233 (1998).
[69] See, e.g., A. Blandin, in *Magnetism - Selected topics*, edited by S. Foner (Gordon and Breach, New York, 1976).
[70] A.V. Andreev and A. Kamenev, Phys. Rev. Lett. **81**, 3199 (1998).
[71] M. Eto amd H. Kamimura, Phys. Rev. Lett. **61**, 2790 (1988).
[72] T. Vojta, F. Epperlein, and M. Schreiber, Phys. Rev. Lett. **81**, 4212 (1998).
[73] P. Fulde, *Electron Correlations in Molecules and Solids*, (Springer, Berlin, 1995).

# INDEX

## A

Anticommutation relations 4
Action 21,25
Artificial atom 293
Aharonov-Bohm flux 275,279
Anderson's theorem 200
Anderson insulator 271
Analytic 57
Anti-analytic 57
Anomaly 65,94
Angle-resolved photoemission
   225,228,234
Anharmonic ratio 73
Angular momentum 237
Angular momentum shell 102
Activation energy 129
Anti-Skyrmion 141
Adiabatic
   switching 174
   transport 203
AND gate 178
Avoided crossing 196
Atomic layer 244
Antiferromagnetic
   coupling 216
   configuration 222,223,232
Antiferromagnetism 252
Asymptotic expansion 265

## B

Bosonization 2,7
Bound state 226
Boltzmann equation 220
Ballistic approach 220,256
Bose operators 10,12
Bose representation 16,19
BCS theory 184
Boundary condition 32,35,275
Band structure 50,242,245,251
Biquadratic exchange 242
Branch cut 75
Brownian motion 171
Brillouin zone 251
Bohr radius 264
Backscattering 36,39,42

## C

Current-voltage relation 1,38,48,270
Carbon nanotubes 2
Conductance 4,5,257,258,272
Conductivity 128
Coherent states 16
Channel 23
Charge density 26,35,41,43
Charged gap 100
Capacitance 31,110
Coulomb interaction 33,119,141,263
   gas 44,47,76
   blockade 301
   glass 301
Chemical potential 41,100,111
Collapsed blip 48
Conformal field 53
Correlation function
   54,58,61,66,70,74,77,78
   differential equation 72
Chiral fields 57
   ground states 135
Complex plane 58
Confining potential 102,103,152
Conformal spins 59
   Dimensions 59
   Weights 59
   Symmetry 59
Curved space 62
Crossing symmetry 74
Critical theory 77
   point 92
   density 269
Central charge 78,82
Casimir operator 82,85,87
Compressible states 99
Controlled-not (C-NOT) gate 178
Currents 89
Current-in-plane (CIP) 216,217,222
Current-perpendicular-to-plane (CPP)
   216,222,256
Computation 169
Cyclotron orbit 104
Composite
   excitation 127
   fermion 156,157
   skyrmion 157
Corobino geometry 107
Commensurate antiferromagnet 218

Coherence 167
Coherent state 186
Condensate 184,185
Cooper pair 184,190,191
Computational complexity 180
Conditional phase shift 203
Continuum 226
Cubic symmetry 251
Compressibility 272,299,301
Crystallization 278
Configuration interaction method 304
Cooperative process 123

**D**

Density operators 7,22
Decoherence 165,166,183,201,205
Density fluctuations 21
Density of states 225,261
Defects 253,266
Dispersion 22
Dielectric function 30
    Constant 30
Density matrix 42
Descendant 71
Dyson equation 240,244
Dissipative quantum mechanics 46
Dual field 57
De Haas-van-Alphen effect 101,115,125
Disorder 101,106,152,160,273,288
Diffusion current 132
Double well potential 170
Drude conductivity 270
Dingle temperature 116
Donor
 Si 104

**E**

Electron-hole pairs 10
Ensemble average 277
Easy-axis 145
Entanglement 179
Electron gas 99,263,265
Energy dispersion 6
Edge
 states 102
 current 105,111
External charge 29,30
Electron-electron interaction 101,144,270
Electrostatics 31,266
Eigenvalue 62,245,246
Eigenenergy 274
Eigenvector 274
Exponents 77
Effective action 95

Energy gap 100
Entropy 172
Exchange coupling 231,243,249,250,255
Exact diagonalization 264,274
Extended plane wave 266
Electroneutrality 30,32

**F**

Fermi liquid 1,263,263
Fermi
 level 226,228,232
 point 6
Fermi surface 233,251,252,253,254
Fermi-Wigner crossover 273
Fractional
 regime 101
 gap 127
Free-electron model 224
Fermi function 4,242
Fermion
 Dirac 84,92
 real 92
 free 81
Fano-Anderson model 4
Fock space 9,11,13
Fermi operators 7
 Spinless 26
Fixed point 28
Friedel oscillations 30,41
Four-terminal voltage 40
Finite size effects 59
Fusion rules 66,67
Filling factor 99,117,120,133,283
Flux
 quantum 127,143
 qubit 194,196
Ferromagnet 237,241
Ferromagnetic
 configuration 222,223,232,250
 layer 249
Field effect device 267

**G**

Gaussian 45,53,56
Ground state 11,21,277,284,285,286
Ginzburg-Landau equation 187
Generating functional 55
Green's function 60,244
Gapless spectrum 62,302
Gapped phase 136
Gauge 94
Green function 226
Geometric phase 203

**H**
Hilbert space 69,166
Heat capacity 151
Hypergeometric function 74
Heterostructure 109,116,153,267,272
Hall plateau 115
Hartree-Fock approximation 121,277,296,302
Hyperfine coupling 146
Heisenberg exchange 242
Hopping matrix 259
Hubbard repulsion 273
Hamiltonian matrix 277
Harmonic trap 293
Hubbard Hamiltonian 303

**I**
Impurity 1,26,30,35,43
Insulator 259
Imaginary part 247
Ideal-model gate 179
Interface 213,216,271
Interfacial roughness 236
Interacting electrons 38
Ising model 75
Integer regime 101,130
Interband transition 146
Incompressible states 99,
    fluid 129
Incompressibility gap 100
Ionization energy 141
Inverse photoemission 228,229
Ion trap 183
Inductance 198
Incommensurate spin density 218
Interstitial site 266

**J**
Josephson
    tunnelling 185
    tunnel junction 187,190,195
    current 187
Junction capacitance 189,191,195
Jullière's formula 260
Jastrow wavefunction 287

**K**
Kinetic energy 263
Keldysh
    contour 42,43,44,46
    formalism 238,239
Kac-Moody algebras 81,82
Knizhnik-Zamolo0dchikov equations 86
Kohn-Sham density functional 132

Knight shift 146
Kubo formula 238
Koopman's theorem 299,300

**L**
Landauer formula 4
Landau level 101,114,116,121,153
Lanczos algorithm 274
Left-moving electrons 5,12
Ladder operators 12,17,24
Log-normal distribution 290
Lagrangian 21,91
Long wavelength response 29
Laplace operator 55
Laplace-Beltrami operator 98
Laurent components 67
Linear chain 204
Lehmann expansion 97
Longitudinal resistivity 100
Lifshitz-Kosevich formula 150
Langevin equation 171
Layer thickness 221
Liquid-solid phase 263,281,287
Liquid helium 267
Lattice
    Hexagonal  265
    Bcc        265
Lattice model 273
Localization 295,297
    length 294

**M**
Macroscopic quantum states 208
Minimal model 76,79
Metal-insulator transition 264,271
    Re-entrant 271
Mean field 189
Moebius transformation 243
Mean free path 221
Mesoscopic 33
MOSFET 109,268
Metric tensor 62
Monodromy 74
Monolayer 216
Magnetoresistance
    Giant 213,214,219,250,256,257
    Normal 213,292
    Anisotropic 214
    Colossal 214
    Tunnelling 237,250,259,260
Marginal operator 77
Magnetization
    99,118,102,126,131,134,213

collinear 102,239
Magnetic field 100,102,269
Magnetic fluctuation 206
Magnetic oscillation 112,114,119
Multiquantum wells 115
Matrix Ricatti equation 248
Magnetometry 113
Magnetoplasmon 136
Melting 266
Mobility 115
  gap 106
Moore's law 172
Measurement 207
Magnetic moment 213,237
Magnetic multilayer 214,221
Molecular beam epitaxy 214
Majority spin 230,235
Minority spin 230,235

## N
Normal order 20
Nesting 252
Noise 205,206
Non-linear Schroedinger equation 186
Non-Abelian bosonization 88
Non-equilibrium
  currents 107,109
  charges 109
  configuration 231
NMR 125,202
Nodal structure 266,287

## O
One-dimension 3,7,9,224
One-band model 241
Optical pumping 146,147
Optical absorption 147
OR gate 178
Order parameter 188
Ohmic dissipation 200
Oscillation 232,237
Organic single crystal 267

## P
Particle current 4
Poisson statistics 276,298
Potential energy 225
Pulse shape 205
Photon 182
Positive energies 11
Partition function 11
Particle operator 12
Persistent current 275,277,278

Periodicity 246
Percolation 76,264
Public-key cryptography 180
Polymers 76
Phase fields 13,14,16,20,24
Particle-hole
  excitations 17
  symmetry 144
Path integral 42
Primary field 63,64,78,85
Potts model 79
Pauli matrices 90
Pressure 100
Poisson equation 120
Polarization 143,146,289
Photoluminescence 148
Particle number operator 12
Phase separation 161
Phonon 117
Phase breaking length 272
Phase separation 121

## Q
Quantum Hall
  effect 100
  integer effect 139
  fractional 268
  ferromagnet 120,121,125,140,144,157
Quantum diffusion 277
Quantum
  well 124,141,224,230
  dot 183,293,301
Quantum Monte Carlo 265
Quasi-particle 126,130,206
Qubit 165,190
Quasi-hole 126
Quantum
    computation 165
    state control 165
    information 165,175,176
    logic gate 177,200
    Turing machine 181
Quantum-classical crossover 263,301
Quantum phase transition 263,301
Quasi-periodic 247

## R
Resonance 227
Right-moving electrons 5,11
Reservoirs 32,33
Representations 70,73,85,86
Riemann curvature 77
Radiative recombination 148

Read head 216
Resistor
   model 221
   network 223
Resistivity 269
Random potential 273
Random matrix theory 276

S
Spectral density 230
Spectral parameter 276
Semi-classical approach 96
Surface Green function 247
Scattering 3
Single electron transistor (SET) 193,200
Shor's algorithm 167
Superfluid 165
Selection rules 147
Scalability 183
String 22
Stoner
   model 120,122,291
   instability 290,303
Semi-infinite substrate 243
Skyrmion 123,139,148,155,159
Schottky nuclear heat capacity 149
Schrœdinger picture 23
   equation 98
Schrœdinger's cat 166
Superposition 166
Second-order phase transition 296
Spin 24
   density 25
   polarization 28,260
   wave 38
   texture 141
   flip 143,219
   valve 219
   gap 147
   exciton 160
   current 238,240
Single-particle excitations 264
Saturation field 216
Superlattices 216
Screening 29,106,134,272
Spin-charge separation 39
Spintronics 213
Scaling
   function 49
   dimension 62
Scaling theory of localization 270
Stress-energy tensor 62,63,65,70,78,84
Strip geometry 67
Superconductor 199

sl(2)
Superconducting
   devices 165
   circuits 186
Spinless fermions 82,263
Sugawara Hamiltonian 83,89
Single qubit gate 201
SU(N) 83
Specific heat 66,134
Skipping orbit 104
Solid 266,276
Spin-orbit coupling 120,144
Scanning capacitance spectroscopy 135
Scanning electron microscope 237
Soliton 139
Site orbitals 282
Skyrme crystal 144
Selection rules 147
Stationary phase approximation
   232,248,249,250,255,257
Single-orbital formula 243
Strong coupling 264
Slater-Jastrow 265
Slater determinant 266,281
Shubnikov-de Haas measurement 268
Semiconductor 2
Subband 119,121

T
Thin film 213
Two dimensions 264
Transport 1
Torque method 231,234
Torus 273,275
Tight-binding model 238
Tomonaga-Luttinger model 1,5,26,36
Transmission amplitude 3,5
   coefficient 3
Transversal motion 26
Turing machine 168,170
Trilayer 224,231,235,237,255
Tricritical 79
Tensor 66
Tower of states 71
Topological invariant 91
   term 97
   excitation 123
Tunnelling microscopy 125
Tunnelling 173,260
Time reversal symmetry 276
Thermodynamic gap 129
Tunable Josephson energy 197
Two qubit device 200
Two Charge Qubit gate 202
Two Flux Qubit gate 203

Transition metal 215
Transistor 268


**U**
Unitary 77
Ungapped phase 136
Universal logarithmic correction 270

**V**
Voltage 26
Van Leuwen theorem 103
Virasoro algebra 63,68
    generators 71,72,86
Variational energy 266

**W**
Wire 1,29
  Nonequilibrium 2
Ward identities 63,68
Weal localization 267
Wess-Zumino-Novikov-Witten model
    83,91,92,93
Wess-Zumino term 93,96,97
Wigner
  solid 263
  lattice 301
  molecule 278,287,292
  glass 301
Wigner-Dyson statistics 280,281,283,298
Weak coupling 264

**X**
XOR gate 178

**Y**

**Z**
Zero bias anomaly 48
Zeeman energy 142,144,152,154,158,276

Printed in the United States
1142100001B/33-119